Modem Theory

At the heart of any modern communication system is the modem, connecting the data source to the communication channel. This first course in the mathematical theory of modem design introduces the theory of digital modulation and coding that underpins the design of digital telecommunications systems. A detailed treatment of core subjects is provided, including baseband and passband modulation and demodulation, equalization, and sequence estimation. The modulation waveforms for communication channels and digital recording channels are treated in a common setting and with unified terminology. A variety of more advanced topics is also covered, such as trellis codes, turbo codes, the Viterbi algorithm, block codes, maximum likelihood and maximum posterior probability, iterative demodulation, and jamming. Numerous end-of-chapter exercises are also included to test the reader's understanding throughout. This insightful book is ideal for senior undergraduate students studying digital communications and is also a useful reference for practicing engineers.

RICHARD E. BLAHUT is the Henry Magnuski Professor of Electrical and Computer Engineering at the University of Illinois, Urbana-Champaign. He is a Life Fellow of the IEEE and the recipient of many awards including the IEEE Alexander Graham Bell Medal (1998) and Claude E. Shannon Award (2005), the Tau Beta Pi Daniel C. Drucker Eminent Faculty Award, and the IEEE Millennium Medal. He was named a Fellow of the IBM Corporation in 1980, where he worked for over 30 years, and was elected to the National Academy of Engineering in 1990.

Modem Theory

An Introduction to Telecommunications

Richard E. Blahut

University of Illinois, Urbana-Champaign

CAMBRIDGE
UNIVERSITY PRESS

CAMBRIDGE
UNIVERSITY PRESS

University Printing House, Cambridge CB2 8BS, United Kingdom

One Liberty Plaza, 20th Floor, New York, NY 10006, USA

477 Williamstown Road, Port Melbourne, VIC 3207, Australia

314-321, 3rd Floor, Plot 3, Splendor Forum, Jasola District Centre, New Delhi - 110025, India

79 Anson Road, #06-04/06, Singapore 079906

Cambridge University Press is part of the University of Cambridge.

It furthers the University's mission by disseminating knowledge in the pursuit of education, learning and research at the highest international levels of excellence.

www.cambridge.org
Information on this title: www.cambridge.org/9780521780148

© Cambridge University Press 2010

First published 2010

A catalogue record for this publication is available from the British Library

Library of Congress Cataloging in Publication data
Blahut, Richard E.
  Modem theory : an introduction to telecommunications / Richard Blahut.
     p. cm.
  Includes bibliographical references and index.
  ISBN 978-0-521-78014-8 (hardback)
  1. Coding theory.   2. Signal processing–Digital techniques–Mathematics.
  3. Modulation (Electronics)–Mathematics.   4. Modems.   I. Title.
  TK5102.92.B64 2010
  621.39´814–dc22   2009031867

ISBN 978-0-521-78014-8 Hardback

In loving memory of
Gayle Jones Blahut (1962–2008)
   — who always had the right touch

Contents

Preface

The field of telecommunication consists of the theory and the practice of communication at a distance, principally electronic communication. Many systems for telecommunication now take the form of large, complex, interconnected data networks with both wired and wireless segments, and the design of such systems is based on a rich theory. Communication theory studies methods for the design of signaling waveforms to transmit information from point to point, as within a telecommunication system. Communication theory is that part of information theory that is concerned with the explicit design of suitable waveforms to convey messages and with the performance of those waveforms when received in the presence of noise and other channel impairments. Digital telecommunication theory, or *modem theory*, is that part of communication theory in which digital modulation and demodulation techniques play a prominent role in the communication process, either because the information to be transmitted is digital or because the information is temporarily represented in digital form for the purpose of transmission.

Digital communication systems are in widespread use and are now in the process of sweeping away even the time-honored analog communication systems, such as those used in radio, television, and telephony. The main task of communication theory is the design of efficient waveforms for the transmission of information over band-limited or power-limited channels. The most sweeping conclusion of information theory is that all communication is essentially digital. The nature of the data that is transmitted is unimportant to the design of a digital communication system. This is in marked contrast to analog communication systems, such as radio or television, in which the properties of the transmitted waveform are inextricably tied up with the properties of the application, and only weakly tied to considerations of the communication channel. To make the point more strongly, we can give a whimsical definition of a digital communication system as a communication system designed to best use a given channel, and an analog communication system as one designed to best fit a given source. The spectrum of a well-designed digital communication waveform is a good match to the passband characteristics of the channel; the only essential way in which the source affects the spectrum of the waveform is by the bit rate. In contrast, the spectrum of an analog communication waveform depends critically on the properties of the source.

The purpose of this book is to give a general introduction to the modulation waveforms and demodulation techniques that are central to the design of digital telecommunication systems. Moreover, because recording is essentially a communication process – from a past time to a future time – modulation techniques are also central to the design of magnetic and optical recording systems. Modulation waveforms for passband channels and for baseband channels, such as magnetic recording channels, are treated in a common setting and with unified terminology.

The topics of this book are confined to the modulation layer of communication theory. The topics at the modulation layer lie above other topics needed for the physical design of communication equipment and lie below topics in other layers of the theory that deal with networking, routing, and application sessions. These are topics for other books. The compaction and compression of source data, including analog or voice, are also topics that are not treated in this book. These, also, are topics for other books.

The material in this book consists, for the most part, of selected chapters from *Digital Transmission of Information* (published in 1990), which have been rewritten and expanded to fit the needs of a course in modem theory and digital telecommunications. Waveforms and modulators are studied in Chapters 2 and 5 for baseband and passband channels, respectively. Basic demodulators are developed in Chapters 3 and 6 for baseband and passband channels, respectively. More advanced methods of demodulation for channels with dispersion are studied in Chapter 4. In Chapter 3, the matched filter is introduced as a filter to maximize signal-to-noise ratio prior to a demodulation decision. A stronger statement of optimality of the matched filter is deferred to Chapter 7, where it is shown to be part of the maximum-likelihood demodulator for both coherent and noncoherent demodulation in gaussian noise. These first seven chapters contain the central ideas of modulation and demodulation, which are at the core of the theory of modems. The final five chapters then go deeper into the subject by developing some of the other topics that are needed to round out the foundations of the theory of modems. Chapter 8 treats methods of synchronizing the transmitter and the receiver so that they have the same time reference. Chapters 9 and 10 discuss methods of coding for communication channels to control errors. Rather than modulate one data bit, or a few data bits, at a time into a communication waveform, a coded representation modulates an entire message into a communication waveform so that cross-checks can eliminate errors. Chapter 9 discusses codes designed for an additive noise channel, Chapter 10 discusses codes designed for a discrete channel, usually binary. Finally, Chapters 11 and 12 advance the theory beyond the simple linear channel studied in most of the book. Chapter 11 studies the robustness of modems in the presence of simple nonlinearities and fading. Chapter 12 discusses techniques for the prevention of intentional disruption of communications by a malicious adversary known as a jammer.

Modern digital telephony, now in widespread use, is an almost miraculous system – partly wireless and partly wired – in which our everywhere environment is filled with a dense but faint electromagnetic fabric that millions of telephone users can tap into to draw out conversations, text, and data almost without end, and certainly without awareness of the immense theory and the extensive industry that make this fabric possible. This is the real miracle of the recent decades, and this technological miracle has changed the world far more than has any political theory or practice. This book studies digital communication theory, which consists of the collection of waveform methods that underlie the telecommunication system. A deeper investigation of the merits of these methods is provided within the subject of information theory, which is only touched on in this book. The study of modems is a first step in understanding this wondrous wireless fabric, as well as the enormous and sophisticated wired backbone that underlies this global system. The study of information theory provides a fuller understanding of the optimality of these methods.

This book evolved within the rich environment of students and faculty at the University of Illinois. I could not have found a better set of colleagues anywhere with which to interact, and no environment more intellectually stimulating. The quality of the book has much to do with the typing skills of Mrs Frances Bridges and the editing skills of Mrs Helen Metzinger. And, as always, Barbara made it possible.

*A man may expresse and signifie the intentions of his minde,
at any distance of place by objects . . . capable of a twofold
difference onely.*

Sir Francis Bacon (1561–1626)

1 Introduction

A point-to-point communication system transfers a message from one point to another through a noisy environment called a *communication channel*. A familiar example of a communication channel is formed by the propagation of an electromagnetic wave from a transmitting antenna to a receiving antenna. The message is carried by the time-varying parameters of the electromagnetic wave. Another example of a communication channel is a waveform propagating through a coaxial cable that connects a jack mounted on an office wall to another such jack on another wall or to a central node. In these examples, the waveform as it appears at the receiver is contaminated by noise, by interference, and by other impairments. The transmitted message must be protected against such impairments and distortion in the channel. Early communication systems were designed to protect their messages from the environment by the simple expedient of transmitting at low data rates with high power. Later, message design techniques were introduced that led to the development of far more sophisticated communication systems with much better performance. Modern message design is the art of piecing together a number of waveform ideas in order to transmit as many bits per second as is practical within the available power and bandwidth. It is by the performance at low transmitted energy per bit that one judges the quality of a digital communication system. The purpose of this book is to develop modern waveform techniques for the digital transmission of information.

1.1 Transmission of information

An overview of a digital communication system is shown in Figure 1.1. A message originating in an information source is to be transmitted to an information user through a channel. The digital communication system consists of a device called a *transmitter*, which prepares the source message for the communication channel, and another device called a *receiver*, which prepares the channel output for the user. The operation of the transmitter is called *modulation* or *encoding*. The operation of the receiver is called *demodulation* or *decoding*. Many point-to-point communication systems are two-way

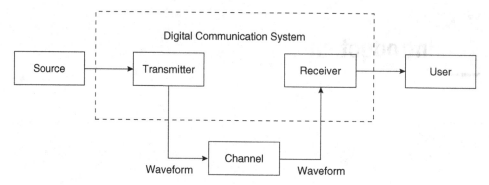

Figure 1.1. Overview of a digital communication system.

systems in which both a modulator and a demodulator are combined in a single package called a *modem*, the theory of which is the topic of this book.

At the physical level, a communication channel is normally an analog channel in that it transmits waveforms. The user information may arise as digital source data or may arise as an analog signal. The source data must eventually be modulated into an analog waveform that suits the analog channel. When the source signal is an analog waveform – perhaps a continuous-time analog waveform such as voice or video – a digital communication system first converts that waveform to a digital representation (which may take the form of a stream of bits), processes that digital representation in some way, but then converts it back into a continuous-time analog waveform for passage through the channel. The analog waveform passing through the channel will be completely different from the analog waveform generated by the source. The analog waveform at the channel output may be converted later by means of sampling and quantization to intermediate digital data for processing and demodulation. The intermediate digital data may be very different in its character from both the digital data that was transmitted and the final digital data produced by the demodulator. Ultimately, if required by the user, the digital data may be reconverted to its original analog form, such as voice or video. This multiple conversion between digital data and analog data may seem to be complicated and expensive, but it is worthwhile for many reasons. The channel waveform is matched to the nature of the channel, not to the nature of the source. Moreover, the digital data can be rerouted through many kinds of digital links or storage devices while it is in digital form, or it can be merged and mingled with other data traffic passing through a network.

Analog modulation is still widely used in radio and television, and until recently, was also used in phonography and voice telephony. Analog modulation techniques make relatively superficial changes to the signal in order to send it through the channel; there is no significant effort to tailor the waveform to suit the channel at any deeper level. Digital communication waveforms are more sophisticated. Digital communication theory endeavors to find waveforms that are closely matched to the characteristics of the channel and that are tolerant of the impairments in the channel so that the reliable flow of information through the channel is ensured. The characteristics of the source are of no interest in designing good waveforms for a channel. Good waveforms for digital communication are designed to match the characteristics of the channel; the source information is then encoded into this channel waveform. A digital communication system might require a considerable amount of electronic circuitry to translate the source waveform into a form more suitable for the channel, but electronics is now cheap. In contrast, most channels are comparatively expensive, and it is important to make the best use of a channel.

Even an application that might appear to be intrinsically an analog application, such as broadcast television, can be partitioned into two tasks – the task of delivering so many bits per second through the channel and the task of representing the video signal by the available number of bits per second. This is the important and (once) surprising separation principle of information theory, which says that the task of transmitting the output of a source through a channel can be separated, without meaningful loss, into the task of forming a binary representation of the source output and the task of sending a binary datastream through the channel. For digital transmission to be effective, both of these tasks must be implemented efficiently. Otherwise, there will be disadvantages such as increased bandwidth or larger transmitted power. The source data must first be compressed, then modulated into a suitable transmission waveform, perhaps a spectrally-efficient waveform that carries multiple bits per second per hertz, or perhaps an energy-efficient waveform that uses low power but occupies a large bandwidth.

The disadvantages that were once cited for digital communications are really not compelling; any validity that these claims once had has crumbled under the progress of technology. Cost was once a significant disadvantage of digital communication systems, but is no longer. Digital modulation of an analog signal was once believed to require larger bandwidth than direct analog modulation. This is not true. By using modern methods of data compression, data compaction, and bandwidth-efficient modulation – conveying multiple bits per second per hertz – a digital communication system will actually use less bandwidth than an analog communication system. Another presumed disadvantage that is sometimes mentioned is that of quantization noise. Quantization noise, however, is completely under the control of the designer of the quantization scheme and will be the only important source of noise in the signal that is presented to the user. The modern view is that quantization noise is a price cheerfully paid for the more important advantage of removing the effects of channel noise and other impairments from the received signal. It is a truism of information theory that, in a well-designed system, the quantization noise will always be less than the channel noise it replaces.

On the other hand, there are numerous and compelling advantages of digital communication. Every link becomes simply a "bit pipe" characterized by its data rate and its probability of bit error. It makes no difference to the communication links whether the transmitted bits represent a digitized voice signal or a computer program. Many kinds of data source can share a common digital communication link, and the many kinds of communication links are suitable for any data source. Errors due to noise and interference are almost completely suppressed by the use of specialized codes for the prevention of error. A digital datastream can be routed through many physically different links in a complex system, and can be intermingled with other digital traffic in the network. A digital datastream is compatible with standardized encryption and antijam equipment.

The digital datastream can be regenerated and remodulated at every repeater that it passes through so that the effect of additive channel noise, or other impairments, does not accumulate in the signal. Analog repeaters, on the other hand, consist of amplifiers that amplify both signal and noise. Noise accumulates in an analog communication waveform as it passes through each of a series of repeaters.

Finally, because digital communication systems are built in large part from digital circuitry, data can be readily buffered in random-access digital memories or on magnetic disks. Many functions of a modem can be programmed into a microprocessor or designed into a special-purpose digital integrated circuit. Thus a digital data format is compatible with the many other digital systems and subsystems of the modern world.

1.2   A brief historical survey

The historical development of modem theory can be divided into several phases: traditional methods such as PSK, PAM, QAM, and orthogonal signaling, which were developed early; the multilevel signal constellation designs of the 1970s; the coded modulation and precoding techniques of the 1980s and 1990s, and the graphical methods of the past decade. It is a curiosity of technological history that the earliest communication systems such as telegraphy (1832,1844) actually can be classified as digital communication systems. Even the time-honored Morse code is a digital communication waveform in that it uses a discrete alphabet. Telegraphy created an early communications industry but lacked the popular appeal of later analog communication systems such as the telephone (1876) and the phonograph (1877). These analog communication systems were dominant for most of the twentieth century.

The earliest broadcast systems for communication were concerned with the transfer of analog continuous signals, first radio signals (1920) and then television signals (1923, 1927). Analog modulation techniques were developed for embedding a continuous-time signal into a carrier waveform that could be propagated through a channel such as an electromagnetic-wave channel. These techniques are still employed in systems that demand low cost or have strong historical roots, such as radio, telephony, and television. There are indications, however, that analog modulation is becoming outdated even for those applications, and only the enormous investment in existing equipment will forestall the inevitable demise of the familiar AM and FM radios. Indeed, the evolution from digital communication to analog communication that began in the 1870s is now being countered by an evolution from analog communication back to digital communication that found its first strength in the 1970s. Even the analog phono-graph record, after 100 years of popularity and dominance, has now been completely superseded by the compact disk.

The earliest radio transmitters used a form of analog modulation called *amplitude modulation*. This method maps a signal $s(t)$ into a waveform $c(t)$ given by

$$c(t) = [1 + ms(t)] \cos 2\pi f_0 t$$

where m is a small constant called the *modulation index* such that $ms(t)$ is much smaller than one, and f_0 is a constant called the *carrier frequency* such that f_0 is large compared to the largest frequency for which $S(f)$, the Fourier transform of $s(t)$, is nonzero. Even though the fidelity of the received signal is not noteworthy, amplitude modulation became very popular early on because the mapping from $s(t)$ to $c(t)$ could be implemented in the transmitter very simply, and the inverse mapping from $c(t)$ to $s(t)$ could be implemented simply in the receiver, though only approximately.

Frequency modulation is an alternative analog modulation technique given by the following map from $s(t)$ to $c(t)$:

$$c(t) = \sin \left(2\pi f_0 t + \int_0^t ms(\xi) d\xi \right)$$

where, again, the carrier frequency f_0 is large compared to the largest frequency for which $S(f)$ is significant. Frequency modulation was naively proposed very early as a method to conserve the radio spectrum. The naive argument was that the term $ms(t)$ is an "instantaneous frequency" perturbing the carrier frequency f_0 and, if the modulation index m is made very small, the bandwidth of the transform $C(f)$ could be made much smaller than the bandwidth of $S(f)$. Carson (1922) argued that this is an ill-considered plan, as is easily seen by looking at the approximation

$$c(t) \approx \sin 2\pi f_0 t + \cos 2\pi f_0 t \left[m \int_0^t s(\xi) d\xi \right]$$

when m is small. The second term has the same Fourier transform as the bracketed component, but translated in frequency by f_0. Because the integral of $s(t)$ has the same frequency components as $s(t)$, the spectral width is not reduced. As a result of this observation, frequency modulation temporarily fell out of favor. Armstrong (1936) reawakened interest in frequency modulation when he realized it had a much different property that was desirable. When the modulation index is large, the inverse mapping from the modulated waveform $c(t)$ back to the signal $s(t)$ is much less sensitive to additive noise in the received signal than is the case for amplitude modulation, – at least when the noise is small. Frequency demodulation implemented with a hardlimiter suppresses noise and weak interference, and so frequency modulation has come to be preferred to amplitude modulation because of its higher fidelity.

The basic methods of analog modulation are also used in modified forms such as single-sideband modulation or vestigial-sideband modulation. These modified forms are attempts to improve the efficiency of the modulation waveform in its use of the spectrum. Other analog methods, such as Dolby (1967) modulation, are used to match the analog source signal more closely to the noise characteristics of the channel. All methods for modifying the techniques of analog modulation are stopgap methods. They do not attack the deficiencies of analog modulation head-on. Eventually, possibly with a few exceptions, such methods will be abandoned in favor of digital modulation.

The superiority of digital communication seems obvious today to any observer of recent technological trends. Yet to an earlier generation it was not obvious at all. Shannon's original development of information theory, which was published in 1948 and implicitly argued for the optimality of digital communications, was widely questioned at the time. Communication theory became much more mathematical when Shannon's view became widely appreciated. His view is that communication is intrinsically a statistical process, both because the message is random and because the noise is random. The message is random because there is little point in transmitting a predetermined message if the receiver already knows it. If there are only a few possible predetermined messages already known to the receiver, one of which must be sent, then there is no need to send the entire message. Only a few prearranged bits need to be transmitted to identify the chosen message to the receiver. But this already implies that there is some uncertainty about which message will be the chosen message. Even this simple example introduces randomness, and as the number of possible messages increases, the randomness increases, and so the number of bits needed in the message increases as well.

Randomness is an essential ingredient in the theory of communication also because of noise in the channel. This statistical view of communication, encompassing both random messages and noisy channels, was promoted by Shannon (1948, 1949). Earlier, Rice (1945) had made extensive study of the effect of channel noise on received analog communication waveforms. Shannon developed the broader and (at that time) counterintuitive view that the waveform could be designed to make the channel noise essentially inconsequential to the quality of the received waveform. He realized that combating noise was a job for both the transmitter and the receiver, not for the receiver alone. In his papers, Shannon laid a firm foundation for the development of digital communication. A different paper dealing with applications that were transitional between analog and digital communication was due to Oliver, Pierce, and Shannon (1948). The period of the 1940s appears to be the time when people began thinking deeply about the fundamental nature of the communication problem and the return to digital signaling began to accelerate. There were, however, many earlier studies and applications of digital signaling as in the work of Nyquist (1924, 1928) and Hartley (1928). Aschoff (1983) gives a good early history of digital signaling.

1.3   Point-to-point digital communication

A simple block diagram of a point-to-point digital communication system is shown in Figure 1.1. The model in Figure 1.1 is quite general and can be applied to a variety of communication systems, and also to magnetic and optical storage systems. The boxes labeled "channel," "source," and "user" in Figure 1.1 represent those parts of the system that are not under the control of the designer. The identification of the channel may be somewhat arbitrary because some of the physical components such as amplifiers might in some circumstances be considered to be part of the channel and in other circumstances might be considered to be part of the modulator and demodulator.

It is the task of the designer to connect the data source to the data sink by designing the boxes labeled "transmitter" and "receiver". These boxes are also called, more simply, the *encoder* and *decoder* or the *modulator* and *demodulator*. The latter names are usually preferred when the channel is a waveform channel while the former are usually preferred for discrete channels. Consequently, the transmitter is also called the *encoder/modulator*, and the receiver is also called the *demodulator/decoder*. Often a single package that can be used either as a transmitter or a receiver (or both simultaneously) is desired. A modulator and demodulator combined into a single box is called a *modem*. The term "modem" might also be used to include the encoder and decoder as well, and usually includes other supporting functions that extend beyond modulation and demodulation but which are needed to make the modem work. The terms "transmitter" and "receiver" are sometimes preferred as the broader terms that include such supporting functions. Figure 1.2 and Figure 1.3 show the functions normally included in the transmitter and receiver.

Modern practice in the design of communication systems is to separate the design tasks associated with the data source and the data user from the design tasks associated with the channel. This leads technology in the direction of greater flexibility in that the source data, when reduced to a stream of bits, might be transmitted through any one of many possible channels or even through a network of different channels. To

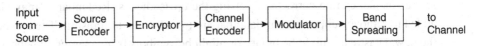

Figure 1.2. Primary transmitter functions.

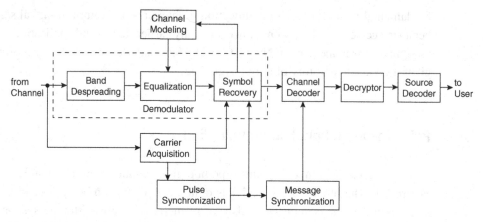

Figure 1.3. Primary receiver functions.

do this, the functions of the transmitter and receiver are broken into more detailed functions as described by the block diagrams of Figures 1.2 and 1.3. The transmitter includes a source encoder, a channel encoder, and a modulator; the receiver includes a demodulator, a channel decoder, and a source decoder. Information theory teaches us that there is no consequential loss in performance because of partitioning the problem in this way. Moreover, for our topics, there is no loss in generality if the interface between the source encoder and the channel encoder, as well as the interface between the channel decoder and source decoder are regarded to be serial streams of binary data.

The source data entering a digital communication system may be analog or digital. Upon entering the transmitter, analog data will first be digitized. In the process of digitization, continuous time may be reduced to discrete time by the process of sampling a source waveform of finite bandwidth. Then the datastream is processed by a source encoder, whose purpose is to represent the source data compactly by a stream of bits called the *source codestream*. At this point, the source data has been reduced to a commonplace stream of bits, superficially displaying no trace of the origin of the source data. Indeed, data from several completely different kinds of sources now may be merged into a single bit stream. The source data might then be encrypted to prevent eavesdropping by an unauthorized receiver. Again, the encrypted bit stream is another commonplace bit stream superficially displaying no trace of its origin.

The datastream is next processed by the channel encoder, which transforms the datastream into a new datastream called the *channel codestream*. The channel codestream has redundancy in the form of elaborate cross checks built into it, so that errors arising in the channel can be corrected by using the cross checks. Redundancy may also be added to remove subpatterns of data that are troublesome to transmit. The symbols of the new datastream might be binary or might be symbols from a larger alphabet called the *channel alphabet*. The stream of channel codewords is passed to the modulator, which converts the sequence of discrete code symbols into a continuous function of time called the *channel waveform*. The modulator does this by replacing each symbol of the channel codeword by the corresponding analog symbol from a finite set of analog symbols composing the *modulation alphabet*. Here, discrete time is also reconverted to continuous time by some form of interpolation. The sequence of analog symbols composes the channel waveform, which is either transmitted through the channel directly or after it is further modified to spread its bandwidth. The reason that bandspreading might be used is to protect the signal from some kinds of fading or interference, possibly intentional interference created by an adversary. The input to the channel is the channel waveform formed by the transmitter. The channel waveform is now a continuous-time waveform. Although the source waveform might also have been a continuous-time waveform, the appearance and properties of the channel waveform will be quite different from those of the source waveform. The channel waveform then passes through the channel where it may be severely attenuated and usually changed in other ways.

The input to the receiver is the output of the channel. Because the channel is subject to various types of noise, dispersion, distortion, and interference, the waveform seen at the channel output differs from the waveform at the channel input. The waveform will always be subjected to thermal noise in the receiver, which is additive gaussian noise, and this is the disturbance that we shall study most thoroughly. The waveform may also be subjected to many kinds of impulsive noise, burst noise, or other forms of nongaussian noise. Upon entering the receiver, if the waveform has been bandspread in the transmitter, it is first despread. The demodulator may then convert the received waveform into a stream of discrete channel symbols based on a best estimate of each transmitted symbol. Sometimes the demodulator makes errors because the received waveform has been impaired and is not the same as the waveform that was transmitted. Perhaps to quantify its confidence, the demodulator may append confidence annotations to each demodulated symbol. The final sequence of symbols from the demodulator is called the *received word* or the *senseword*. It is called a *soft senseword* if it is not reduced to the channel input alphabet or if it includes confidence annotations. The symbols of the senseword need not match those of the transmitted channel codeword; the senseword symbols may take values in a different alphabet.

The function of the channel decoder is to use the redundancy in the channel codeword to correct the errors in the senseword and then to produce an estimate of the datastream that appeared at the input to the channel encoder. If the datastream has been encrypted, it is now decrypted to produce an estimate of the sequence of source codewords. Possibly at this point the datastream contains source codewords from more than one source, and these source codewords must be demultiplexed. If all errors have been corrected by the channel decoder, each estimated source codeword matches the original source codeword. The source decoder performs the inverse operation of the source encoder and delivers its output to the user.

The modulator and the demodulator are studied under the term *modulation theory*. Modulation theory is the core of digital communications and the subject of this book. The methods of baseband modulation and baseband demodulation are studied in Chapters 2 and 3, while passband modulation and passband demodulation are studied in Chapter 5 and Chapter 6. The formal justification for the structure of the optimal receiver will be given with the aid of the maximum-likelihood principle, which is studied in Chapter 7.

The receiver also includes other functions, such as equalization and synchronization, shown in Figure 1.3, that are needed to support demodulation. The corresponding structure of the optimal receiver will not fully emerge until these functions are developed as a consequence of the maximum-likelihood principle in Chapter 7. The receiver may also include intentional nonlinearities, perhaps meant to clip strong interfering signals, or to control dynamic range. These are studied in Chapter 11.

1.4   Networks for digital communication

Digital point-to-point communication systems can be combined to form digital communication networks. Modern communication networks were developed relatively recently, so they are mostly digital. The main exception to this is the telephone network which began early as an analog network, and was converted piecemeal to a digital network through the years, but still with some vestiges of analog communication. The telephone network is a kind of network in which switching (*circuit switching*) is used to create a temporary communication channel (a *virtual channel*) between two points. A broadcast system, shown in Figure 1.4, might also be classified as a kind of communication network. However, if the same waveform is sent to all receivers, the design of a broadcast system is really quite similar to the design of a point-to-point communication system. Therefore both the early telephone network and a broadcast system, in their origins, can be seen as kinds of analog point-to-point communication systems. Other, more recent, communication networks are digital.

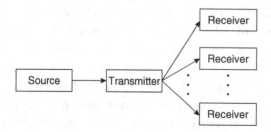

Figure 1.4. Broadcast communications.

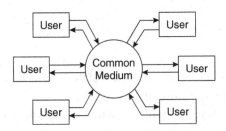

Figure 1.5. A multiaccess communication network.

Figure 1.5 shows a modern form of a communication network, known as a *multiaccess network*. It consists of a single channel, called a *multiaccess channel*, to which is attached a multitude of users who exchange messages. The rules by which the users access the shared channel are called *channel protocols*. The protocol is one new element that makes a digital communication network more complicated than a digital point-to-point communication system.

The simplest multiaccess channel is the *fixed-assignment multiaccess channel* of which *time-division multiaccess signaling* is a good example. In a time-division multiaccess system, each user is preassigned a fixed time interval called a *time slot* during which it can transmit. In this way, the channel is available to each user in turn. Similarly, for *frequency-division multiaccess signaling* each user is assigned a fixed frequency slot. One disadvantage of the time-division or frequency-division protocol is that it is inefficient; if a user has nothing to transmit, its time slot or frequency slot is wasted. Another disadvantage is that the set of users needs to be fairly well specified so that time slots or frequency slots can be assigned. Both transmitter and receiver need to know the slot assignments. Moreover, each user of a time-division system needs to have its clock synchronized to system time. If it is possible for new or dormant users to become active, then there must be a control channel to assign time slots or frequency slots.

At the other extreme are networks for which the set of users is very large, not very well defined, and each individual user wants to transmit only rarely but at random times and at its own convenience. Then the signaling method is known as *demand-assignment multiaccess signaling*. One kind of protocol for this case is the kind known as a *contention-resolution algorithm*. Users transmit whenever it suits them, but the collision of simultaneous messages may cause those messages to be lost or damaged. In the event of such a conflict, all users in the collision retransmit their lost messages at a later time. Each user chooses a later time to retransmit by a common algorithm that chooses a random, or effectively random, time so that another collision is unlikely. Contention-resolution algorithms have seen wide use, but the behavior and optimality of these algorithms is not fully understood.

1.5   The Fourier transform

The design of communication waveforms makes extensive use of the Fourier transform. We shall review the theory of the Fourier transform in this section.

A signal $s(t)$ is a real-valued or complex-valued function of time that has finite energy

$$E_p = \int_{-\infty}^{\infty} |s(t)|^2 dt \; < \; \infty.$$

We shall also call the signal $s(t)$ a *pulse*, usually when we can regard it as a simple signal. The *support* of $s(t)$ is the set of t for which $s(t)$ is nonzero. The pulse $s(t)$ has a *finite duration* – or has finite *timewidth* – if its support is contained in a finite interval. Because we regard t to be time, the pulse $s(t)$ is called a *time-domain signal*.

The *Fourier transform* $S(f)$ of the signal $s(t)$ is defined as

$$S(f) = \int_{-\infty}^{\infty} s(t) e^{-j2\pi ft} dt.$$

The function $S(f)$ is sometimes called the *frequency-domain signal*. The relationship between a function and its Fourier transform is denoted in summary form by the notation

$$s(t) \leftrightarrow S(f).$$

The two-way arrow implies that $s(t)$ can be converted to $S(f)$, and $S(f)$ can be converted to $s(t)$. Sometimes this relationship is abbreviated by the functional notation

$$S(f) = \mathcal{F}[s(t)]$$

and

$$s(t) = \mathcal{F}^{-1}[S(f)].$$

We also call the Fourier transform $S(f)$ the *spectrum* or the *amplitude spectrum* of the pulse $s(t)$, and we call $|S(f)|^2$ the *power spectrum* of the pulse $s(t)$. The *support* of $S(f)$ is the set of f for which $S(f)$ is nonzero. The spectrum $S(f)$ has *finite bandwidth* if its support is contained in a finite interval.

The Fourier transform is a linear operation. This means that if

$$s(t) = as_1(t) + bs_2(t)$$

for any complex constants a and b, then

$$S(f) = aS_1(f) + bS_2(f).$$

If $S(f)$ is the Fourier transform of the signal $s(t)$, then it is not the Fourier transform of any other signal.[1] Consequently, the Fourier transform can be inverted. This is proved in the following theorem, which gives an explicit formula for the inverse Fourier transform.

Theorem 1.5.1 (Inverse Fourier transform) *If $S(f)$ is the Fourier transform of $s(t)$, then*

$$s(t) = \int_{-\infty}^{\infty} S(f)e^{j2\pi ft}df$$

for all t at which $s(t)$ is continuous.

Proof We shall sketch only an informal "proof" here, because the symbolism of the impulse function $\delta(t)$ is used. (The impulse function is defined in Section 1.6.) By definition

$$S(f) = \int_{-\infty}^{\infty} s(\xi)e^{-j2\pi f\xi}d\xi.$$

Therefore

$$\int_{-\infty}^{\infty} S(f)e^{j2\pi ft}df = \int_{-\infty}^{\infty}\left[\int_{-\infty}^{\infty} s(\xi)e^{-j2\pi f\xi}d\xi\right]e^{j2\pi ft}df$$

$$= \int_{-\infty}^{\infty} s(\xi)\left[\int_{-\infty}^{\infty} e^{-j2\pi(\xi-t)f}df\right]d\xi.$$

The inner integral does not have a proper meaning. The integral is infinite if $\xi = t$ and otherwise is the integral of a sine wave over all time, which is undefined. Without further justification, we replace the inner integral by a delta function to write

$$\int_{-\infty}^{\infty} S(f)e^{j2\pi ft}df = \int_{-\infty}^{\infty} s(\xi)\delta(\xi - t)d\xi$$

$$= s(t).$$

This completes the informal proof.   ∎

The formula for the inverse Fourier transform has the same structure as the formula for the Fourier transform itself except for a change in sign in the exponent. Consequently, there is a duality in Fourier transform pairs. Specifically, if

$$s(t) \leftrightarrow S(f)$$

---

[1] This statement requires the definition that two pulses $s_1(t)$ and $s_2(t)$ are equivalent if and only if the difference pulse $s_1(t) - s_2(t)$ has zero energy.

is a Fourier transform pair, then

$$\dot{S}(t) \leftrightarrow s(-f)$$

is also a Fourier transform pair.

Theorem 1.5.2 *Scaling of the time axis by the real constant a changes s(t) to s(at) and changes the transform $S(f)$ to $|a|^{-1}S(f/a)$.*

Proof If a is positive, set $at = t'$, so that

$$\int_{-\infty}^{\infty} s(at)e^{-j2\pi ft}dt = \int_{-\infty}^{\infty} s(t')e^{-j2\pi ft'/a}dt'/a$$

$$= \frac{1}{a}S(f/a).$$

If a is negative, the sign reversal can be accommodated by writing $|a|$. ∎

In the next several theorems we develop a general *shift property*, which is known as the *delay theorem* when used to shift the time origin, and as the *modulation theorem* when used to shift the frequency origin.

Theorem 1.5.3 (Delay theorem) *If s(t) has Fourier transform $S(f)$, then $s(t - t_0)$ has Fourier transform $S(f)e^{-j2\pi t_0 f}$.*

Proof By a simple change in variable,

$$\int_{-\infty}^{\infty} s(t - t_0)e^{-j2\pi ft}dt = \int_{-\infty}^{\infty} s(\tau)e^{-j2\pi f(\tau + t_0)}d\tau$$

$$= S(f)e^{-j2\pi t_0 f}$$

as was to be proved. ∎

The delay theorem is illustrated in Figure 1.6.

Theorem 1.5.4 (Modulation theorem) *If s(t) has Fourier transform $S(f)$, then $s(t)e^{j2\pi f_0 t}$ has Fourier transform $S(f - f_0)$.*

Proof By a simple change in variable,

$$\int_{-\infty}^{\infty} s(t)e^{j2\pi f_0 t}e^{-j2\pi ft}dt = \int_{-\infty}^{\infty} s(t)e^{-j2\pi(f - f_0)t}dt$$

$$= S(f - f_0),$$

as was to be proved. ∎

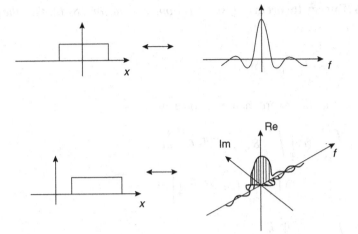

Figure 1.6. Illustrating the delay theorem.

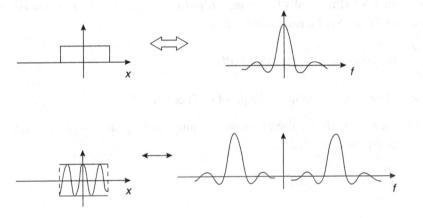

Figure 1.7. Illustrating the modulation theorem.

Corollary 1.5.5 (Modulation theorem) *If $s(t)$ has Fourier transform $S(f)$, then $s(t)\cos 2\pi f_0 t$ has Fourier transform $\frac{1}{2}[S(f+f_0)+S(f-f_0)]$, and $s(t)\sin 2\pi f_0 t$ has Fourier transform $\frac{1}{2}j[S(f+f_0)-S(f-f_0)]$.*

Proof   The proof follows immediately from the theorem and the relations

$$\cos 2\pi f_0 t = \frac{1}{2}\left(e^{j2\pi f_0 t}+e^{-j2\pi f_0 t}\right)$$

$$\sin 2\pi f_0 t = \frac{1}{2j}\left(e^{j2\pi f_0 t}-e^{-j2\pi f_0 t}\right).$$ ∎

The modulation theorem is illustrated in Figure 1.7.

Theorem 1.5.6 (Energy theorem)   *If $s(t)$ has Fourier transform $S(f)$, then the pulse energy E_p of $s(t)$ satisfies*

$$E_p = \int_{-\infty}^{\infty} |s(t)|^2 dt = \int_{-\infty}^{\infty} |S(f)|^2 df.$$

Proof   This is a straightforward manipulation as follows:

$$\int_{-\infty}^{\infty} |s(t)|^2 dt = \int_{-\infty}^{\infty} s(t) \left[\int_{-\infty}^{\infty} S(f) e^{j2\pi ft} df \right]^* dt$$

$$= \int_{-\infty}^{\infty} S^*(f) \left[\int_{-\infty}^{\infty} s(t) e^{-j2\pi ft} dt \right] df$$

$$= \int_{-\infty}^{\infty} |S(f)|^2 df.$$

∎

The energy theorem is a special case of the following theorem.

Theorem 1.5.7 (Parseval's theorem)   *If pulses $s_1(t)$ and $s_2(t)$ have Fourier transforms $S_1(f)$ and $S_2(f)$, respectively, then*

$$\int_{-\infty}^{\infty} s_1(t) s_2^*(t) dt = \int_{-\infty}^{\infty} S_1(f) S_2^*(f) df.$$

Proof   The proof is similar to the proof of Theorem 1.5.6.   ∎

A linear filter with impulse response $g(t)$ and input signal $s(t)$ has an output signal $r(t)$ given by the convolution

$$r(t) = \int_{-\infty}^{\infty} s(\xi) g(t - \xi) d\xi$$

$$= \int_{-\infty}^{\infty} s(t - \xi) g(\xi) d\xi.$$

The next theorem, a fundamental theorem for the study of the effect of a linear filter on a signal, says that a convolution in the time domain becomes a product in the frequency domain. Because multiplication is simpler than convolution, it often is easier to think about a filtering problem in the frequency domain.

Theorem 1.5.8 (Convolution theorem)   *If $s(t)$, $g(t)$, and $r(t)$ have Fourier transforms $S(f)$, $G(f)$, and $R(f)$, respectively, then*

$$r(t) = \int_{-\infty}^{\infty} g(t - \xi) s(\xi) d\xi$$

if and only if

$$R(f) = G(f)S(f).$$

Proof Expressing $s(t)$ as an inverse Fourier transform gives

$$r(t) = \int_{-\infty}^{\infty} g(t-\xi)\left[\int_{-\infty}^{\infty} S(f)e^{j2\pi f\xi}df\right]d\xi$$

$$= \int_{-\infty}^{\infty} S(f)e^{j2\pi ft}\left[\int_{-\infty}^{\infty} g(t-\xi)e^{-j2\pi f(t-\xi)}d\xi\right]df.$$

Let $\eta = t-\xi$. Then

$$r(t) = \int_{-\infty}^{\infty} S(f)e^{j2\pi ft}\left[\int_{-\infty}^{\infty} g(\eta)e^{-j2\pi f\eta}d\eta\right]df$$

$$= \int_{-\infty}^{\infty} S(f)G(f)e^{j2\pi ft}df.$$

Consequently, by the uniqueness of the inverse Fourier transform,

$$R(f) = S(f)G(f).$$

Because the argument can be read in either direction, the theorem is proved in both directions. ∎

A special case of the convolution theorem says that the convolution of a pulse $s(t)$ with the pulse $s^*(-t)$ has Fourier transform $|S(f)|^2$ because the Fourier transform of $s^*(-t)$ is $S^*(f)$. Similarly, the pulse $|s(t)|^2$ has Fourier transform equal to the convolution of $S(f)$ with $S^*(-f)$.

Theorem 1.5.9 (Differentiation) *If pulse $s(t)$ has Fourier transform $S(f)$, then the derivative $ds(t)/dt$, if it exists, has Fourier transform $j2\pi fS(f)$.*

Proof Allowing the interchange of differentiation and integration gives

$$\frac{ds(t)}{dt} = \frac{d}{dt}\int_{-\infty}^{\infty} S(f)e^{j2\pi ft}df$$

$$= \int_{-\infty}^{\infty} [j2\pi fS(f)]e^{j2\pi ft}df.$$

Then

$$\frac{ds(t)}{dt} \leftrightarrow j2\pi fS(f)$$

by the uniqueness of the Fourier transform. ∎

Theorem 1.5.10 (Poisson sum formula)

$$\sum_{\ell=-\infty}^{\infty} s(t+\ell T) = \frac{1}{T}\sum_{\ell=-\infty}^{\infty} e^{j2\pi \ell t/T}S\left(\frac{\ell}{T}\right).$$

Proof   The expression

$$\sum_{\ell=-\infty}^{\infty} s(t + \ell T) = s(t) * \sum_{\ell=-\infty}^{\infty} \delta(t + \ell T)$$

has the Fourier transform

$$\sum_{\ell=-\infty}^{\infty} s(t + \ell T) \leftrightarrow S(f)\frac{1}{T} \sum_{\ell=-\infty}^{\infty} \delta\left(f + \frac{\ell}{T}\right)$$

(as derived in the next section). Expressing the left side as an inverse Fourier transform of the right side gives

$$\sum_{\ell=-\infty}^{\infty} s(t + \ell T) = \int_{-\infty}^{\infty} e^{j2\pi ft}\left[\frac{1}{T} \sum_{\ell=-\infty}^{\infty} S(f)\delta\left(f + \frac{\ell}{T}\right)\right] df$$

$$= \frac{1}{T} \sum_{\ell=-\infty}^{\infty} \int_{-\infty}^{\infty} e^{j2\pi ft} S(f)\delta\left(f + \frac{\ell}{T}\right) df$$

$$= \frac{1}{T} \sum_{\ell=-\infty}^{\infty} e^{j2\pi \ell t/T} S\left(\frac{\ell}{T}\right)$$

as was to be proved.   ∎

1.6   Transforms of some useful functions

A list of some useful one-dimensional Fourier transform pairs is given in Table 1.1. Some of the entries in this table are developed in this section. Other entries will arise later in the book.

The most elementary pulse is the *rectangular pulse*, defined as

$$\text{rect}(t) = \begin{cases} 1 & \text{if } |t| \leq 1/2 \\ 0 & \text{if } |t| > 1/2. \end{cases}$$

The Fourier transform of a rectangular pulse is readily evaluated. Let

$$s(t) = \text{rect}\left(\frac{t}{T}\right).$$

Table 1.1. *A table of one-dimensional Fourier transform pairs*

$s(t)$	$S(f)$
rect(t)	sinc(f)
sinc(t)	rect(f)
$\delta(t)$	1
1	$\delta(f)$
$e^{-\pi t^2}$	$e^{-\pi f^2}$
$e^{j\pi t^2}$	$\frac{1+j}{\sqrt{2}} e^{-j\pi f^2}$
$\cos \pi t$	$\frac{1}{2}\delta\left(f - \frac{1}{2}\right) + \frac{1}{2}\delta\left(f + \frac{1}{2}\right)$
$\sin \pi t$	$\frac{j}{2}\delta\left(f - \frac{1}{2}\right) - \frac{j}{2}\delta\left(f + \frac{1}{2}\right)$
$e^{-|t|}$	$\frac{2}{1+(2\pi f)^2}$
$\text{comb}_N(t)$	$\text{dirc}_N(f)$
$\text{comb}(t)$	$\text{comb}(f)$

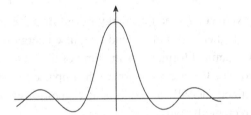

Figure 1.8. The sinc function.

Then

$$S(f) = \int_{-T/2}^{T/2} e^{-j2\pi ft}\, dt$$

$$= \frac{1}{-j2\pi f}\left[e^{-j\pi fT} - e^{j\pi fT} \right]$$

$$= T\,\text{sinc}(fT)$$

where the *sinc function* or *sinc pulse*, illustrated in Figure 1.8, is defined as

$$\text{sinc}(x) = \frac{\sin \pi x}{\pi x}.$$

This Fourier transform pair is written concisely as

$$\text{rect}\left(\frac{t}{T}\right) \leftrightarrow T\text{sinc}(fT).$$

By the duality property of the Fourier transform, we can immediately write

$$W\text{sinc}(tW) \leftrightarrow \text{rect}\left(\frac{f}{W}\right).$$

The next example is the *impulse function* $\delta(t)$. Though the impulse function is not a true function,[2] it is very useful in formal manipulations. It can be defined in a symbolic way as follows. A rectangular pulse of unit area and width T is given by

$$s(t) = \frac{1}{T}\text{rect}\left(\frac{t}{T}\right)$$

$$= \begin{cases} \frac{1}{T} & |t| \leq T/2 \\ 0 & |t| > T/2. \end{cases}$$

This pulse has a unit area for every value of T, and the energy E_p of this pulse is $1/T$. As T decreases, the pulse becomes higher and thinner, and the pulse energy goes to infinity. The impulse function $\delta(t)$ is defined formally as the limit of this sequence of rectangular pulses as T goes to zero. The Fourier transform of the impulse function of $\delta(t)$ does not exist, but it can be defined formally as the limit as T goes to zero of the sequence of Fourier transforms of rectangular pulses.

Thus, we write

$$\delta(t) \leftrightarrow 1$$

to indicate that

$$S(f) = \lim_{T \to 0} \frac{1}{T}(T\text{sinc}\,fT)$$

$$= 1.$$

Because the energy E_p of $\delta(t)$ is not finite, nor is $\delta(t)$ a proper function, this is not a proper Fourier transform pair in the sense of the original definition. However, because this improper pair is frequently useful in formal manipulations, the notion of a Fourier transform is enlarged by appending this pair to the list of Fourier transform pairs.

---

[2] The impulse function is an example of a *generalized function*. Generalized functions are created to enlarge the notion of a function so that certain converging sequences of functions do have a limit. A formal theory of generalized functions has been developed. The impulse function, and its properties, properly belong to that theory.

A doublet of impulses, given by $\delta(t-\frac{1}{2})+\delta(t+\frac{1}{2})$ has Fourier transform $e^{-j\pi f}+e^{j\pi f}$, as follows from the delay property. More generally, a finite train of N regularly spaced impulses, called a *finite comb function*, and denoted

$$\text{comb}_N(t) = \sum_{\ell=0}^{N-1} \delta(t - \ell + \tfrac{1}{2}(N-1))$$

has a Fourier transform

$$S(f) = \sum_{\ell=0}^{N-1} e^{-j2\pi(\ell-\frac{1}{2}(N-1)f)}.$$

Using the identity $\sum_{\ell=0}^{N-1} x^\ell = \frac{1-x^N}{1-x}$, the summation can be reduced to

$$S(f) = \frac{e^{j\pi Nf} - e^{-j\pi Nf}}{e^{j\pi f} - e^{-j\pi f}}$$

$$= \frac{\sin \pi Nf}{\sin \pi f}.$$

The right side is known as a *dirichlet function*, which is denoted $\text{dirc}_N(x)$ and defined by

$$\text{dirc}_N(x) = \frac{\sin \pi Nx}{\sin \pi x}.$$

For integer values of x, the dirichlet function has value $\pm N$ (the sign depending on whether N is odd or even). It has its first zeros at $x = \pm 1/N$. The dirichlet function is small if x differs from an integer by more than a few multiples of $1/N$. For any integer i, $\text{dirc}_N(x + i) = \pm \text{dirc}_N(x)$.

Informally, as N goes to infinity, the train of impulses is known as the *comb function*, which is defined by

$$\text{comb}(x) = \sum_{\ell=-\infty}^{\infty} \delta(t - \ell).$$

The Fourier transform of the improper function $\text{comb}(t)$ is $\text{comb}(f)$. This improper Fourier transform pair is informally justified by regarding $\text{comb}(t)$ as the limit as N goes to infinity, N odd, of $\text{comb}_N(t)$. Then the Fourier transform of $\text{comb}(t)$ is the limit of $\text{dirc}_N(f)$, N odd, as N goes to infinity. For integer values of f, this is the limit of N as N goes to infinity. For other values of f, the limit is zero. Thus, in the limit, the Fourier transform is the comb function.

Other simple pulses are shown in Figure 1.9. Some of these pulses are concentrated in a time interval, say between $-T/2$ and $T/2$, and are equal to zero outside of this

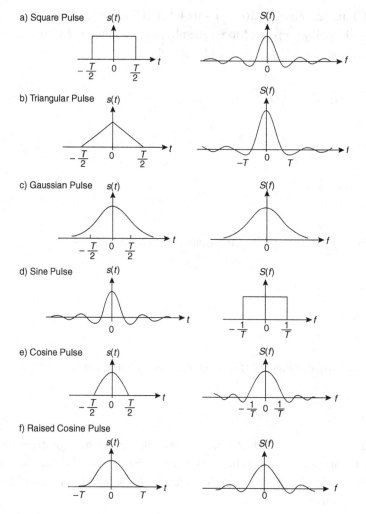

a) Square Pulse $s(t)$ $S(f)$

b) Triangular Pulse $s(t)$ $S(f)$

c) Gaussian Pulse $s(t)$ $S(f)$

d) Sine Pulse $s(t)$ $S(f)$

e) Cosine Pulse $s(t)$ $S(f)$

f) Raised Cosine Pulse $s(t)$ $S(f)$

Figure 1.9. Some simple pulses and their spectra.

interval. For example, the rectangular pulse and the triangular pulse are of this form. Other pulses, such as the sinc pulse, have tails that go on indefinitely. The gaussian pulse

$$s(t) = e^{-\pi t^2}$$

also has tails that go on indefinitely. The Fourier transform of the gaussian pulse is

$$S(f) = \int_{-\infty}^{\infty} e^{-\pi t^2} e^{-j2\pi ft} dt.$$

Because $s(t)$ is an even function, this integral can be written

$$S(f) = \int_{-\infty}^{\infty} e^{-\pi t^2} \cos 2\pi f t \, dt$$

$$= e^{-\pi f^2}$$

where the last line follows from consulting a table of definite integrals. Thus we have the following elegant Fourier transform pair

$$e^{-\pi t^2} \leftrightarrow e^{-\pi f^2}.$$

More generally, using the scaling property of the Fourier transform, we have

$$e^{-at^2} \leftrightarrow \sqrt{\frac{\pi}{a}} e^{-\pi^2 f^2/a}.$$

The rectangular pulse, the sinc pulse, and the gaussian pulse all have finite energy, and so each has a Fourier transform. Other pulses, such as the impulse, have infinite energy and so do not have a Fourier transform. Often, however, (as for the comb function) it is possible to enlarge the definition of the Fourier transform by appending appropriate limits of sequences. Thus, the *chirp pulse*

$$s(t) = e^{j\pi t^2}$$

has infinite energy. However, the finite chirp pulse,

$$s_T(t) = e^{j\pi t^2} \text{rect}\left(\frac{t}{T}\right),$$

has finite energy and so has a Fourier transform $S_T(f)$ for every value of T. The Fourier transform of $s(t)$ can be defined informally as the limit of the transform $S_T(f)$ as T goes to infinity. This limit can be computed directly by writing

$$S(f) = \int_{-\infty}^{\infty} e^{j\pi t^2} e^{-j2\pi f t} \, dt.$$

To evaluate the integral complete the square in the exponent

$$S(f) = e^{-j\pi f^2} \int_{-\infty}^{\infty} e^{j\pi(t-f)^2} \, dt.$$

By a change of variables in the integral, and using Euler's formula, this becomes

$$S(f) = e^{-j\pi f^2} \left[\int_{-\infty}^{\infty} \cos \pi t^2 dt + j \int_{-\infty}^{\infty} \sin \pi t^2 dt \right].$$

The integrals are now standard tabulated integrals, and each is equal to $1/\sqrt{2}$. Therefore the Fourier transform of the chirp pulse is

$$S(f) = \frac{1+j}{\sqrt{2}} e^{-j\pi f^2}$$

as is listed in Table 1.1.

Notice that in the foregoing derivation, as in the derivation of the Fourier transform of an impulse, the Fourier transform of a limit of a sequence of functions is defined as the limit of the sequence of Fourier transforms. This technique, used with care, is a very useful way to enlarge the set of Fourier transform pairs.

1.7   Gaussian random variables

The analysis of the performance of a digital communication system involves the computation of the probability of bit error when the received signal is corrupted by noise, interference, and other impairments. Thus, the methods of probability theory are essential, as is the theory of random processes. The most important random variables in this book are gaussian random variables, and the most important random processes are gaussian random processes. In this section, we shall briefly review the fundamental principles of the theory of gaussian random variables.

Any continuous real random variable X can be described by its probability density function $p(x)$. This function is defined so that the probability that the random variable X lies in the interval $[a, b]$ is given by

$$\Pr[X \in [a, b]] = \int_a^b p(x) dx.$$

A *gaussian random variable* X is a real random variable that is described by a gaussian probability density function. The gaussian probability density function of mean $\bar{x} = \mathrm{E}[x]$ and variance $\sigma^2 = \mathrm{E}[(x - \bar{x})^2]$ is given by

$$p(x) = \frac{1}{\sqrt{2\pi\sigma^2}} e^{-(x-\bar{x})^2/2\sigma^2}.$$

We shall want to know the probability that a gaussian random variable Y exceeds a value x. This is expressed in terms of a function[3] $Q(x)$, defined as

$$Q(x) = \int_x^\infty \frac{1}{\sqrt{2\pi}} e^{-y^2/2} dy.$$

This integral cannot be expressed in closed form. It can be integrated numerically and is widely tabulated.

For positive values of x, the function $Q(x)$ can be bounded as

$$Q(x) \leq \frac{e^{-x^2/2}}{x\sqrt{2\pi}}$$

which is sometimes helpful in understanding the behavior of expressions involving $Q(x)$.

The probability that the random variable X with mean $\bar{x}$ and variance σ^2 exceeds the threshold Θ is given by

$$p = \int_\Theta^\infty \frac{1}{\sqrt{2\pi\sigma^2}} e^{-(x-\bar{x})^2/2\sigma^2}$$

$$= Q\left(\frac{\Theta - \bar{x}}{\sigma}\right).$$

A pair of jointly gaussian random variables (X, Y) is described by the bivariate gaussian probability density function

$$p(x, y) = \frac{1}{2\pi|\Sigma|} e^{-(x-\bar{x}, y-\bar{y})\Sigma^{-1}(x-\bar{x}, y-\bar{y})^T}$$

with mean $(\bar{x}, \bar{y}) = E[(x, y)]$ and covariance matrix $E[(x-\bar{x}, y-\bar{y})^T(x-\bar{x}, y-\bar{y})] = \Sigma$, where $|\Sigma|$ denotes the determinant of Σ.

To find the probability that a bivariate gaussian random variable (X, Y) with mean $(\bar{x}, \bar{y})$ and diagonal covariance matrix $\Sigma = \sigma^2 I$ lies in the half plane defined to be on the side of the line $ax + by = c$ that does not include the point $(\bar{x}, \bar{y})$, simply translate and rotate the coordinate system so that, in the new coordinate system, the mean $(\bar{x}, \bar{y})$

---

[3] The function $Q(x)$ is closely related to a function known as the *complementary error function*

$$\text{erfc}(x) = \frac{2}{\sqrt{\pi}} \int_x^\infty e^{-y^2} dy.$$

The *error function* $\text{erf}(x) = 1 - \text{erfc}(x)$ is a widely tabulated function.

is at the origin, and the straight line becomes the vertical line at $x = d/2$. Then

$$p = \int_{d/2}^{\infty} \int_{-\infty}^{\infty} \frac{1}{2\pi\sigma^2} e^{-x^2/2\sigma^2} e^{-y^2/2\sigma^2} \, dx \, dy$$

$$= Q\left(\frac{d}{2\sigma}\right)$$

where $d/2$ is the perpendicular distance from the straight line to the point $(\bar{x}, \bar{y})$. (This useful fact will be restated in Section 6.10 as Theorem 6.10.1 at which time the plane is partitioned into two half planes with respect to two points separated by distance d.)

1.8   Circular random processes

The study of noise in a receiver requires the introduction of the topic of random processes. Most of our needs regarding random processes will be met by an understanding of the important class of random processes known as stationary, complex baseband random processes. These are largely studied by means of their correlation functions and by the Fourier transform of the correlation function, known as the power density spectrum.

A stationary, real baseband random process $X(t)$ has an autocorrelation function (or, more simply, correlation function) defined as

$$R_{XX}(\tau) = E[X(t)X(t + \tau)].$$

The correlation function is independent of t, and always exists for stationary random processes. It also exists for some other random processes that are not stationary. Accordingly, a random process $X(t)$ for which $E[X(t)X(t + \tau)]$ is independent of t is called a *covariance-stationary* random process.

The cross-correlation function between two stationary random processes $X(t)$ and $Y(t)$ is given by

$$R_{XY}(\tau) = E[X(t)Y(t + \tau)].$$

It is clear that $R_{YX}(\tau) = R_{XY}(-\tau)$.

A stationary, complex baseband, random process, denoted $Z(t) = X(t) + jY(t)$, where $X(t)$ and $Y(t)$ are real random processes, has an autocorrelation function given by

$$R_{ZZ}(\tau) = E[Z^*(t)Z(t + \tau)]$$

$$= R_{XX}(\tau) + R_{YY}(\tau) + j[R_{XY}(\tau) - R_{XY}(-\tau)]$$

using the elementary fact that $R_{YX}(\tau) = R_{XY}(-\tau)$.

It is generally not possible to recover $R_{XX}(\tau)$, $R_{YY}(\tau)$, and $R_{XY}(\tau)$ from $R_{ZZ}(\tau)$. Accordingly, define

$$\tilde{R}_{ZZ}(\tau) = E[Z(t)Z(t+\tau)]$$
$$= R_{XX}(\tau) - R_{YY}(\tau) + j[R_{XY}(\tau) + R_{XY}(-\tau)].$$

Then

$$R_{XX}(\tau) = \tfrac{1}{2}\mathrm{Re}\,[R_{ZZ}(\tau) + \tilde{R}_{ZZ}(\tau)]$$
$$R_{YY}(\tau) = \tfrac{1}{2}\mathrm{Re}\,[R_{ZZ}(\tau) - \tilde{R}_{ZZ}(\tau)]$$
$$R_{XY}(\tau) = \tfrac{1}{2}\mathrm{Im}\,[R_{ZZ}(\tau) + \tilde{R}_{ZZ}(\tau)].$$

The most common random processes that occur in the study of modems are the random processes caused by thermal noise in the receiver. Thermal noise is always gaussian, and ordinarily stationary. We shall have frequent occasion to encounter gaussian noise at baseband, at passband, and at complex baseband.

Definition 1.8.1 *A circular random process is a complex baseband, covariance stationary random process for which*

$$R_{XX}(\tau) = R_{YY}(\tau)$$
$$R_{XY}(\tau) = 0.$$

It is straightforward to verify that the circular property of a stationary random process is preserved under the operations of filtering, modulation, and sampling. In particular, the circular property is preserved under multiplication by $e^{j\theta}$. We shall often use complex, circular, and stationary gaussian noise in this book because complex thermal noise is complex, circular, and stationary gaussian noise.

Problems for Chapter 1

1.1.   a. Derive the Fourier transform of the trapezoidal pulse given in the following illustration.
       (**Hint:** The trapezoidal pulse can be obtained by convolving two rectangular pulses.)

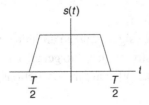

b. Derive the Fourier transform of the cosine pulse

$$s(t) = \begin{cases} \cos \pi t/T & |t| \le T/2 \\ 0 & \text{otherwise} \end{cases}$$

by using the modulation theorem combined with the formula for the Fourier transform of a rectangular pulse.

1.2. Prove that

$$\frac{2}{\pi^2} \int_0^\infty \frac{\sin(ax)\sin(bx)}{x^2} dx = \min(a,b).$$

1.3. a. A die has six distinguishable faces. How many bits are needed to specify the outcome of a roll of a die? How many bits are needed to specify the outcome of a roll of a pair of dice? Does the answer depend on whether the two dice are identical or not?

b. How many bits are needed to specify a single letter of English (from an alphabet of 26 letters)? How many bits are needed to specify an ordered pair of letters of English?

c. A deck of cards is a set of 52 distinguishable elements. How many bits does it take to describe a selection of one card from a deck? How many bits does it take to describe a selection of two cards from a deck? How many bits does it take to describe a selection of 26 cards from a deck?

1.4. If $S(f)$ and $R(f)$ are the Fourier transforms of $s(t)$ and $r(t)$, what is the Fourier transform of $s(t)r(t)$?

1.5. A sequence of binary digits is transmitted in a certain communication system. Any given digit is received in error with probability p and received correctly with probability $1 - p$. Errors occur independently from digit to digit. Out of a sequence of n transmitted digits, what is the probability that no more than j of these are received in error?

1.6. The operation known as "mixing" is a fundamental operation of modulation and demodulation. It consists of multiplying a passband signal

$$s(t) = a(t) \cos 2\pi f_0 t$$

by $\cos 2\pi (f_0 - f_1)t$ followed by a filtering operation. Show that, under suitable conditions, mixing changes the carrier frequency f_0 to f_1. What happens if $s(t)$ is instead multiplied by $\sin 2\pi (f_0 - f_1)t$?

1.7. Given a device that squares its input signal, and an unlimited number of adders and filters, show how one can construct a mixer to generate $a(t) \cos 2\pi f_1 t$ from $a(t) \cos 2\pi f_0 t$ and $\cos 2\pi (f_0 - f_1)t$. What are the restrictions on f_0, f_1, and the spectrum of $a(t)$?

1.8. Find a set of sixteen binary words of blocklength seven such that every word of the set differs from every other word in at least three bit positions. Explain how this code can be used to transmit four bits at a time through a binary channel that never makes more than one bit error in a block of seven bits.

1.9. Show that the amplitude and phase of a passband waveform

$$\tilde{v}(t) = v_R(t) \cos 2\pi f_0 t - v_I(t) \sin 2\pi f_0 t$$

are the same as the amplitude and phase of the complex representation

$$v(t) = v_R(t) + j v_I(t)$$

of that passband waveform. If the phase of the passband waveform is offset by ϕ, how does the phase of the complex representation change? If the reference frequency of the passband waveform is changed by Δf, what happens to the complex representation?

1.10. a. Show that $Q(x) < \frac{1}{2} e^{-x^2/2}$ for $x > 0$.

   Hint: First show that $t^2 - x^2 > (t - x)^2$ for $t > x > 0$. Then apply this inequality to

$$e^{x^2/2} Q(x) = \frac{1}{\sqrt{2\pi}} \int_x^\infty e^{-(t^2 - x^2)/2} dt.$$

   b. Show that $Q(x) < e^{-x^2/2}/x\sqrt{2\pi}$ for $x > 0$.

   Hint: First show that

$$\int_x^\infty \frac{x - t}{\sqrt{2\pi}} e^{-t^2/2} dt \le 0.$$

   c. Estimate the tightness of the bound in part b when x is large.

1.11. a. Prove that if random variables X and Y are independent, they are uncorrelated.

   b. Prove that if random variables X and Y are uncorrelated and gaussian, they are independent.

1.12. a. Prove that for any vector random variable X^n, the covariance matrix Σ is positive definite. That is, show that

$$a^T \Sigma a \ge 0$$

   for any vector a of blocklength n.

   b. Prove, for any random variables X, Y, and Z, that

$$\mathrm{cov}(aX + bY + c, dZ + e) = ad\,\mathrm{cov}(X, Z) + bd\,\mathrm{cov}(Y, Z).$$

1.13. (Function of a random variable.) Let $y = f(x)$ where $f(x)$ is a monotonic function. Let g be the inverse of $f(x)$ so that $g(f(x)) = x$. If X is a real-valued random variable with probability density function $p_X(x)$, prove that Y is a random variable with probability density function $p_Y(y)$ given by

$$p_Y(y) = p_X(g(y)) \frac{dg(y)}{dx}.$$

1.14. Prove that the circular property of a stationary, complex baseband random process is preserved under the operations of filtering, modulation, and sampling.

1.15. Let $s(t)$ be a waveform, possibly complex, whose energy is finite. Let

$$S(f) = \int_{-\infty}^{\infty} s(t) e^{-j2\pi ft} dt$$

be the Fourier transform of $s(t)$. The sampled version of $s(t)$ is

$$s_\delta(t) = \sum_{\ell=-\infty}^{\infty} \delta(t - \ell T) s(t)$$

where $\delta(t)$ is the delta function. Prove that $s_\delta(t)$ has the Fourier transform

$$S_\delta(f) = \frac{1}{T} \sum_{\ell=-\infty}^{\infty} S(f - \ell/T).$$

1.16. The *central limit theorem* says that if $X_1, X_2, \ldots, X_n$ are independent, identically distributed random variables with zero mean and finite variance σ^2, then $Z_n = \frac{1}{\sqrt{n}}(X_1 + X_2 + \cdots + X_n)$ has a probability density function that approaches the gaussian distribution with zero mean and variance σ^2.

a. Explain why this suggests that the gaussian distribution is the (only) fixed point of the following recursion of probability density functions:

$$p_{2^n}(\sqrt{2}x) = p_{2^{n-1}}(x) * p_{2^{n-1}}(x).$$

b. Compute and graph the terms of the recursion starting with $p_1(x) = \text{rect}(tx)$. How many iterations does it take to make your graph visually identical to a gaussian?

2   Baseband Modulation

A *waveform channel* is a channel whose inputs are continuous functions of time. A *baseband channel* is a waveform channel suitable for an input waveform that has a spectrum confined to an interval of frequencies centered about the zero frequency. In this chapter, we shall study the design of waveforms and modulators for the baseband channel.

The function of a digital modulator is to convert a digital datastream into a waveform representation of the datastream that can be accepted by the waveform channel. The waveform formed by the modulator is designed to accommodate the spectral characteristics of the channel, to obtain high rates of data transmission, to minimize transmitted power, and to keep the bit error rate small.

A modulation waveform cannot be judged independently of the performance of the demodulator. To understand how a baseband communication system works, it is necessary to study both the baseband modulation techniques of this chapter and the baseband demodulation techniques of Chapter 3. The final test of a modem is in the ability of the demodulator to recover the symbols of the input datastream from the channel output signal when received in the presence of noise, interference, distortion, and other impairments.

2.1   Baseband and passband channels

A waveform channel is a channel whose input is a continuous function of time, here denoted $c(t)$, and whose output is another function of time, here denoted $v(t)$. A *linear channel* is one that satisfies the *superposition principle*: if input $c(t)$ causes output $v(t)$ and input $c'(t)$ causes output $v'(t)$, then for any real numbers a and b, input $ac(t) + bc'(t)$ causes output $av(t) + bv'(t)$.

Every linear, time-invariant channel can be described as a linear filter. A linear baseband channel is depicted in Figure 2.1 as a linear filter with impulse response $h(t)$ and transfer function $H(f)$, which is the Fourier transform of $h(t)$. The output of the linear channel $h(t)$ is given by the convolution

$$v(t) = \int_{-\infty}^{\infty} h(\xi)c(t - \xi)d\xi.$$

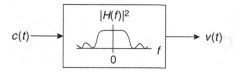

Figure 2.1. A baseband transfer function.

In the frequency domain, the convolution becomes the product

$$V(f) = H(f)C(f)$$

as stated by the convolution theorem.

A *baseband channel* is a linear channel $h(t)$ for which the support of $H(f)$ is a finite interval of the frequency axis that contains the origin. In practice, this condition is relaxed to require only that $H(f)$ is large at frequencies in the vicinity of the zero frequency, and negligible at frequencies far away from zero. A *passband channel* is a linear channel for which the support of $H(f)$ is confined to two finite intervals centered at frequencies $\pm f_0$, where f_0 is large compared to the width of each of the intervals. In practice, this condition is relaxed to require only that $H(f)$ be negligible outside of these two intervals centered at $\pm f_0$.

In this chapter, we shall usually suppress consideration of the channel by setting $H(f) = 1$. This assumption is the same as assuming that $H(f)$ is at least as wide as $C(f)$ and equal to one whenever $C(f)$ is significantly different from zero. Then $V(f)$ is equal to $C(f)$, and the precise shape of $H(f)$ at other frequencies is not important. On occasion, as in Chapter 9, we will impose other requirements on $H(f)$, and $H(f)$ will play a larger role in the discussion.

2.2   Baseband waveforms for binary signaling

The purpose of the modulator is to convert a stream of digital data into a waveform $c(t)$ that is suitable to pass through the channel. Usually, we shall think of the input to the modulator as a serial stream of binary data flowing at a constant rate, say one input data bit every T_b seconds.

A pulse, $s(t)$, is a real function of time whose energy, defined by

$$E_p = \int_{-\infty}^{\infty} s^2(t)dt$$

is finite. Often we prefer to display a separate amplitude, expressed as a free parameter A, in addition to the pulse shape $s(t)$. Then the pulse will be denoted $As(t)$ and the energy is

$$E_p = \int_{-\infty}^{\infty} A^2 s^2(t)dt.$$

Although it may seem simpler to simply use $s(t)$ for the pulse, it will prove to be useful to have the pulse amplitude A displayed separately so that the pulse $s(t)$ can be chosen to be without physical units. The physical units such as volts or amperes can be attached to the amplitude A. When we choose to use the latter form, we may choose to normalize $s(t)$ so that $\int_{-\infty}^{\infty} s^2(t)dt = 1$. Then $E_p = A^2$.

A simple way to modulate the datastream into a channel waveform is to map the datastream into a concatenation of pulses, called a *modulation waveform*, as described by the equation

$$c(t) = \sum_{\ell=-\infty}^{\infty} a_\ell s(t - \ell T)$$

where a_ℓ depends on the ℓth data bit. The equation for $c(t)$ can be regarded as a prescription for converting discrete time to continuous time. The discrete-time sequence a_ℓ for $\ell = \ldots, -1, 0, 1, \ldots$ representing the sequence of data bits is converted into a continuous-time waveform $c(t)$ using the pulse shape $s(t)$ to fill in the time axis.

Whenever we prefer, we may instead write the one-sided expression

$$c(t) = \sum_{\ell=0}^{\infty} a_\ell s(t - \ell T)$$

to show the datastream beginning at time zero.

One way to define a_ℓ is

$$a_\ell = \begin{cases} 1 & \text{if the } \ell\text{th data bit is a 1} \\ -1 & \text{if the } \ell\text{th data bit is a 0.} \end{cases}$$

Alternatively, when $s(t)$ is normalized so that $\int_{-\infty}^{\infty} |s(t)|^2 dt = 1$, we may define

$$a_\ell = \begin{cases} A & \text{if the } \ell\text{th data bit is a 1} \\ -A & \text{if the } \ell\text{th data bit is a 0.} \end{cases}$$

This waveform $c(t)$ is called an *antipodal signaling waveform* or, when at passband, a *binary phase-shift keyed signaling waveform* (BPSK). An example of an antipodal waveform using triangular pulses is shown in Figure 2.2.

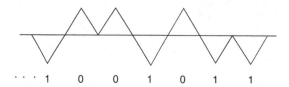

Figure 2.2. Example of an antipodal waveform.

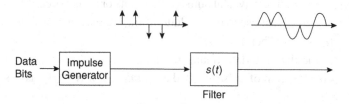

Figure 2.3. Functional description of a modulator for antipodal signaling.

The energy per transmitted bit, denoted E_b, is equal to the energy E_p in the pulse $As(t)$ because there is one bit for every pulse. If A is the amplitude of the transmitted pulse and $s(t)$ has unit energy, then

$$E_b = \int_{-\infty}^{\infty} A^2 s^2(t) dt = A^2.$$

We shall usually prefer to use this convention in which the energy in the pulse $s(t)$ is equal to one and the bit energy is carried by the term A^2.

One can choose any convenient pulse shape for an antipodal signaling waveform. The distinction between the role of the pulse shape and the decision to use the sign of the pulse as the method of modulating data bits is emphasized by writing an equivalent mathematical expression for the waveform as

$$c(t) = \left[\sum_{\ell=-\infty}^{\infty} a_\ell \delta(t - \ell T) \right] * s(t).$$

With this description, the signs of the impulses $a_\ell \delta(t - \ell T)$ convey the data bits; the stream of impulses is then passed through a filter with impulse response $s(t)$ in order to create a practical waveform formed using pulse $s(t)$. A functional description of an antipodal modulator is shown in Figure 2.3.

An alternative way to relate the pulse amplitude a_ℓ to the ℓth data bit is as follows:

$$a_\ell = \begin{cases} A & \text{if the } \ell\text{th data bit is a 1} \\ 0 & \text{if the } \ell\text{th data bit is a 0.} \end{cases}$$

1   1   0   1   0   0   1   ···

Figure 2.4. Example of an OOK waveform.

1   1   0   1   0   0   1   ···

Figure 2.5. Example of an NRZ waveform.

This waveform is called a *binary on–off-keyed* (OOK) *signaling waveform*. An example of an OOK waveform is shown in Figure 2.4. The energy needed to transmit a one is E_p, while no energy is needed to transmit a zero. If the data bits are half zeros and half ones, the average energy per transmitted bit is

$$E_b = \frac{1}{2} \int_{-\infty}^{\infty} A^2 s^2(t)dt + \frac{1}{2}0$$

$$= \frac{1}{2}A^2.$$

It might appear from this simple analysis that OOK uses less energy than antipodal signaling. This is so if A is held fixed, but we shall see in Chapter 3 that, in order to combat noise, a larger value of A must be used for OOK than for antipodal signaling.

An on–off-keyed waveform using rectangular pulses is shown in Figure 2.5. In this example, the pulse width is equal to the pulse spacing T. This special case of OOK is called a *nonreturn-to-zero* (NRZ) signaling waveform, which is often used in applications in which waveform simplicity is of first importance. The disadvantage of NRZ is the same as any waveform using a rectangular pulse; the spectrum $S(f)$ has sidelobes that fall off very slowly, as $1/f$. A large bandwidth is needed to support the leading and trailing edges of the pulses. Accordingly, to transmit NRZ requires a large bandwidth in comparison to the data rate. This requirement is not acceptable in most modern communication systems.

2.3   Baseband waveforms for multilevel signaling

Instead of modulating one bit at a time at each signaling instant ℓT, as in antipodal signaling, one can modulate several bits at a time at each signaling instant. Thus in place of binary signaling, we shall introduce M-ary signaling. There are two main variations

Figure 2.6. Some signal constellations on the real line.

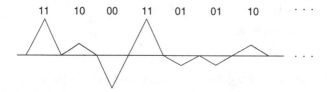

Figure 2.7. Example of a multilevel baseband waveform.

of M-ary signaling. In Section 2.8, we shall study the M-ary orthogonal pulse families. This method of signaling requires bandwidths that grow larger with M; increasing M is paid for with more bandwidth. In the present section, we shall study the opposite case in which bandwidth is constrained. Only one pulse shape $s(t)$ is used, and there are M pulse amplitudes. Figure 2.6 shows some alphabets of pulse amplitudes. These alphabets are called *signal constellations* and are denoted $\mathcal{S}$. The signal constellation $\mathcal{S} = \{c_0, c_1, \ldots, c_{M-1}\}$ on the real line is a finite set of M real numbers; usually $M = 2^k$ for some integer k. The uniformly-spaced four-ary signal constellation, for example, is $\mathcal{S} = \{-3A, -A, A, 3A\}$.

Each point of the signal constellation $\mathcal{S}$ represents one of the allowable amplitudes of the transmitted pulse. The modulation waveform, called a *multilevel signaling waveform*, is

$$c(t) = \sum_{\ell=-\infty}^{\infty} a_\ell s(t - \ell T) \qquad a_\ell \in \mathcal{S}$$

where $\int_{-\infty}^{\infty} s(t)^2 dt = 1$. To use the signal constellation, one must choose a pulse $s(t)$ and assign a k-bit binary number to each of the 2^k points of the signal constellation. The k data bits at the ℓth time are then represented by a pulse of the amplitude specified by those k bits.

An example of a four-ary waveform, using a triangular pulse and the signal constellation $\mathcal{S} = \{-3A, -A, A, 3A\}$, is shown in Figure 2.7. This waveform $c(t)$ can be expressed as

$$c(t) = \sum_{\ell=-\infty}^{\infty} a_\ell s(t - \ell T)$$

where

$$a_\ell = \begin{cases} -3A & \text{if the } \ell\text{th pair of data bits is 00} \\ -A & \text{if the } \ell\text{th pair of data bits is 01} \\ A & \text{if the } \ell\text{th pair of data bits is 10} \\ 3A & \text{if the } \ell\text{th pair of data bits is 11.} \end{cases}$$

This waveform, called an *amplitude-shift keyed* (ASK) signaling waveform or a *pulse-amplitude modulation* (PAM) signaling waveform, transmits two data bits in time T by using the four-ary signal constellation on the real line from Figure 2.6 and a triangular pulse. In the same way, one could design a four-ary ASK waveform using this same signal constellation, but using a different pulse for $s(t)$.

In this example, the data patterns have been assigned to points of the signal constellation in the natural counting order. This means that 01 and 10 are assigned to adjacent points. If an error confusing these two points is made, it results in two bit errors. An alternative assignment is to use a *Gray code*

$$a_\ell = \begin{cases} -3A & \text{if the } \ell\text{th pair of data bits is 00} \\ -A & \text{if the } \ell\text{th pair of data bits is 01} \\ A & \text{if the } \ell\text{th pair of data bits is 11} \\ 3A & \text{if the } \ell\text{th pair of data bits is 10.} \end{cases}$$

Now a single symbol error rarely results in two bit errors. The use of a Gray code gives only a trivial performance improvement, but it does so for very little cost.

The average energy per symbol of the signal constellation $\{c_0, c_1, \ldots, c_{M-1}\}$ with $M = 2^k$ is

$$E_s = \frac{1}{M} \sum_{m=0}^{M-1} c_m^2$$

under the assumption that each symbol is used with the same probability. Because there are $k = \log_2 M$ bits in a symbol, the average energy per bit of the signal constellation is

$$E_b = \frac{1}{k} E_s.$$

The expected energy E_L contained in a block of L symbols is the expectation

$$
E_L = \int_{-\infty}^{\infty} \mathrm{E} \left[\sum_{\ell=0}^{L} a_\ell s(t - \ell T) dt \right]^2
$$

$$
= \int_{-\infty}^{\infty} \sum_{\ell=0}^{L} \sum_{\ell'=0}^{L} \mathrm{E}[a_\ell a_{\ell'}] s(t - \ell T) s(t - \ell' T) dt
$$

$$
= \sum_{\ell=0}^{L} \sum_{\ell'=0}^{L} \mathrm{E}[a_\ell a_{\ell'}] \int_{-\infty}^{\infty} s(t - \ell T) s(t - \ell' T) dt.
$$

We may assume that the data bits are maximally random. This means that every data pattern is equally probable, and the symbols are independent. Therefore

$$
\mathrm{E}[a_\ell a_{\ell'}] = \begin{cases} E_s & \text{if } \ell = \ell' \\ 0 & \text{if } \ell \neq \ell'. \end{cases}
$$

Finally, we have

$$
E_L = \sum_{\ell=0}^{L} E_s \int_{-\infty}^{\infty} s^2(t - \ell T) dt = L E_s \int_{-\infty}^{\infty} s^2(t) dt.
$$

Using the convention that the pulse $s(t)$ has energy equal to one, the average energy in a block of L symbols is LE_s. The average energy E_s of the k bits resides in the energy of the signal constellation.

For example, for the four-point signal constellation defined above,

$$
E_s = \tfrac{1}{4}[(-3A)^2 + (-A)^2 + (A)^2 + (3A)^2]
$$

$$
= 5A^2.
$$

The expected energy per block is LE_s. Because there are two bits in each symbol and L symbols in the block, the average energy per bit is $5A^2/2$.

2.4   Nyquist pulses

For any choice of signaling waveform, the pulse shape $s(t)$ must be chosen so that the sequence of modulation amplitudes a_ℓ representing the sequence of data bits can be recovered from the signaling waveform

$$
c(t) = \sum_{\ell'=-\infty}^{\infty} a_{\ell'} s(t - \ell' T).
$$

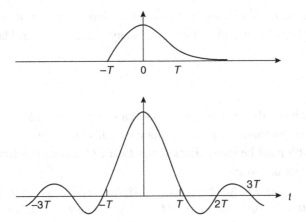

Figure 2.8. Some pulses displaying intersymbol interference.

Let $t = \ell T$. Then

$$c(\ell T) = \sum_{\ell'=-\infty}^{\infty} a_{\ell'} s(\ell T - \ell' T).$$

If $s(\ell T)$ is nonzero for some value of ℓ other than zero, as illustrated by two examples in Figure 2.8, then the pulse $s(t)$ is said to have *intersymbol interference* because $c(\ell T)$ will then depend on $a_{\ell'}$ for some ℓ' other than ℓ. There will be no intersymbol interference if $s(\ell T) = 0$ for all ℓ not equal to zero because then

$$c(\ell T) = \sum_{\ell'=-\infty}^{\infty} a_{\ell'} s(\ell T - \ell' T)$$

$$= a_{\ell}.$$

Definition 2.4.1 *A Nyquist pulse for a signaling interval of duration T is any pulse* $s(t)$ *that satisfies*

$$s(\ell T) = \begin{cases} 1 & \ell = 0 \\ 0 & \ell \neq 0. \end{cases}$$

A Nyquist pulse must be used for signaling whenever one wants to avoid intersymbol interference at the sampling instants ℓT. These *Nyquist samples* are samples of the pulse as it is received, not as it is transmitted. This remark is important whenever the pulse changes shape within the channel. In Chapter 3, we shall require the Nyquist property to hold for a filtered version of $s(t)$ rather than for $s(t)$ itself. This will be appropriate because our concern will be with intersymbol interference in the signal after it is filtered in the receiver and not with intersymbol interference in the transmitted signal.

The rectangular pulse $\text{rect}(t/T)$ is one example of a Nyquist pulse; it has no intersymbol interference because $\text{rect}(\ell T/T) = 0$ if ℓ is any nonzero integer. Then

$$s(\ell T) = \begin{cases} 1 & \ell = 0 \\ 0 & \ell \neq 0. \end{cases}$$

The rectangular pulse can be used only when the channel can support the wide spectrum of a rectangular pulse. The spectrum of the rectangular pulse falls off as $1/f$ – so slowly that the channel bandwidth must be many times larger than $1/T$ in order to transmit the pulse without significant distortion.

The sinc pulse, $\text{sinc}(t/T)$, is another example of a Nyquist pulse. The sinc pulse has no intersymbol interference because $\text{sinc}(\ell T/T) = 0$ for integer $\ell \neq 0$. However, the sinc pulse suffers from the fact that it has infinite duration. More seriously, the amplitude of the sinc pulse falls off only as $1/t$. The amplitude falls off so slowly that in the modulated waveform

$$c(t) = \sum_{\ell=-\infty}^{\infty} a_\ell s(t - \ell T),$$

the distant sidelobes of many pulses will occasionally reinforce each other to produce very large amplitudes in $c(t)$ at times t that are not integer multiples of T, possibly exceeding the linear region of the transmitter or receiver. The possibility of large amplitudes also implies that the signal $c(t)$ may pass through its sample points at ℓT very steeply, which in turn suggests sensitivity to timing errors. Therefore the sinc pulse is not often a useful pulse shape in practice.

The rectangular pulse and the sinc pulse may be regarded as the two extreme cases of pulses, one of which is too spread on the frequency axis, and one of which is too spread on the time axis. Good pulses avoid each of these disadvantages.

To design a practical pulse $s(t)$ that has no intersymbol interference, one must design a Nyquist pulse whose time-domain sidelobes fall off rather quickly. This means that the spectrum $S(f)$ should not have sharp edges. The next theorem, known as the *Nyquist criterion*, shows how the pulse $s(t)$ may be designed in the frequency domain so that there is no intersymbol interference.

Theorem 2.4.2 *The pulse $s(t)$ is a Nyquist pulse if and only if the transform $S(f)$ satisfies*

$$\frac{1}{T} \sum_{n=-\infty}^{\infty} S\left(f + \frac{n}{T}\right) = 1 \quad |f| \leq 1/2T.$$

Proof The inverse Fourier transform is

$$s(t) = \int_{-\infty}^{\infty} S(f)e^{j2\pi ft} df.$$

Therefore

$$s(\ell T) = \int_{-\infty}^{\infty} S(f)e^{j2\pi f \ell T} df.$$

We break up the frequency axis into contiguous intervals of length $1/T$, with the nth interval running from $(2n-1)/2T$ to $(2n+1)/2T$, and then express $s(\ell T)$ as a sum of pieces

$$s(\ell T) = \sum_{n=-\infty}^{\infty} \int_{(2n-1)/2T}^{(2n+1)/2T} S(f)e^{j2\pi f \ell T} df.$$

In the nth term of the sum, replace f by $f + \frac{n}{T}$ and notice that $e^{j2\pi \ell n} = 1$ to write

$$s(\ell T) = \sum_{n=-\infty}^{\infty} \int_{-1/2T}^{1/2T} S\left(f + \frac{n}{T}\right) e^{j2\pi f \ell T} df$$

$$= \int_{-1/2T}^{1/2T} \left[\sum_{n=-\infty}^{\infty} S\left(f + \frac{n}{T}\right)\right] e^{j2\pi f \ell T} df$$

$$= \int_{-1/2T}^{1/2T} S'(f)e^{j2\pi f \ell T} df$$

where $S'(f)$ is equal to the bracketed term in the previous line. Suppose that $S'(f) = \text{rect}\left(\frac{f}{T}\right)$. Then $s(\ell T) = \text{sinc}(\ell)$. Therefore

$$s(\ell T) = \begin{cases} 1 & \text{if } \ell = 0 \\ 0 & \text{if } \ell \neq 0. \end{cases}$$

Moreover, the uniqueness of the Fourier series expansion says that the Fourier coefficients of $S'(f)$ uniquely specify $S'(f)$. Therefore $S'(f)$ can be equal to $\text{rect}\left(\frac{f}{T}\right)$ only if $s(\ell T) = \text{sinc}(\ell)$. Thus there is only one function, namely $S'(f)$, on the stated interval that corresponds to the stated sequence of samples $s(\ell T)$.   ■

Nyquist pulses are pulses that satisfy Theorem 2.4.2. Figure 2.9 illustrates the meaning of the theorem. It shows how translated copies of the spectrum must add to a constant. We may describe this more vividly by saying that "out-of-band" tails of the spectrum "fold" back into the band to form a virtual rectangular spectrum. Often, as in Figure 2.9, the infinite sum on n appearing in Theorem 2.4.2 can be replaced by a finite sum because $S(f)$ is nonzero only on a finite interval of the frequency axis.

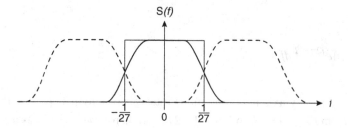

Figure 2.9. Illustrating the Nyquist criterion.

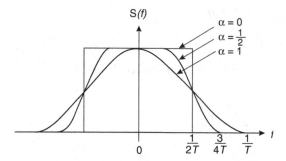

Figure 2.10. Raised-cosine spectra.

Specifically, let W be such that $S(f) = 0$ if $|f| > W$. Then the condition can be written as the finite sum

$$\frac{1}{T} \sum_{n=-N}^{N} S\left(f + \frac{n}{T}\right) = 1 \quad \text{for } |f| \le \frac{1}{2T}$$

where N is the integer $\lfloor 2TW \rfloor$. If

$$\frac{W}{2} < \frac{1}{2T} < W,$$

then N equals 1 in the sum.

Some common examples of Nyquist pulses are those whose Fourier transforms are defined as

$$S(f) = \begin{cases} T & 0 \le |f| \le \frac{1-\alpha}{2T} \\ \frac{T}{2}\left[1 - \sin\frac{\pi T(|f|-1/2T)}{\alpha}\right] & \frac{1-\alpha}{2T} \le |f| \le \frac{1+\alpha}{2T} \\ 0 & |f| \ge \frac{1+\alpha}{2T} \end{cases}$$

where α is a parameter between 0 and 1. These spectra, sometimes called *raised-cosine spectra*, are shown in Figure 2.10. The figure makes it obvious that the raised-cosine

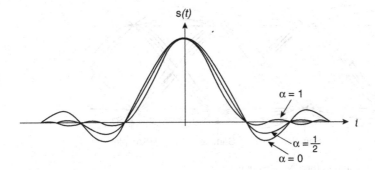

Figure 2.11. Some Nyquist pulses.

spectra satisfy the Nyquist criterion. The corresponding Nyquist pulses $s(t)$, shown in Figure 2.11, are given by

$$s(t) = \frac{\sin \pi t/T}{\pi t/T} \frac{\cos \alpha \pi t/T}{1 - 4\alpha^2 t^2/T^2}.$$

For fixed nonzero α, the tails of the pulse $s(t)$ decay as $1/t^3$ for large $|t|$. Although the pulse tails persist for an infinite time, they are eventually small enough so they can be truncated with only negligible perturbations of the zero crossings.

2.5 Eye diagrams

For an antipodal signaling waveform using the pulse $s(t)$, the set of all possible data-streams corresponds to the set of all possible sequences of the values $a_\ell = \pm A$ for $\ell = \ldots, -1, 0, +1, \ldots$. Therefore, corresponding to the infinite number of possible datastreams, one obtains an infinite number of possible waveforms, each of the form

$$c(t) = \sum_{\ell=-\infty}^{\infty} a_\ell s(t - \ell T).$$

Figure 2.12 shows some of these waveforms superimposed on a common graph using the Nyquist pulse

$$s(t) = \frac{\sin \pi t/T}{\pi t/T} \frac{\cos \pi t/2T}{1 - t^2/T^2}.$$

The superimposed waveforms in Figure 2.12 form what is called an *eye diagram*. Every waveform forming the eye diagram neatly passes through either $+A$ or $-A$ at

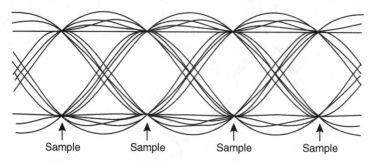

Figure 2.12. Eye diagram for a Nyquist pulse.

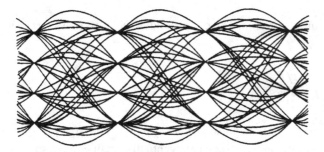

Figure 2.13. Eye diagram for a four-level baseband waveform.

every sampling instant according to the values of its data sequence. Between sampling instants, however, the waveforms take values that depend on the pulse shape $s(t)$ and the specific sequence of bits. Figure 2.12 should make it visually apparent why a Nyquist pulse other than a sinc pulse should be used. The waveforms in the figure do not have unreasonably large values between the sampling instants, but if a sinc pulse were used, some very large excursions of the waveforms would occasionally occur because of the constructive superposition of sidelobes.

Another observation from Figure 2.12 is that the waveforms do not cross the zero line at the same time. If the system uses the zero crossings to synchronize time – as some systems do – there will be some data-dependent time jitter in the local clock.

The eye diagram for a multilevel signaling waveform is defined in the same way. It is the set of all possible modulated waveforms $c(t)$ superimposed on the same graph. The eye diagram for a four-ary ASK waveform, shown in Figure 2.13, consists of the superposition of many signaling waveforms, each of which passes through one of the four values $-3A, -A, A,$ or $3A$ at every sampling instant. As before, the data is found by the value of $c(t)$ at the sampling instants. The values of $c(t)$ between the sampling instants are related to the shape of the spectrum of the Nyquist pulse $s(t)$.

2.6   Differential modulation

An alternative modulation convention for converting a user datastream into a baseband waveform that is sometimes preferred is described in this section. Figure 2.14 shows a waveform in which the data bits are used to define the transitions in the pulse amplitude rather than the amplitude itself. This waveform, which looks like NRZ, is called *NRZ inverse* (NRZI). It can be defined mathematically as

$$c(t) = \sum_{\ell=-\infty}^{\infty} a_\ell s(t - \ell T)$$

where $a_\ell = 0$ if the ℓth data bit is equal to the $(\ell - 1)$th data bit, and otherwise $a_\ell = A$. The pulse $s(t)$ is a rectangular pulse of width T. The NRZI waveform requires a reference bit at the start that does not convey a data bit. If the datastream begins at time zero, say, then set $a_{-1} = 0$.

The NRZI waveform may be used in conjunction with a demodulator that uses a transition detector. This is because a transition detector does not reproduce a transmitted NRZ bitstream. Instead, at each bit time, it produces a zero bit if the current channel waveform level is the same as the previous channel waveform level; otherwise it produces a one. Therefore, this demodulator does recover the datastream modulated into the NRZI waveform.

The behavior of the transition demodulator has something of the nature of a differentiation; indeed, this demodulator is often implemented with a differentiator and a threshold. This kind of demodulator is frequently used in applications where a differentiation occurs naturally as part of the receiver, as in many read heads for magnetic tapes and disks. Since a differentiation is inevitable in such applications, one might choose a waveform that can be so demodulated.

If for an NRZI waveform, however, a demodulator reconstructs instead the bit sequence a_ℓ/A, then the demodulator output datastream is not equal to the input datastream; it has been changed. In this case, the modulator sends (or writes) the data as NRZI data, but the demodulator senses (or reads) the data instead as NRZ data. One might also have the opposite situation where the modulator sends the data as NRZ and the demodulator senses the data as NRZI. In either case, an additional device must be

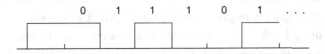

Figure 2.14. Example of an NRZI waveform.

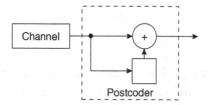

Figure 2.15. Precoder or Postcoder for NRZI.

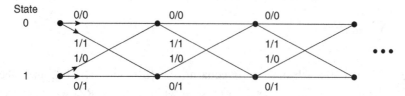

Figure 2.16. Trellis for NRZI.

used to reconvert the data between the two forms because the user will want the data returned to its original form.

A waveform with data represented in one form, such as NRZI, can be made to look like a waveform with data represented in another form by the use of an operation prior to the modulator that prepares the data. A *precoder* is a transformation on a bitstream prior to entering a channel that cancels another transformation that will take place on the bitstream within the modulator/demodulator or within the channel. The cascade of the precoder and the modulator/demodulator returns the bitstream to its original form. Figure 2.15 shows a precoder for an NRZI waveform that is demodulated as NRZ. The plus sign denotes a modulo-two adder, or an exclusive-or gate, and the box is a delay of one time unit. The precoder requires only that the initial reference level for the NRZI modulation is the zero level. Alternatively, one can use a device following the demodulator. Then it is known as a *postcoder*. The precoder shown in Figure 2.15 works equally well as a postcoder for an NRZI waveform demodulated as NRZ. It does not matter whether the logic circuit is placed prior to the modulator or following the demodulator.

The NRZI waveform is our first example of a channel waveform with memory within the waveform. In later chapters, we shall encounter channel waveforms with much more elaborate memory. To keep track of these kinds of waveforms, a kind of graph known as a *trellis* is very useful. Figure 2.16 gives a trellis for the NRZI waveform.

In the trellis of Figure 2.16, there are two nodes in a column, denoting the states 0 and 1. These nodes represent the two levels that the waveform could have had at the previous time instant. The column of states is replicated horizontally to represent the possible states at successive time instants. The horizontal direction represents time, and the column of nodes is replicated at each time instant. The nodes are connected by branches that show how the situation might change. Each branch is labeled by two bits. The bit on the left is the new data bit that takes the system to a new state. The bit on the right is the new channel bit. In this simple example, the new channel bit and the new state are equal. This need not be true in other labeled trellises.

The notion of a trellis will be used later in more general situations. A general trellis consists of a set of nodes arranged in a vertical column used to represent the state of a memory at a single time instant; each of the nodes represents one of the possible states of the memory. Each step through the trellis is called a *frame* of the trellis. Permissible transitions in the state of the memory are depicted by lines connecting the nodes at successive time instants. One can see how the state changes with time by following a path through the trellis. The branches are labeled with data bits, and in this way, the data bits specify the path. The branches are also labeled with the channel bits to the right of the data bit labels. The sequence of channel bits on the path taken through the trellis forms the channel sequence.

We shall examine many trellises in later chapters.

2.7   Binary orthogonal signaling at baseband

A much different kind of binary signaling waveform can be defined that uses two different pulses to represent the two bit values. Two such pulses, $s_0(t)$ and $s_1(t)$, are called an *orthogonal* pair of pulses[1] if they satisfy

$$\int_{-\infty}^{\infty} s_0(t)s_1^*(t)dt = 0.$$

For baseband signaling, both $s_0(t)$ and $s_1(t)$ are real-valued pulses. Usually there is also an implied requirement that both pulses are Nyquist pulses and that, for all integer ℓ,

$$\int_{-\infty}^{\infty} s_0(t)s_1^*(t - \ell T)dt = 0.$$

This condition will hold trivially if both pulses are supported only on the interval $[-T/2, T/2]$.

To modulate the datastream into a channel waveform, let

$$c(t) = \sum_{\ell=-\infty}^{\infty} [a_\ell s_0(t - \ell T) + \bar{a}_\ell s_1(t - \ell T)]$$

[1] Although we use only real pulses for baseband signaling, we regard it as good practice here to write a complex conjugate here because it would be required if the pulse were complex. For real pulses, the complex conjugation operator $*$ is completely superfluous.

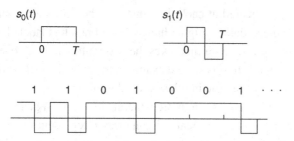

Figure 2.17. An example of a binary orthogonal waveform at baseband.

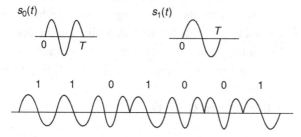

Figure 2.18. Another example of a binary orthogonal waveform at baseband.

where

$$(a_\ell, \bar{a}_\ell) = \begin{cases} (A,0) & \text{if the } \ell\text{th data bit is a } 0 \\ (0,A) & \text{if the } \ell\text{th data bit is a } 1. \end{cases}$$

This waveform is called a *binary orthogonal signaling waveform*.

One example of a binary orthogonal signaling waveform is shown in Figure 2.17. Clearly, that pair of pulses, $s_0(t)$ and $s_1(t)$, is an orthogonal pair. The waveform has the appearance of a binary antipodal waveform with a bit duration of $T/2$. In fact, however, it is something different. This example shows that one cannot always identify a signaling waveform from its appearance alone.

A second example of a waveform for binary orthogonal signaling is the pair of pulses

$$s_0(t) = \sin 2\pi f_0 t \qquad 0 \le t \le T$$
$$s_1(t) = \sin 2\pi f_1 t \qquad 0 \le t \le T$$

with

$$\int_0^T \sin 2\pi f_0 t \, \sin 2\pi f_1 t \, dt = 0,$$

as shown in Figure 2.18. To ensure that the pulses are orthogonal, the frequencies f_0 and f_1 are chosen to have the harmonic relationship $f_0 = m/(2T)$ and $f_1 = m'/(2T)$ where

m and m' are distinct positive integers. When f_0 and f_1 are very large in comparison to $1/T$, this signaling waveform is better treated as a passband waveform. Then it is called (binary) *frequency-shift keying* (FSK). Moreover, any pair of orthogonal signaling waveforms often is loosely referred to as binary FSK, the general case then taking its name from the special case.

2.8   *M*-ary orthogonal signaling at baseband

In general, we are concerned with both the power and the bandwidth used by a wave-form. A channel for which the power constraint is the more important constraint is called a *power-limited channel*. A channel for which the bandwidth constraint is the more important constraint is called a *bandwidth-limited channel*. Of course, both con-straints are important in every application, but usually one or the other is of primary importance.

In Section 2.3, we studied a signaling method known as *M*-ary pulse amplitude modulation, which is suitable for channels with a tight bandwidth constraint. In this section, we shall study *M*-ary orthogonal signaling at baseband, which is suitable for channels whose bandwidth constraint is so weak that it can be ignored, but for which power must be limited. In contrast to the *M*-ary orthogonal signaling alphabets, which use *M* distinct pulses, are the *M*-ary signal constellations, which are used with a single pulse.

The pulse alphabets for *M*-ary signaling without a bandwidth constraint that we shall study are the *orthogonal pulse alphabets* and the *simplex pulse alphabets*. These signal-ing alphabets are the most important by far, both in practice and in theory, whenever the bandwidth constraint is not a factor. Nonorthogonal pulse alphabets may also be used when there is no bandwidth constraint, but orthogonal pulses are generally preferred. The reason for the orthogonality condition is to make the pulses easy to distinguish in the presence of noise.

Definition 2.8.1   *An M-ary orthogonal pulse alphabet is a set of M pulses $s_m(t)$ for $m = 0, \ldots, M - 1$ having the properties of equal energy and orthogonality, given by*

$$\int_{-\infty}^{\infty} s_m(t)s_n^*(t)dt = \begin{cases} 1 & m = n \\ 0 & m \neq n \end{cases}$$

and

$$\int_{-\infty}^{\infty} s_m(t)s_n^*(t - \ell T)dt = 0 \qquad \ell \neq 0.$$

In this section we consider only pulse alphabets of real pulses. Waveforms that form an orthogonal pulse alphabet are usually designed to have a finite interval of duration

T as their common support and this property means that the second condition of the theorem is trivial.

Usually for applications, $M = 2^k$ for some integer k. To use the M-ary pulse alphabet, one must assign one k-bit binary number to each of the 2^k pulses of the pulse alphabet. A k-bit number is transmitted at time ℓT by transmitting the pulse $s_m(t - \ell T)$ assigned to that k-bit number.

An M-ary modulator is a device that, upon receiving discrete symbol m, forms pulse $s_m(t)$ and passes $As_m(t)$ to the waveform channel. To transmit a stream of binary data, the datastream is broken into k-bit words; the ℓth such word defines a k-bit number m_ℓ that is mapped into the pulse $s_{m_\ell}(t)$. The transmitted waveform is

$$c(t) = \sum_{\ell=-\infty}^{\infty} As_{m_\ell}(t - \ell T).$$

An M-ary demodulator is a device that, based on the noisy output of the waveform channel, forms an estimate, $\widehat{m}_\ell$, of the integer m_ℓ at the input to the modulator at the ℓth input time. The waveform channel, combined with an M-ary modulator and an M-ary demodulator, constitutes a finite-input, discrete-time channel. When $\widehat{m}_\ell$ is not equal to m_ℓ the demodulator has made a symbol error at time ℓT.

Figure 2.19 shows examples of signaling waveforms using three different sets of four-ary orthogonal pulses. They are known as four-ary frequency-shift keying (FSK), four-ary pulse-position modulation (PPM), and four-ary code-shift keying (CSK). A four-ary CSK waveform is shown in Figure 2.20. Larger alphabets of M-ary orthogonal pulses, perhaps with M equal to 32 or 64, are in common use. All of these kinds of

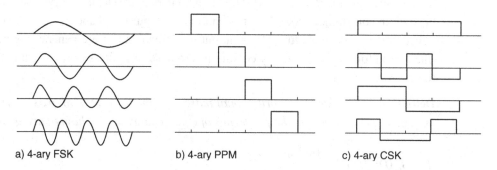

a) 4-ary FSK   b) 4-ary PPM   c) 4-ary CSK

Figure 2.19. Some examples of baseband four-ary waveforms.

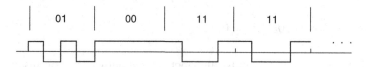

Figure 2.20. Example of a four-ary CSK waveform.

orthogonal pulse alphabets are sometimes referred to collectively as M-ary FSK (or MFSK), where now the term FSK is used in the looser sense to mean orthogonal signaling.

The total transmitted energy $E_p = A^2$ is used to transmit a symbol that conveys $k = \log_2 M$ bits. To express the energy in a standardized way so that performance can be readily compared for various values of M, the energy per bit is defined as

$$E_b = \frac{E_p}{k} = \frac{E_p}{\log_2 M}.$$

The energy per bit E_b is a more useful way to state performance; the energy per pulse E_p is the more tangible quantity.

Any orthogonal set of M pulses can be used to construct another set of M pulses known as a *simplex pulse alphabet*.[2] The average of all the pulses is subtracted from each individual pulse. Thus

$$q_m(t) = s_m(t) - \frac{1}{M} \sum_{m=0}^{M-1} s_m(t).$$

The simplex pulses have less energy than the orthogonal pulses. This is shown by the simple calculation

$$E_q = \int_{-\infty}^{\infty} A^2 |q_m(t)|^2 dt = \left(1 - \frac{1}{M}\right) E_p.$$

The simplex pulses are not orthogonal. Rather, they have the stronger property of negative correlation. For $m \neq m'$, the correlation is

$$\int_{-\infty}^{\infty} q_m(t) q_{m'}(t) dt = -\frac{1}{M}.$$

We shall see in Chapter 3 that a simplex family of pulses constructed from an orthogonal family has the same probability of error as the orthogonal family of pulses even though the energy is less.

Any $M/2$-ary orthogonal set of pulses can be used to construct another set of M pulses, known as an M-ary *biorthogonal pulse alphabet*. The biorthogonal alphabet of M pulses consists of the $M/2$ orthogonal pulses and the negatives of the $M/2$ orthogonal pulses. The biorthogonal pulses either have a correlation equal to zero or to the negative of the pulse energy. The latter case occurs only when a pulse is correlated with its own negative. Of the $M(M-1)/2$ distinct cross-correlations, $M/2$ of them will be negative and the rest will be zero.

---

[2] More directly, we could define an M-ary simplex pulse alphabet, renormalized, as any set of M pulses $q_m(t)$ of unit energy such that the correlation of any two distinct pulses is $-1/(M-1)$.

2.9   Power density spectrum of baseband waveforms

When modulated with a deterministic datastream of finite duration, the channel signaling waveform

$$c(t) = \sum_{\ell=0}^{n-1} a_\ell s(t - \ell T)$$

has a Fourier transform that can be calculated by using the delay theorem as

$$C(f) = S(f) \sum_{\ell=0}^{n-1} a_\ell e^{-j2\pi f \ell T}$$

$$= S(f)A(f).$$

The spectrum $C(f)$ factors into a *pulse-shape factor* $S(f)$ and an *array factor* $A(f)$. The array factor depends on the actual values taken by the data, so we cannot actually determine $A(f)$ without knowing the data. The array factor $A(f)$ is random if the data is random.

When modulated with a datastream that is infinitely long, the channel signaling waveform

$$c(t) = \sum_{\ell=-\infty}^{\infty} a_\ell s(t - \ell T)$$

has infinite energy, so it does not have a Fourier transform. Moreover, we do not know the actual data sequence. Indeed, it would be artificial to study only one data sequence. Instead, we usually prefer a description that is actually more useful: a probabilistic description. Loosely, we may expect that if n is very large and the data is random, then the particular $A(f)$ will not be important and $|C(f)|^2$ will have nearly the same shape as $|S(f)|^2$. This heuristic picture can be made rigorous by going immediately to the case of an infinitely long and random datastream. Then $c(t)$ becomes a random process; the properties of this random process are suggested by the eye diagrams that are shown in Figures 2.12 and 2.13.

To describe the random process $c(t)$, one wants to deal with some form of the power density spectrum. However, $c(t)$ is not stationary, so as it stands, it does not have a power density spectrum. This difficulty can be handled in two ways – either by treating $c(t)$ as a kind of random process known as a cyclostationary random process, or alternatively, by introducing a random time delay so that the waveform $c(t)$ now has no preferred time instants. We will choose the latter approach. The antipodal signaling waveform now becomes the stationary random process

$$c(t) = \sum_{\ell=-\infty}^{\infty} a_\ell s(t - \ell T - \alpha)$$

where α is uniformly distributed over $[0, T]$, and the a_ℓ are independent random variables taking values in the signal constellation. The autocorrelation function of the random waveform is

$$R(\tau) = E[c(t)c(t + \tau)]$$

$$= E\left[\sum_{\ell=-\infty}^{\infty} \sum_{\ell'=-\infty}^{\infty} a_\ell a_{\ell'} s(t - \ell T - \alpha)s(t + \tau - \ell'T - \alpha)\right].$$

Move the expectation inside the sum and recall that the random variables a_ℓ and α are independent to obtain

$$R(\tau) = \sum_{\ell=-\infty}^{\infty} \sum_{\ell'=-\infty}^{\infty} E[a_\ell a_{\ell'}]E[s(t - \ell T - \alpha)s(t + \tau - \ell'T - \alpha)]$$

$$= \overline{a^2} \sum_{\ell=-\infty}^{\infty} E[s(t - \ell T - \alpha)s(t + \tau - \ell T - \alpha)]$$

because $E[a_\ell a_{\ell'}] = \overline{a^2}\delta_{\ell\ell'}$. Finally,

$$R(\tau) = \frac{\overline{a^2}}{T} \sum_{\ell=-\infty}^{\infty} \int_0^T s(t - \ell T - \alpha)s(t + \tau - \ell T - \alpha)d\alpha$$

$$= \frac{\overline{a^2}}{T} \sum_{\ell=-\infty}^{\infty} \int_{\ell T}^{(\ell+1)T} s(t - \xi)s(t + \tau - \xi)d\xi$$

$$= \frac{\overline{a^2}}{T} \int_{-\infty}^{\infty} s(t)s(t + \tau)dt.$$

The power density spectrum $\Phi_c(f)$ of the random waveform $c(t)$ is the Fourier transform of $R(\tau)$, and so it is proportional to $|S(f)|^2$. Thus

$$\Phi_c(f) = \frac{\overline{a^2}}{T}|S(f)|^2.$$

The power density spectrum of $c(t)$ is related to the transform of a single pulse. The bandwidth occupied by the modulated waveform $c(t)$ is the same as the bandwidth occupied by the pulse $s(t)$.

Problems for Chapter 2

2.1. The following two sequences of blocklength four are orthogonal

$$
\begin{array}{cccc}
1 & 1 & 1 & 1 \\
1 & 1 & -1 & -1
\end{array}
$$

because $\sum_i a_i b_i = 0$.

a. Construct four sequences of blocklength four that are orthogonal.

b. Can you construct eight sequences of blocklength 4, with "chips" taking the value $1, -1, j, -j$, $(j = \sqrt{-1})$, that are pairwise orthogonal using the definition $\sum_i a_i b_i^* = 0$ for orthogonality of complex sequences?

c. Repeat part b by using the definition $\sum_i a_i b_i = 0$ for orthogonality.

d. Repeat part b by using the definition $\text{Re}[\sum_i a_i b_i] = 0$ for orthogonality.

2.2. Let $q_m(t)$ for $m = 0, \ldots, M-1$ denote the elements of a simplex pulse alphabet. Define the squared euclidean distance between two pulses as

$$
d^2(q_m, q_{m'}) = \int_{-\infty}^{\infty} |q_m(t) - q_{m'}(t)|^2 dt.
$$

For a simplex alphabet, show that the squared euclidean distance $d^2(q_m, q_{m'})$, where m is not equal to m', is a constant that is independent of m and m'.

2.3. Prove that an antipodal signaling waveform $c(t)$ based on sinc pulses will occasionally take on values larger than any constant, no matter how large the constant.

2.4. a. Show that when M equals 2, M-ary biorthogonal signaling reduces to binary antipodal signaling (BPSK).

b. Show that when M equals 2, M-ary simplex signaling reduces to binary antipodal signaling (BPSK).

2.5. A four-level ASK waveform is used on a channel with a possible sign inversion ("gain" of ± 1). Design a differential encoding scheme to make this sign change "transparent".

2.6. Sketch a family of eight-ary biorthogonal pulse alphabets.

2.7. The legacy telephone channel can be grossly characterized as an ideal passband channel from 300 Hz to 2700 Hz.

a. Choose a signal constellation and a symbol rate to obtain a 9600 bits/second telephone line modem.

b. Choose a pulse shape $s(t)$ that will pass through the channel without causing intersymbol interference. Describe the transmitted waveform.

2.8. A filter whose impulse response is a Nyquist pulse is called a *Nyquist filter*. Does the cascade of two Nyquist filters, in general, equal a Nyquist filter? Find the set of all Nyquist filters that will produce another Nyquist filter when cascaded with itself.

2.9. Prove that a pulse $s(t)$ with a raised-cosine spectrum is a Nyquist pulse.

2.10. Let

$$s(t) = \cos(\pi t/T)\text{rect}(t/T).$$

a. Find the bandwidth B containing 90 percent of the pulse energy. That is,

$$\int_{-B/2}^{B/2} |S(f)|^2 df = .9E_p.$$

b. Find the bandwidth B containing 90 percent of the power of a binary antipodal signaling waveform that uses $s(t)$.

2.11. Prove that if $s_0(t)$ and $s_1(t)$ are orthogonal baseband pulses whose spectra are confined to bandwidth W, then $s_0(t) \cos 2\pi f_0 t$ and $s_1(t) \cos 2\pi f_0 t$ are orthogonal pulses as well, provided f_0 is larger than W. Give an example for which the statement fails when f_0 is smaller than W.

2.12. A binary orthogonal signaling waveform at baseband is given by

$$c(t) = \sum_{\ell=-\infty}^{\infty} [a_\ell s_0(t - \ell T) + \bar{a}_\ell s_1(t - \ell T)]$$

where $s_0(t)$ and $s_1(t)$ are orthogonal with transforms $S_0(f)$ and $S_1(f)$, and

$$(a_\ell, \bar{a}_\ell) = \begin{cases} (1, 0) & \text{if the } \ell\text{th data bit is a 0} \\ (0, 1) & \text{if the } \ell\text{th data bit is a 1.} \end{cases}$$

Find the power density spectrum of $c(t)$ when made stationary by including a random time offset.

2.13. A baseband antipodal signaling waveform using the pulse shape of width T_b, described by

$$s(t) = \begin{cases} 1 & -T_b/2 \leq t < 0 \\ -1 & 0 \leq t \leq T_b/2, \end{cases}$$

is called a *Manchester signaling waveform*. An *inverse Manchester signaling waveform* is defined as

$$c(t) = \sum_{\ell=0}^{\infty} a_\ell s(t - \ell T_b)$$

where $a_\ell = 1$ if the ℓth data bit is the same as $a_{\ell-1}$, and otherwise $a_\ell = -1$.

Design an encoder and decoder to use with a baseband antipodal modulator and demodulator with bit duration $T_b/2$ to obtain an inverse Manchester waveform with bit duration T_b.

2.14. Suppose that a channel $h(t)$ and a pulse $s(t)$ are given with transforms as illustrated.

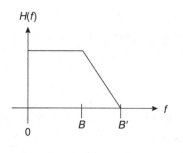
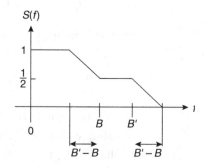

a. Show that the pulse can be used for binary antipodal signaling transmitted at a rate of $2B$ bits per second without intersymbol interference at the output of the channel. Is there intersymbol interference at the input to the channel?

b. What might be the purpose of including nonzero energy in the transmitted pulse at frequencies that are not received at the channel output?

c. Find $s(t)$ and the time-domain pulse shape at the filter output.

d. Is $s(t)$ a Nyquist pulse for any signaling interval?

2.15. Let

$$s(t) = \frac{4}{\pi\sqrt{T}} \frac{\cos(2\pi t/T)}{1 - (4t/T)^2}.$$

a. Find $s(t) * s(-t)$.

b. Contrast the zeros of $s(t)$ with the zeros of $s(t) * s(-t)$.

c. Is $s(t)$ a Nyquist pulse? Is $s(t) * s(-t)$ a Nyquist pulse?

2.16. A baseband pulse has a transfer function $S(f)$ that is a trapezoid as illustrated.

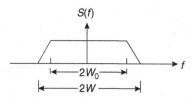

Prove that, for some T, the pulse $s(t)$ is a Nyquist pulse. At what spacing are its zeros? Prove that, when the pulse $s(t)$ is convolved with itself, the result of the convolution is not a Nyquist pulse.

Notes for Chapter 2

The design of pulses that are free of intersymbol interference was studied by Nyquist (1928). The construction of waveforms without intersymbol interference is the converse of the sampling problem, so it is not surprising that Nyquist's work deals with both topics. The M-ary orthogonal communication systems were first derived and analyzed by Kotel'nikov (1959), and studied further by Viterbi (1961). The simplex codes were studied by Balakrishnan (1961), who showed that simplex codes are locally optimal. It has never been proved that simplex codes are globally optimal. A long-standing conjecture states that, of all M-ary pulse alphabets of energy E_p, none has a smaller probability of error when used on a white gaussian-noise channel than a simplex family. The NRZI waveform is also called *differential binary*. When used as a passband waveform, it is called *differential phase-shift keying* (DPSK).

3 Baseband Demodulation

The function of a digital demodulator is to reconvert a waveform received in noise back into the stream of data symbols from the discrete data alphabet. We usually regard this datastream as a binary datastream. A demodulator is judged by its ability to recover the user datastream with low probability of bit error even when the received channel waveform is contaminated by distortion, interference, and noise. The probability of symbol error or bit error at the demodulator output is also called the symbol error rate or the bit error rate. The demodulator is designed to make these error rates small. We shall concentrate our discussion on optimum demodulation in the presence of additive noise because additive noise is the most fundamental disturbance in a communication system.

The energy per data bit is the primary physical quantity that determines the ability of the communication system to tolerate noise. For this purpose, energy (or power) always refers to that portion of the energy in the waveform that reaches the receiver. This will be only a small fraction of the energy sent by the transmitter.

The study of the optimum demodulation of a waveform in additive noise is an application of the statistical theory of hypothesis testing. The basic principles are surprisingly simple and concise. The most basic principle, and the heart of this chapter, is the principle of the matched filter. A very wide variety of modulation waveforms are demodulated by the same general method of passing a received signal through a matched filter and sampling the output of that matched filter. In this chapter, we shall study the matched filter and apply it to the demodulation of baseband signaling waveforms. The demodulators studied in this chapter deal with only one symbol at a time. Sequence demodulation, which is necessary whenever the channel output exhibits intersymbol interference, either intentionally or unintentionally, will be studied in Chapter 4. The matched filter will be applied to passband waveforms in Chapter 6.

3.1 The matched filter

Let $As(t)$ be a pulse of finite energy E_p:

$$E_p = \int_{-\infty}^{\infty} |As(t)|^2 dt = \int_{-\infty}^{\infty} |AS(f)|^2 df \leq \infty$$

where the constant A is included so that we may set the energy of $s(t)$ equal to one whenever convenient. Equality of the integrals in the time domain and the frequency domain is a consequence of Parseval's formula. In this chapter, we shall restrict $s(t)$ to be a real function of time in order to ensure clarity and because baseband waveforms, which we treat in this chapter, are real. However, with very little change, the ideas of this section also apply to complex functions of time, as we shall see in Chapter 6. Therefore, although this chapter deals only with baseband (or real) signals the derivation of the matched filter is written so that it can be read for the passband (or complex) case as well.

Let $n(t)$ be zero mean, stationary baseband noise with variance σ^2, correlation function $\phi(\tau)$, and power density spectrum $N(f)$. It is not necessary in this section to specify the noise in any further detail. In later sections of this chapter, the noise will be made more specific by giving its probability density function, usually a gaussian probability density function.

If the received signal consists of the pulse $As(t)$ in additive stationary noise, it is

$$v(t) = As(t) + n(t).$$

If the received signal consists of noise only, it is

$$v(t) = n(t).$$

Our task for the moment is to design a procedure that will detect whether the received signal $v(t)$ consists of a pulse in noise or of noise only. We choose at this time to consider only the class of *linear detectors*. That detection procedure is to pass the received signal $v(t)$ through a filter with impulse response $g(t)$, as shown in Figure 3.1, and then to test the amplitude of the signal at the output of the filter at time instant t_0. If the amplitude at time t_0 is larger than a fixed number, Θ, called a *threshold*, the decision is that the pulse $As(t)$ is present in the received signal $v(t)$. If the amplitude is smaller than the threshold, the decision is that the pulse is not present in $v(t)$. It remains only to choose the filter impulse response $g(t)$ and the decision threshold Θ.

The output of a filter with impulse response $g(t)$ and input $As(t) + n(t)$ is

$$u(t) = \int_{-\infty}^{\infty} g(\xi)[As(t - \xi) + n(t - \xi)]d\xi.$$

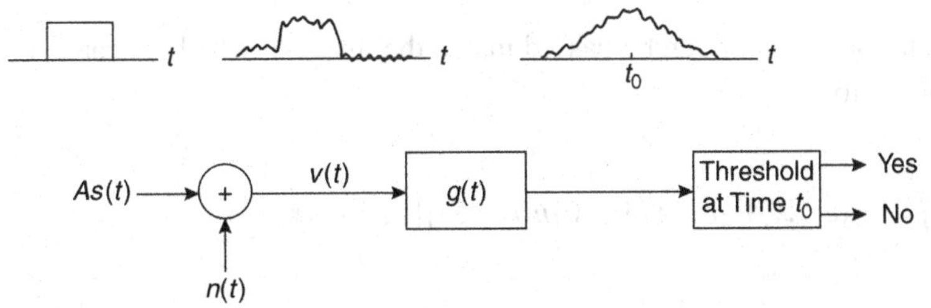

Figure 3.1. Detection of a pulse in noise.

We do not yet restrict $g(t)$ to be a causal filter, so we have set the lower limit to $-\infty$. Because $n(t)$ is a random process, the filter output is a random process. We denote the expected value of $u(t)$ by $Ar(t)$, which is the noise-free output of the filter, and the variance of $u(t)$ by σ^2. That is,

$$Ar(t) = E[u(t)]$$
$$\sigma^2 = \text{var}[u(t)].$$

The variance σ^2 does not depend on t because $n(t)$ is stationary.

The filter $g(t)$ should be chosen to pass the signal $s(t)$ and to reject the noise $n(t)$. Because it cannot fully achieve both of these goals simultaneously, the filter is chosen as the best compromise between the two goals. Specifically, we choose to maximize the ratio of signal power to noise power at the single time instant $t = 0$. The signal-to-noise ratio at time zero is simply the ratio $S/N = [Ar(0)]^2/\sigma^2$. We choose $g(t)$ to maximize the signal-to-noise ratio S/N. The filter $g(t)$ that achieves this maximum is called the *matched filter* for pulse $As(t)$ in noise $n(t)$. We shall derive the matched filter in this section.

The matched-filter demodulator is optimal in the restricted class consisting of demodulators with this linear structure, although it need not be optimal in the larger class of all possible demodulators. It is not at all obvious at this point that restricting the search to demodulators of this form is the optimum thing to do. Indeed, in general, it is not. It is true, however, that if the noise is gaussian, the matched filter is the first step in any optimum demodulation procedure. This is a consequence of the maximum-likelihood principle, which will not be studied until Chapter 7. Even when the noise is not gaussian, it is still a very practical and widely-used demodulation procedure. We shall study demodulation more fully in the next section, and there we shall see the role of the matched filter in digital communication.

To proceed, first determine $Ar(t)$ as follows:

$$Ar(t) = E[u(t)]$$
$$= E\left[\int_{-\infty}^{\infty} g(\xi)[As(t-\xi) + n(t-\xi)]d\xi\right].$$

The expectation operator can be passed inside the integral, which, because $g(t)$ is not random, leads to

$$Ar(t) = \int_{-\infty}^{\infty} g(\xi)[E[As(t-\xi)] + E[n(t-\xi)]]d\xi.$$

But $E[As(t)] = As(t)$ because the pulse is not random, and $E[n(t)] = 0$ because $n(t)$ has zero mean. Therefore the expected value of the output $Ar(t)$ at time zero is the signal component at the output of the filter at time zero. In particular,

$$r(0) = \int_{-\infty}^{\infty} g(\xi)s(-\xi)d\xi.$$

Next, Parseval's formula

$$\int_{-\infty}^{\infty} a(\xi)b^*(\xi)d\xi = \int_{-\infty}^{\infty} A(f)B^*(f)df$$

can now be used on the right side with $a(\xi) = g(\xi)$ and $b^*(\xi) = s(-\xi)$. Because the Fourier transform of $b(t) = s^*(-t)$ is easily found to be $B(f) = S^*(f)$, and $(S^*(f))^* = S(f)$, we conclude that

$$r(0) = \int_{-\infty}^{\infty} G(f)S(f)df,$$

and

$$|Ar(0)|^2 = \left|A \int_{-\infty}^{\infty} G(f)S(f)df\right|^2$$

is the signal power S at time zero.

Next, we find an expression for the noise power. The variance of $u(0)$ is

$$\text{var}[u(0)] = E[|u(0) - E[u(0)]|^2]$$

$$= E\left[\int_{-\infty}^{\infty} g(\xi_1)n(-\xi_1)d\xi_1 \int_{-\infty}^{\infty} g(\xi_2)n(-\xi_2)d\xi_2\right]$$

$$= \int_{-\infty}^{\infty}\int_{-\infty}^{\infty} g(\xi_1)g(\xi_2)E[n(-\xi_1)n(-\xi_2)]d\xi_1 d\xi_2.$$

Now recall that $E[n(-\xi_1)n(-\xi_2)]$ is the correlation function $\phi(\xi_1 - \xi_2)$. Thus:

$$\sigma^2 = \int_{-\infty}^{\infty}\int_{-\infty}^{\infty} g(\xi_1)g(\xi_2)\phi(\xi_1 - \xi_2)d\xi_1 d\xi_2.$$

Make the change of variables $\eta = \xi_1 - \xi_2$ and $\xi = \xi_1$. Then $d\xi_1 d\xi_2 = d\xi d\eta$ and

$$\sigma^2 = \int_{-\infty}^{\infty} \phi(\eta) \int_{-\infty}^{\infty} g(\xi)g(\xi - \eta)d\xi d\eta.$$

The inner integral is the convolution of $g(t)$ with $g(-t)$, which has the Fourier transform $|G(f)|^2$. The first term $\phi(\eta)$ has the Fourier transform $N(f)$. Therefore, by Parseval's formula,

$$\sigma^2 = \int_{-\infty}^{\infty} N(f)|G(f)|^2 df.$$

This is the noise power N at the output of $g(t)$. Our task is to maximize the ratio S/N by the choice of $G(f)$.

The development of the matched filter makes use of a fundamental inequality of functional analysis, known as the *Schwarz inequality*, which we shall derive first.

Theorem 3.1.1 (Schwarz inequality)   *Let $r(t)$ and $s(t)$ be finite energy pulses, real-valued or complex-valued. Then*

$$\int_{-\infty}^{\infty} |r(t)|^2 dt \int_{-\infty}^{\infty} |s(t)|^2 dt \geq \left| \int_{-\infty}^{\infty} r^*(t)s(t)dt \right|^2$$

with equality if and only if $r(t)$ is a constant (real or complex) multiple of $s(t)$.

Proof   Let a and b be any constants, real or complex. Then, because $|ar(t) - bs(t)|^2$ is never negative, we have the inequality

$$0 \leq \int_{-\infty}^{\infty} |ar(t) - bs(t)|^2 df$$

with equality if and only if $ar(t) - bs(t)$ is zero for all t. Therefore expanding the square,

$$0 \leq |a|^2 \int_{-\infty}^{\infty} |r(t)|^2 dt - ab^* \int_{-\infty}^{\infty} r(t)s^*(t)dt$$
$$- a^*b \int_{-\infty}^{\infty} r^*(t)s(t)dt + |b|^2 \int_{-\infty}^{\infty} |s(t)|^2 dt$$

with equality if and only if $ar(t) - bs(t)$ is zero for all t. Now choose the constants a and b as follows:

$$a = \int_{-\infty}^{\infty} r^*(t)s(t)dt$$

$$b = \int_{-\infty}^{\infty} |r(t)|^2 dt.$$

For these constants, the inequality becomes

$$0 \leq |a|^2 b - aba^* - a^*ba + |b|^2 \int_{-\infty}^{\infty} |s(t)|^2 dt.$$

The constant b is real and positive. It can be canceled to give

$$0 \leq -|a|^2 + b \int_{-\infty}^{\infty} |s(t)|^2 dt,$$

which is equivalent to the statement of the theorem.   ∎

We are now ready to state the matched filter. In proving the following theorem, we make use of the fact that the noise power density spectrum $N(f)$ is real and nonnegative at each f and so has a square root at each f.

Theorem 3.1.2 *Suppose that the input pulse $As(t)$ is received in additive gaussian noise whose power density spectrum $N(f)$ is nonzero at all f. The signal-to-noise power ratio S/N at the output of the filter $g(t)$ satisfies*

$$\frac{S}{N} \leq A^2 \int_{-\infty}^{\infty} \frac{|S(f)|^2}{N(f)} df,$$

and equality is achieved when the filter is specified as

$$G(f) = \frac{S^*(f)}{N(f)}.$$

Proof The signal power and the noise power were each calculated earlier in the section. The signal-to-noise power ratio satisfies the following:

$$\frac{S}{N} = \frac{\left| A \int_{-\infty}^{\infty} G(f)S(f)df \right|^2}{\int_{-\infty}^{\infty} N(f)|G(f)|^2 df}$$

$$= \frac{|A|^2 \left| \int_{-\infty}^{\infty} N(f)^{\frac{1}{2}} G(f) \frac{S(f)}{N(f)^{\frac{1}{2}}} df \right|^2}{\int_{-\infty}^{\infty} N(f)|G(f)|^2 df}$$

$$\leq \frac{|A|^2 \int_{-\infty}^{\infty} N(f)|G(f)|^2 df \int_{-\infty}^{\infty} \frac{|S(f)|^2}{N(f)} df}{\int_{-\infty}^{\infty} N(f)|G(f)|^2 df}$$

$$= |A|^2 \int_{-\infty}^{\infty} \frac{|S(f)|^2}{N(f)} df$$

where the inequality is the Schwarz inequality. This bound does not depend on the filter $G(f)$. Choosing

$$G(f) = C\frac{S^*(f)}{N(f)}$$

for any (real or complex) constant C makes the signal-to-noise ratio equal to the bound. Hence this filter provides the maximum signal-to-noise ratio. This completes the proof of the theorem. ∎

The filter given in Theorem 3.1.2 is known as the matched filter, and sometimes, when $N(f)$ is not a constant, as the *whitened matched filter*. Any constant multiple of the matched filter is also a matched filter. Notice that we did not assume that the noise is gaussian. The matched filter maximizes signal-to-noise ratio at time zero for any covariance-stationary noise. More generally, the filter

$$G(f) = C\frac{S^*(f)}{N(f)}e^{-j2\pi ft_0}$$

maximizes the signal-to-noise ratio at time t_0.

The signal-to-noise ratio is maximized by the matched filter. To decide whether the input to the matched filter consists of a pulse in noise or of noise only, the output signal at time t_0 is tested by comparing it to a threshold Θ. If the output $u(t_0)$ at time t_0 exceeds the threshold Θ, then a pulse is declared to be present. If the threshold Θ is set low, then a pulse will rarely be missed, but noise will often be mistaken for a pulse. If the threshold is set high, then noise is unlikely to be mistaken for a pulse, but a pulse is more likely to be missed. By choice of Θ, one can trade off the probability of falsely declaring a pulse when there is none against the probability of missing a pulse when it is there.

If the power density spectrum of the noise is constant, then the noise is called *white noise*. By convention, the power density spectrum of white noise is expressed as $N(f) = N_0/2$ and

$$G(f) = C\frac{S^*(f)}{N_0/2}.$$

In this case, it is convenient to choose for the constant $C = N_0/2$ so that

$$G(f) = S^*(f)$$

and, as stated in the next theorem,

$$g(t) = s^*(-t).$$

The filter $s^*(-t)$ is then simply known as the *matched filter*. The complex conjugate is superfluous in this chapter because $s(t)$ is real, but it is included because we will encounter complex $s(t)$ in later chapters.

Corollary 3.1.3 *Suppose that the pulse $As(t)$ satisfies $\int |s(t)|^2 dt = 1$. In additive white noise of power density spectrum $N_0/2$ watts per hertz, the maximum signal-to-noise power ratio at the output of a filter $g(t)$ is achieved by the matched filter $g(t) = s^*(-t)$, and is*

$$\frac{|A|^2}{\sigma^2} = \frac{2E_p}{N_0}$$

where $E_p = |A|^2$ is the total energy in the signal pulse $As(t)$.

Proof When $N(f) = N_0/2$ and $G(f) = S^*(f)$, the output noise variance is

$$\sigma^2 = \int_{-\infty}^{\infty} N(f)|S(f)|^2 df$$

$$= \frac{N_0}{2},$$

and the output signal-to-noise power ratio becomes

$$\frac{S}{N} = \int_{-\infty}^{\infty} \frac{|A|^2 |S(f)|^2}{N_0/2} dt$$

$$= \frac{2E_p}{N_0}$$

$$= \frac{|A|^2}{\sigma^2}.$$

Moreover, because

$$s(-t) = \int_{-\infty}^{\infty} S(f)e^{-j2\pi ft} df,$$

the inverse Fourier transform of $S^*(f)$ is $s^*(-t)$, and so $g(t) = s^*(-t)$, as was to be proved. ∎

Notice that, when the noise is white, the shape of the signal pulse plays no role in calculating the output signal-to-noise ratio. Only the energy of the pulse matters. This means that one is free to use as a signaling pulse any pulse shape based on criteria other than signal-to-noise ratio.

The output of the matched filter at the sampling instant

$$u(0) = \int_{-\infty}^{\infty} s^*(-\xi)v(-\xi)d\xi$$

$$= \int_{-\infty}^{\infty} v(t)s^*(t)dt$$

is the correlation between the received signal $v(t)$ and the transmitted pulse shape $s(t)$. Thus one can take the view that the function of the matched filter is to compute this correlation.

The expected value of the matched-filter output is

$$Ar(t) = A \int_{-\infty}^{\infty} s(\xi)s^*(\xi - t)d\xi.$$

The integral here is sometimes called the autocorrelation function of the pulse $s(t)$.

The matched filter $s^*(-t)$ will usually be a noncausal filter. To remedy this if $s^*(-t)$ is zero for times t less than $-t_0$, offset the impulse response by a delay time t_0, so that

$$g(t) = s^*(t_0 - t),$$

which is now a causal filter with a delay t_0, the signal-to-noise ratio will be maximized at time t_0 rather than at time zero. An example of a matched filter that includes a time delay to make it causal is shown in Figure 3.2. If, instead, the pulse $s(t)$ has tails that go on indefinitely, as does the gaussian pulse, then it will not be possible to make $s^*(t_0 - t)$ causal. Figure 3.3 shows the situation for which the matched filter is not causal for any finite t_0. In this case, one can approximate the matched filter as closely as desired by choosing t_0 large enough.

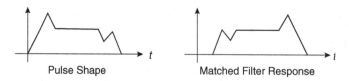

Pulse Shape   Matched Filter Response

Figure 3.2. Example of a matched filter.

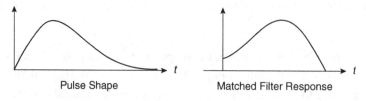

Pulse Shape   Matched Filter Response

Figure 3.3. Example of an approximate matched filter.

There is a slight problem with units whenever we write the pulse as $s(t)$ and the impulse response for the matched filter as $s^*(-t)$ because then the output signal of the matched filter at time t_0 is E_p, and the output signal power is E_p^2. This is dimensionally unsatisfactory because the output power has the units of joules-squared. This is why we find it convenient to redefine the input pulse as $As(t)$ and the matched filter as $s^*(-t)$, where now the normalized pulse $s(t)$ has energy equal to one and $As(t)$ has energy $|A|^2$. Then the expected output of the matched filter $s^*(-t)$ is

$$\int_{-\infty}^{\infty} A|s(t)|^2 dt = A.$$

Similarly, with the filter $s(-t)$, the output noise power now is

$$\sigma^2 = \int_{-\infty}^{\infty} N(f)|S(f)|^2 df$$
$$= \frac{N_0}{2}.$$

This is summarized in the following theorem.

Theorem 3.1.4 *The signal output $Ar(t)$ at time zero of the filter $s^*(-t)$ matched to pulse $As(t)$, with energy $E_p = |A|^2$, is equal to A. If the input noise is white, the output noise variance is $N_0/2$.*

If there are two or more pulses in a discussion, we may have a matched filter for each of them, with all matched filters fed by the identical noisy signal $v(t)$. We are then interested in the correlation between the output noise samples.

Theorem 3.1.5 *Filters with common input that are matched to real orthogonal pulses have uncorrelated noise outputs at time zero if the common noise input is white noise, and independent noise outputs if the common noise input is white and gaussian.*

Proof Consider the outputs of two filters matched to real pulses $s_i(t)$ and $s_j(t)$. Let n_i and n_j be the noise outputs of the real filters $s_i(-t)$ and $s_j(-t)$ at time zero. Then

$$E[n_i n_j] = E\left[\int_{-\infty}^{\infty} n(\xi)s_i(-\xi)d\xi \int_{-\infty}^{\infty} n(\xi')s_j(-\xi')d\xi'\right]$$
$$= \int_{-\infty}^{\infty}\int_{-\infty}^{\infty} s_i(-\xi)s_j(-\xi')E[n(\xi)n(\xi')]d\xi d\xi'.$$

Because the noise is white, $E[n(\xi)n(\xi')] = (N_0/2)\delta(\xi - \xi')$. Therefore

$$E[n_i n_j] = \int_{-\infty}^{\infty} s_i(-\xi)s_j(-\xi')\frac{N_0}{2}\delta(\xi - \xi')d\xi d\xi'.$$

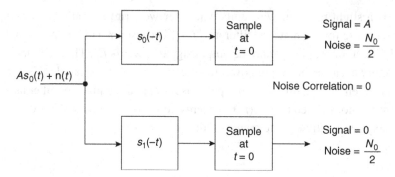

Figure 3.4. Matched filters for orthogonal pulses.

Carrying out one integration gives

$$E[n_i n_j] = \frac{N_0}{2} \int_{-\infty}^{\infty} s_i(-\xi)s_j(-\xi)d\xi.$$

Because when $i \neq j$, the pulses $s_i(t)$ and $s_j(t)$ are orthogonal, the final integral equals zero, so

$$E[n_i n_j] = 0$$

when $i \neq j$. The noise outputs n_i and n_j are uncorrelated at $t = 0$ when $i \neq j$. If the noise inputs are gaussian, then the outputs are also gaussian and so they are independent because uncorrelated gaussian random variables are independent.   ∎

Theorems 3.1.4 and 3.1.5 contain some of the more important properties of matched filters. These properties are portrayed in Figure 3.4. Theorem 3.1.5 also implies several other important facts. Suppose that the real pulse $s(t)$ satisfies

$$\int_{-\infty}^{\infty} s(t)s(t - \ell T)dt = 0.$$

Then $s(t)$ and $s(t - \ell T)$ are orthogonal. When excited by $As(t)$, the real filter $s(-t)$ will have output signal $Ar(t)$ equal to zero at time ℓT for $\ell \neq 0$, and equal to A at time zero. Thus the output $r(t)$ is a Nyquist pulse. Theorem 3.1.5 tells us further that with white noise at the input of the matched filter $s(-t)$, samples at the output of the matched filter will be uncorrelated at times separated by ℓT.

3.2   Demodulation of binary baseband waveforms

In the remainder of this chapter, we shall develop demodulators for the various waveforms described in Chapter 2 when received in additive stationary noise. These demodulators do not require that the noise is gaussian noise. Any additive stationary noise is permitted. However, we calculate the probability-of-error performance of these demodulators only for gaussian noise. We do not make any statement in this chapter that the demodulators are optimal. In general, they are not optimal, but we shall prove in Chapter 7 that these demodulators are optimal whenever the additive noise is gaussian noise.

Figure 3.5 shows the demodulator for binary antipodal signaling. The received signal is passed through a matched filter to maximize the signal-to-noise ratio at the particular time instant at which the filter output is to be sampled. The data bit is declared to be a zero or a one based on whether the sample is positive or negative. Each sample of the matched-filter output is an example of a *decision statistic*. A decision statistic is any function of the received signal upon which a decision is based. The probability that the decision is wrong is called the *probability of bit error* or the *bit error rate*.

The linear demodulator for binary on–off keying is similar. The received signal $v(t)$, which is either equal to $s(t) + n(t)$ or equal to $n(t)$ only, is passed through the matched filter $s(-t)$ and sampled at ℓT. If the ℓth sample is larger than the threshold $\Theta = A/2$, then the ℓth data bit is declared to be a one. Otherwise, it is declared to be a zero. It should be obvious that the threshold Θ should be set midway between 0 and A.

The demodulator for binary orthogonal signaling (binary FSK) has two matched filters, as shown in Figure 3.6. One filter is matched to $s_0(t)$ and the other is matched to $s_1(t)$. The received signal $v(t)$, which is either equal to $s_0(t) + n(t)$ or to $s_1(t) + n(t)$, is

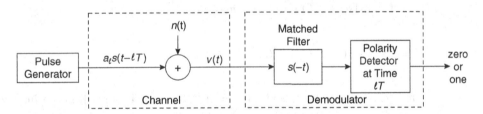

Figure 3.5. Matched-filter demodulation of antipodal signaling.

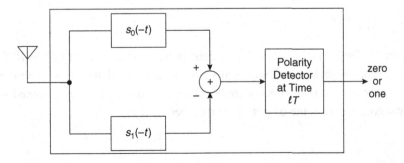

Figure 3.6. Matched-filter demodulation of binary FSK.

applied to both filters. The filter outputs are both sampled at ℓT and the ℓth data bit is declared to be a zero or a one based on which filter output sample is larger. To decide which of the two outputs is larger, subtract the two outputs, one from the other, to form the decision statistic. The sign of the decision statistic determines the demodulated data bit. The decision statistic has noise from each matched-filter output so it has double the noise power of BPSK.

Many of the other demodulators that we shall study also recover the datastream by applying the signal to the input of a matched filter and forming a decision statistic from the filter output at each sampling instant ℓT. The use of the matched-filter output to demodulate one bit at a time implies that there is no intersymbol interference. The output of the matched filter will have no intersymbol interference if the signal pulses at the output of the matched filter are Nyquist pulses. This means that we need no longer require that the pulse at the transmitter be a Nyquist pulse, only that the pulse out of the matched filter is a Nyquist pulse. Therefore we may require that the transmitted pulse $s(t)$ be redefined accordingly so that the pulse at the output of the matched filter is the Nyquist pulse $r(t)$.

For antipodal signaling, the received signal

$$v(t) = \sum_{\ell=-\infty}^{\infty} a_\ell s(t - \ell T) + n(t)$$

is passed through the matched filter $s(-t)$. The filter output is

$$u(t) = \int_{-\infty}^{\infty} a_\ell [s(t - \ell T) * s(-t)] + n(t) * s(-t)$$

$$= \sum_{\ell=-\infty}^{\infty} a_\ell r(t - \ell T) + n'(t)$$

where $r(t) = s(t) * s(-t)$ and $n'(t) = n(t) * s(-t)$. There will be no intersymbol interference at the output of the matched filter if $r(t)$ is a Nyquist pulse. Let $R(f)$ be any function of frequency that is real, nonnegative, and satisfies Theorem 2.4.2. Then $r(t)$ is a Nyquist pulse. Let $s(t)$ have the transform

$$S(f) = R^{\frac{1}{2}}(f)e^{j\theta(f)}$$

for any $\theta(f)$. Then $s(t) * s(-t)$ is the Nyquist pulse $r(t)$ because $|S(f)|^2 = R(f)$.

Henceforth, we shall commonly deal with the received signal at the output of the matched filter and we shall suppose that the transmitted signal $s(t)$ is designed so that the signal component at the output of the matched filter is

$$c(t) = \sum_{\ell=-\infty}^{\infty} a_\ell r(t - \ell T)$$

where $r(t)$ is a Nyquist pulse. Moreover, for antipodal signaling,

$$
a_\ell =
\begin{cases}
A & \text{if the } \ell\text{th data bit is a 1} \\
-A & \text{if the } \ell\text{th data bit is a 0.}
\end{cases}
$$

The amplitude A now represents the amplitude at the output of the matched filter. The ℓth sample of the matched-filter output, given by

$$
u_\ell = u(\ell T)
$$

$$
= \sum_{\ell'=-\infty}^{\infty} a_{\ell'} r(\ell T - \ell' T) + n'(\ell T)
$$

$$
= a_\ell + n'_\ell,
$$

depends on the ℓth data bit through a_ℓ. The ℓth data bit of the antipodal signaling waveform is estimated to be a zero or a one depending on whether the ℓth sample $u(\ell T)$ is negative or positive.

Figure 3.7 shows how the generation of the Nyquist pulse $r(t)$ is now distributed between the transmitter and the receiver. By distributing the pulse generation in this way, we can have both a matched filter and a Nyquist pulse. The pulse as it enters the channel $s(t)$ is no longer a Nyquist pulse. (It might be called a pre-Nyquist pulse.) Therefore, the modulated waveform $c(t)$ entering the channel can look quite complicated and separate data bits will interfere. The matched filter in the receiver, however, will render the waveform more understandable by forming the Nyquist pulse $r(t)$.

For example, let

$$
s(t) = \frac{4}{\pi \sqrt{T}} \frac{\cos(2\pi t/T)}{1 - (4t/T)^2},
$$

which is not a Nyquist pulse. Indeed, for this pulse, $s(\ell T)$ is never equal to zero for integer ℓ. However, at the output of the matched filter

$$
r(t) = s(t) * s(-t)
$$

$$
= \frac{\sin(\pi t/T)}{\pi t/T} \frac{\cos(\pi t/T)}{1 - (4t/T)^2},
$$

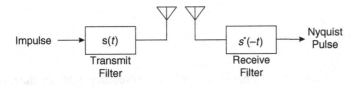

Figure 3.7. Partitioning the generation of a Nyquist pulse.

which is a Nyquist pulse. Often, we speak carelessly of transmitting the Nyquist pulse $r(t)$ when we really mean that the pulse $s(t)$ is actually transmitted; the existence and role of the matched filter is implied.

3.3   Error rates for binary signaling

The performance of a binary signaling scheme is given by the probability of bit error at the demodulator output. An error can be made in either of two ways: the demodulator can detect a zero bit when a one bit was actually transmitted, or the demodulator can detect a one bit when a zero bit was actually transmitted. The probabilities of these two error events need not be equal, but usually one prefers to design the demodulator so that they are equal, and this is the only case we analyze. In order to calculate the probability of demodulation error, the noise probability density function must be specified.

In this section, we shall derive the demodulated bit error rate E_b for each elementary binary signaling waveform transmitted through a distortionless, additive white gaussian-noise channel. The additive white gaussian-noise channel occurs often in practice, and is a standard channel to judge the performance of a communication system. Often the actual channel noise is not known.

The received signal is

$$v(t) = c(t) + n(t)$$

where $c(t)$ is the transmitted waveform, and $n(t)$ is white gaussian noise. Because the noise is white, its power density spectrum is given by

$$N(f) = \frac{N_0}{2}.$$

We shall see that the performance of binary signaling in white gaussian noise depends on N_0 and E_b only through the ratio E_b/N_0.

The decision made by the matched-filter demodulator is based on using the matched-filter output as a decision statistic. When only one pulse shape is used in constructing the waveform, there is only one matched filter and only one decision statistic at time ℓT. There will be no intersymbol interference if the signaling pulse $r(t)$ at the output of the matched filter is a Nyquist pulse. When $v(t)$ is passed through a filter matched to $s(t)$, the filter output is

$$v(t) = \sum_{\ell=-\infty}^{\infty} a_\ell r(t - \ell T) + n'(t).$$

The decision statistic at time ℓT, which is equal to the correlation between the received signal and the transmitted signal, is compared to the threshold Θ. If the decision statistic is above the threshold, a one is declared to be the data bit; if the decision statistic is below the threshold, a zero is declared to be the data bit. The value of the threshold is chosen so that the two kinds of error probability are equal.

When unbiased gaussian noise is applied to the input of any filter, the output of the filter is also gaussian noise and any sample of the filter output is a gaussian random variable with probability density function

$$p(x) = \frac{1}{\sqrt{2\pi}\sigma} e^{-x^2/2\sigma^2}$$

where σ^2 is the variance of that random variable.

Let x denote the decision statistic, consisting of the sampled output of the matched filter. Let $p_0(x)$ denote the probability density function on x when a zero is transmitted; let $p_1(x)$ denote the probability density function on x when a one is transmitted. Figure 3.8 shows $p_0(x)$ and $p_1(x)$ for the case of OOK in gaussian noise. The area of the hatched region of $p_0(x)$ in Figure 3.8 is the probability that the threshold will be exceeded when a zero is transmitted; the area of the hatched region of $p_1(x)$ is the probability that the threshold will not be exceeded when a one is transmitted. These error probabilities are denoted $p_{e|0}$ and $p_{e|1}$ and are given by

$$p_{e|0} = \int_{\Theta}^{\infty} p_0(x)dx$$

$$p_{e|1} = \int_{-\infty}^{\Theta} p_1(x)dx;$$

where Θ is to be chosen so that $p_{e|0} = p_{e|1}$. This value of Θ clearly should be where the two gaussian distributions cross because then the area under the two tails above and below Θ are equal. The error probabilities are the areas of the two tails.

To compute the probability of error for each binary signaling scheme that we have studied, we need to carry out the integration of the corresponding $p_0(x)$ and $p_1(x)$ over the appropriate interval. For some probability density functions, $p_0(x)$ and $p_1(x)$, the

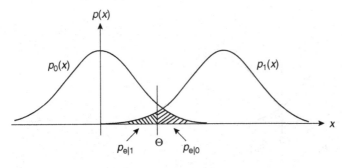

Figure 3.8. Error probabilities for binary OOK.

integrals can be evaluated in closed form, but for gaussian probability density functions, numerical integration is necessary. This is described by the standard function $Q(x)$ that was defined in Section 1.7.

The first case we analyze is antipodal signaling in gaussian noise. Then

$$p_0(x) = \frac{1}{\sqrt{2\pi}\sigma} e^{-(x+A)^2/2\sigma^2}$$

$$p_1(x) = \frac{1}{\sqrt{2\pi}\sigma} e^{-(x-A)^2/2\sigma^2}$$

where A is the mean sampled output of the matched filter, and σ^2 is the noise variance at the output of the matched filter. By symmetry of this case, it is clear that $p_{e|0} = p_{e|1}$ if the threshold Θ equals zero. Therefore

$$p_{e|0} = \int_0^\infty \frac{1}{\sqrt{2\pi}\sigma} e^{-(x+A)^2/2\sigma^2} dx$$

$$p_{e|1} = \int_{-\infty}^0 \frac{1}{\sqrt{2\pi}\sigma} e^{-(x-A)^2/2\sigma^2} dx.$$

These two integrals are illustrated in Figure 3.9. Such integrals cannot be evaluated in terms of elementary functions, but are expressed in terms of the function $Q(x)$.

Theorem 3.3.1 *The average error probability for antipodal signaling (or binary phase-shift keying) with no intersymbol interference, received in additive white gaussian noise is*

$$p_e = \tfrac{1}{2}p_{e|0} + \tfrac{1}{2}p_{e|1}$$

$$= Q\left(\sqrt{\frac{2E_b}{N_0}}\right).$$

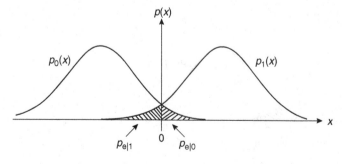

Figure 3.9. Error probabilities for antipodal signaling.

Proof Because the two integrals for $p_{e|0}$ and $p_{e|1}$ are equal, we need to evaluate only one of them. Make the change in variables

$$y = \frac{(x+A)}{\sigma}.$$

Then

$$p_e = p_{e|0} = \int_{A/\sigma}^{\infty} \frac{1}{\sqrt{2\pi}} e^{-y^2/2} dy$$

$$= Q\left(\frac{A}{\sigma}\right)$$

as was to be proved. ∎

The signal-to-noise ratio at the output of the matched filter is given by

$$S/N = \frac{A^2}{\sigma^2} = \frac{2E_p}{N_0} = \frac{2E_b}{N_0}$$

because $E_p = E_b$. Therefore we can write

$$p_e = Q\left(\sqrt{\frac{2E_b}{N_0}}\right).$$

The error probability for antipodal signaling is shown in Figure 3.10. For large values of E_b/N_0, we can use the approximation $Q(x) \approx e^{-x^2/2}/x\sqrt{2\pi}$ to write

$$p_e \approx \frac{e^{-E_b/N_0}}{\sqrt{4\pi E_b/N_0}}$$

or, more coarsely and more simply, the bound that $Q(x) < \frac{1}{2}e^{-x^2/2}$ to write

$$p_e < \frac{1}{2}e^{-E_b/N_0}.$$

Approximations of this kind are useful for rough, order-of-magnitude comparisons.

The demodulator for on–off keying uses a single matched filter with a threshold Θ set at $A/2$, so the probability of error for OOK in gaussian noise, shown in Figure 3.10, can be obtained in the same way as for antipodal signaling. However, with just a bit of thought, we can avoid repeating the work while gaining a little insight. Clearly, the output of the matched filter has the same noise as in the previous case, and the signal output is either zero or A depending on whether the data bit is a zero or a one. This means that the difference in the two signals is A – half as much as in the case of antipodal signaling, which has a difference of $2A$ between the two signals.

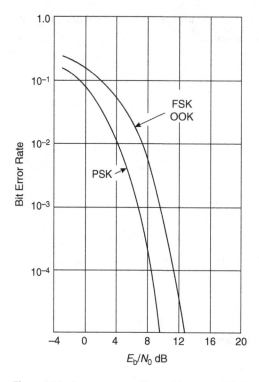

Figure 3.10. Performance of basic binary modulation methods.

Theorem 3.3.2   *The average bit error probability for binary OOK with no intersymbol interference in white gaussian noise is*

$$p_e = Q\left(\sqrt{\frac{E_b}{N_0}}\right).$$

Proof   The detection problem is equivalent to antipodal signaling with the two signal amplitudes $-A/2$ and $A/2$ because the difference between these two numbers is also A, and the noise is the same as before. In that case, the average energy per bit would be proportional to $(A/2)^2$, while in the case of OOK, the average energy per bit is proportional to

$$\tfrac{1}{2}0^2 + \tfrac{1}{2}A^2 = \tfrac{1}{2}A^2$$

because only half the pulses, on average, are nonzero. Therefore the performance of OOK is the same as the performance of antipodal signaling using half as much energy. To achieve the same probability of bit error, baseband OOK requires exactly twice as much power as antipodal signaling. The statement of the theorem follows.   ■

The probability of error for binary frequency-shift keying, or binary orthogonal signaling, will be given next. We will see that, like baseband OOK, binary FSK requires twice the power of antipodal signaling, but for a different reason.

Theorem 3.3.3 *The average bit error probability for a demodulator of binary orthogonal signaling (or binary FSK) with no intersymbol interference in additive white gaussian noise is*

$$p_e = Q\left(\sqrt{\frac{E_b}{N_0}}\right).$$

Proof To decide which output is larger, subtract the two outputs and determine the sign of the difference. The decision statistic then is the difference between the two matched-filter outputs, only one of which has a pulse response in its output. Therefore the expected value of the decision statistic is the same as antipodal signaling with the same pulse energy. The noise power, however, is twice as large because, by Theorem 3.1.5, the noise output is the difference of two independent noise terms of equal variance, and so the noise variances add. Consequently,

$$p_e = Q\left(\sqrt{\frac{2E_p}{2N_0}}\right).$$

Because $E_b = E_p$, the theorem is proved. ∎

3.4   Demodulators for multilevel signaling

An M-ary multilevel signaling waveform, as was described in Section 2.3, is used to convert a waveform channel into a discrete channel with M input symbols corresponding to the M input amplitudes of a signal constellation $S = \{c_0, c_1, \ldots, c_{M-1}\}$. We shall require that $M = 2^k$ for some integer k. The modulator at time ℓT, upon receiving the input data symbol c_m representing k data bits, sets $a_\ell = c_m$ and passes $a_\ell s(t - \ell T)$ through the channel. Therefore, the received waveform in additive white noise $n(t)$ is

$$v(t) = \sum_{\ell=-\infty}^{\infty} a_\ell s(t - \ell T) + n(t)$$

where $s(t)$ is the pulse shape and a_ℓ is the point from the signal constellation chosen at the ℓth time instant to represent k data bits. The task of the demodulator is to recognize that $a_\ell = c_m$. If the demodulator decides that $a_\ell = c_{m'}$ for some $m' \neq m$, then the data symbol demodulated at time ℓT is in error.

Because the same pulse shape, $s(t)$, is used for every point of the signal constellation, there is only one matched filter. We shall suppose again that the pulse shape and the interval T are chosen so that the pulse $r(t)$ at the output of the matched filter is a Nyquist pulse. When $v(t)$ is passed through a filter matched to $s(t)$, the filter output is

$$u(t) = \sum_{\ell=-\infty}^{\infty} a_\ell r(t - \ell T) + n'(t)$$

where $n'(t) = n(t) * s(-t)$, and $r(t)$ is a Nyquist pulse. The ℓth time sample $u_\ell = u(\ell T)$ of the matched-filter output is

$$u_\ell = a_\ell + n'_\ell$$

where a_ℓ is the element of the signal constellation $\mathcal{S} = \{c_0, c_1, \ldots, c_{M-1}\}$ transmitted at time ℓT, and $n'_\ell = n'(\ell T)$ is a sequence of independent noise samples, each of variance σ^2. These samples are gaussian random variables if the channel noise is gaussian. The task of the demodulator is to estimate a_ℓ from u_ℓ. This should be done to minimize the probability of decision error. Evidently, the estimate $\widehat{a}_\ell$ of a_ℓ is the value $c_m \in \mathcal{S}$ that is closest to u_ℓ. That is, the data estimate at time ℓT is

$$\widehat{m}_\ell = \text{argmin}_m |u_\ell - c_m|$$
$$= \text{argmin}_m d(u_\ell, c_m).$$

This decision rule partitions the real line representing the values u_ℓ into regions. The region corresponding to c_m is called the mth *decision region*. The decision region corresponding to c_m is denoted $\mathcal{D}_m$.

For example, the four-ary multilevel signal constellation $\{-3A, -A, A, 3A\}$ is shown in Figure 3.11 together with the decision regions, $\mathcal{D}_0, \mathcal{D}_1, \mathcal{D}_2$, and $\mathcal{D}_3$. The intervals $(-\infty, -2A), (-2A, 0)\ (0, 2A), (2A, \infty)$ are the decision regions. If $u_\ell \in \mathcal{D}_m$, then m is chosen as the estimate of the transmitted data symbol. It should be noted that with this signal constellation and this choice of decision regions, the four points are not equally vulnerable to noise-induced errors. The point denoted $3A$, for example, can only be demodulated incorrectly when the noise is negative, not when the noise is positive.

Figure 3.11. Decision regions.

3.5   Error rates for multilevel signaling

Bit error rates for multilevel signaling can be calculated in nearly the same way as they were for binary antipodal signaling. For example, if the signal constellation is the four-ary constellation $\{-3A, -A, A, 3A\}$ transmitted in additive gaussian noise, then the probability of error when $a_\ell = -A$ is transmitted is the probability that a gaussian random variable with mean $-A$ takes a value smaller than $-2A$ or greater than zero. Altogether, for the four decision regions, there are six probabilities to be tabulated. Each is a similar tail of a gaussian distribution as illustrated in Figure 3.12. Each of these tails has the same probability, which is $Q\left(\frac{A}{\sigma}\right)$, of confusing a point with its nearest neighbor. If each of the four symbols is transmitted with equal probability, equal to $1/4$, then the probability of symbol error is easily seen to be

$$
\begin{aligned}
p_{es} &= \tfrac{1}{4}Q\left(\tfrac{A}{\sigma}\right) + \tfrac{1}{4}2Q\left(\tfrac{A}{\sigma}\right) + \tfrac{1}{4}2Q\left(\tfrac{A}{\sigma}\right) + \tfrac{1}{4}Q\left(\tfrac{A}{\sigma}\right) \\
&= \tfrac{3}{2}Q\left(\tfrac{A}{\sigma}\right) \\
&= \tfrac{3}{2}Q\left(\sqrt{\tfrac{2E_p}{N_0}}\right)
\end{aligned}
$$

because the two middle gaussians each have two error tails, and the other two gaussians each have one error tail. The average energy per symbol is given by

$$
\begin{aligned}
E_s &= \tfrac{1}{4}(-3A)^2 + \tfrac{1}{4}(-A)^2 + \tfrac{1}{4}(A)^2 + \tfrac{1}{4}(3A)^2 \\
&= 5A^2.
\end{aligned}
$$

Because each pulse conveys two bits, the average energy per bit is $E_b = E_s/2 = 5A^2/2 = 5E_p/2$. Therefore the probability of symbol error is

$$
p_{es} = \tfrac{3}{2}Q\left(\sqrt{\tfrac{4\,E_b}{5\,N_0}}\right).
$$

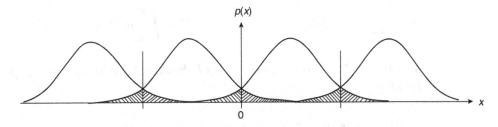

Figure 3.12. Error probabilities for four-ary ASK.

If the data is Gray-coded, then a symbol error will almost always contain only one bit error. Therefore the probability of bit error is

$$p_{eb} \approx \tfrac{3}{4} Q \left(\sqrt{\frac{4}{5} \frac{E_b}{N_0}} \right).$$

This is the bit error rate for four-ary amplitude shift keying. This should be compared to

$$p_{eb} = Q \left(\sqrt{\frac{2E_b}{N_0}} \right)$$

which is the bit error rate for antipodal signaling. Four-ary ASK requires more than twice as much energy per bit to achieve the same bit error rate as antipodal signaling, but it has double the data rate.

3.6   Demodulators for *M*-ary orthogonal signaling

An M-ary orthogonal waveform is used to convert the waveform channel into a channel with M discrete input symbols; usually $M = 2^k$ for some integer k. Upon receiving the mth input value, the modulator transmits the pulse $s_m(t)$ through the channel. The demodulator must then determine m, the index of the transmitted symbol, based on an observation of the received signal $v(t)$ and its knowledge of the M possible choices for $s_m(t)$. The demodulator is based on a bank of matched filters, one filter for each $s_m(t)$.

Demodulation of an M-ary orthogonal signaling waveform proceeds much the same as the demodulation of a binary signaling waveform, but generalized to use multiple matched filters. The received signal

$$v(t) = \sum_{\ell=-\infty}^{\infty} As_{m_\ell}(t - \ell T) + n(t)$$

is passed through a bank of matched filters: $s_m(-t)$ for $m = 0, \ldots, M-1$, one matched filter for each pulse in the M-ary orthogonal family. Each of the M filter outputs

$$u_m(t) = \sum_{\ell'=-\infty}^{\infty} \int_{-\infty}^{\infty} As_{m_{\ell'}}(\xi - \ell'T)s_m(\xi - t)d\xi + \int_{-\infty}^{\infty} n(\xi)s_m(\xi - t)d\xi$$

for $m = 0, \ldots, M-1$, is sampled at $t = \ell T$. Because the pulses are orthogonal, the relationship

$$\int_{-\infty}^{\infty} s_m(\xi - \ell'T)s_{m'}(\xi - \ell T)d\xi = \delta_{\ell\ell'}\delta_{mm'}$$

simplifies the expression for the matched-filter output. If the pulses out of the matched filters are Nyquist pulses, then there is no intersymbol interference. Thus

$$u_m(\ell T) = A\delta_{m_\ell m} + n_{m\ell}$$

where the output noise samples $n_{m\ell}$ are independent, identically distributed random variables. They are uncorrelated if $n(t)$ is white noise and they are independent if the noise is white and gaussian. Each transmitted pulse affects only one matched-filter sample at one time. By finding the index of the matched filter with the largest output at time ℓT, one has an estimate $\widehat{m}_\ell$ of which pulse was transmitted at time ℓT.

3.7   Error rates for *M*-ary orthogonal signaling

A demodulator for an M-ary orthogonal signaling waveform passes the received signal through a bank of M matched filters, one filter matched to $s_m(t)$ for $m = 0, \ldots, M-1$, and samples the output at time ℓT. The value of m corresponding to the filter output $u_m(\ell T)$ with the largest output value is demodulated as the estimated channel output symbol $\widehat{m}$ at time ℓT. Usually, the output of the correct filter will have the largest value, so m will be correctly demodulated, but sometimes the noise will be such that the wrong filter will have a larger output than the correct filter. Then a demodulation error occurs.

We saw in Section 3.6 that the bank of matched filters has output samples given by

$$u_m(\ell T) = A\delta_{m_\ell m} + n_{m\ell}$$

where the noise samples $n_{m\ell}$ are identically distributed random variables with variance σ^2. Suppose that the channel noise $n(t)$ is white gaussian noise. Then the $n_{m\ell}$ are independent gaussian random variables and, by Theorem 3.1.4, we have $E[n_{m\ell}]^2 = \sigma^2 = N_0/2$.

The discrete channel makes an error at time ℓT if the mth symbol is the input to the waveform channel and if the output of some other matched filter exceeds the output of the mth matched filter. The probability of error is independent of m if all of the pulses in the pulse alphabet have the same energy. We will evaluate the probability of error in two ways: first in an approximate way using the union bound,[1] then in an exact way. Each method of analysis has its virtues. An exact analysis is preferred for actual calculations. The approximate analysis, however, is valuable for the insight it provides.

---

[1] For any set of events $\mathcal{E}_i$, the union bound is

$$\Pr[\cup_i \mathcal{E}_i] \le \sum_i \Pr[\mathcal{E}_i].$$

Let $p_{m'|m}$ be the probability that the output sample of the m'th matched filter is larger than the output sample of the mth matched filter under the condition that the mth symbol was transmitted. Then by the union bound,

$$p_e \leq \sum_{m' \neq m} p_{m'|m}.$$

But a formula for $p_{m'|m}$ is already known. Because it is the probability of incorrectly separating two orthogonal pulses, it is the same as the probability of error of binary FSK:

$$p_{m'|m} = Q\left(\sqrt{\frac{E_p}{N_0}}\right).$$

For M-ary orthogonal signaling, because $E_p = E_b \log_2 M$ and there are $M-1$ identical terms in the sum, the union bound immediately gives

$$p_e \leq (M-1)Q\left(\sqrt{\log_2 M \frac{E_b}{N_0}}\right).$$

This inequality is actually an asymptotic equality as E_b/N_0 grows large. The reason for this can be understood by considering why the union bound yields an inequality rather than an equality. There will be occasions for which two incorrect matched filters both have output samples that exceed the output sample of the correct matched filter and such events are counted as two errors in computing the union bound even though there is only one error. This double counting occurs with negligible probability for large E_b/N_0, but for small E_b/N_0 this double counting overestimates p_e and makes the bound useless. An exact analysis is necessary to understand the error quantitatively. Fortunately, an exact analysis is possible.

An exact expression for the probability of error is given by the following theorem, which is a generalization of Theorem 3.3.3.

Theorem 3.7.1 *The probability of symbol error p_e of a matched-filter demodulator for an M-ary orthogonal waveform alphabet used without intersymbol interference on an additive white gaussian-noise channel is given as a function of E_b/N_0 by*

$$p_e = 1 - \int_{-\infty}^{\infty} \frac{1}{\sqrt{2\pi}} e^{-x^2/2} \left[\int_{-\infty}^{x+\sqrt{(2E_b/N_0)\log_2 M}} \frac{1}{\sqrt{2\pi}} e^{-y^2/2} dy\right]^{M-1} dx.$$

Proof Without loss of generality, suppose that $s_m(t)$ is the transmitted pulse. Let z be the sampled output of the mth matched filter. Let σ^2 denote the noise variance at the output of each matched filter, and let A denote the signal at the mth filter output. Let $\gamma = E_b/N_0$. Recall that

$$\frac{A^2}{\sigma^2} = \frac{2E_p}{N_0} = \frac{2E_b \log_2 M}{N_0} = 2\gamma \log_2 M .$$

The probability that the m'th filter output, with $m' \neq m$, exceeds z is

$$P_{e|z,m'} = \int_z^\infty \frac{1}{\sqrt{2\pi}\sigma} e^{-w^2/2\sigma^2} dw.$$

Hence

$$1 - P_{e|z,m'} = \int_{-\infty}^z \frac{1}{\sqrt{2\pi}\sigma} e^{-w^2/2\sigma^2} dw.$$

The probability that every other filter output is smaller than the mth filter output is

$$1 - P_{e|z} = \left[\int_{-\infty}^z \frac{1}{\sqrt{2\pi}\sigma} e^{-w^2/2\sigma^2} dw \right]^{M-1},$$

because the filter outputs are independent random variables. This is true for each value of z. Taking the expected value of this expression with respect to z

$$1 - P_e = \int_{-\infty}^\infty \frac{1}{\sqrt{2\pi}\sigma} e^{-(z-A)^2/2\sigma^2} \left[\int_{-\infty}^z \frac{1}{\sqrt{2\pi}\sigma} e^{-w^2/2\sigma^2} dw \right]^{M-1} dz$$

gives the average probability of error p_e. Finally, make the following changes in the variables of integration

$$x = \frac{z - \sqrt{2\gamma \log_2 M}}{\sigma} \qquad y = \frac{w}{\sigma}$$

to complete the proof of the theorem. ■

The probability of symbol error in Theorem 3.7.1 can be re-expressed in terms of the function $Q(x)$ as

$$p_e = 1 - \int_{-\infty}^\infty \frac{1}{\sqrt{2\pi}} e^{-x^2/2} \left[1 - Q(x + \sqrt{2\gamma \log_2 M}) \right]^{M-1} dx$$

where $\gamma = E_b/N_0$.

The probability of symbol error in Theorem 3.7.1 depends only on M and E_b/N_0. The equation can be numerically integrated for each value of M and E_b/N_0. The result

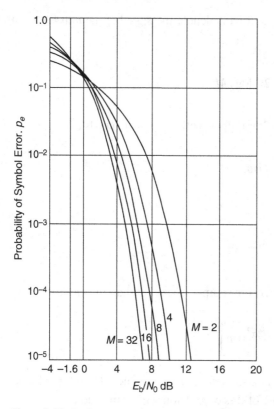

Figure 3.13. Performance of M-ary orthogonal signaling.

is shown in Figure 3.13. The expression of the theorem is unattractive for computing very small p_e because it finds p_e indirectly by first computing a number that is close to one, then subtracting from one. Integration by parts (left as an exercise) can be used to put the expression into a more satisfactory form.

Now we will find the probability of symbol error for a simplex family. Because a simplex family is closely related to an orthogonal family, it is not surprising that this probability can be stated as a corollary.

Corollary 3.7.2 *An M-ary simplex pulse alphabet with pulse energy* $\left(1 - \frac{1}{M}\right)E_p$ *can be demodulated with the same probability of symbol error as an M-ary orthogonal family with pulse energy* E_p.

Proof Let the simplex family be given by

$$q_m(t) = s_m(t) - \frac{1}{M} \sum_{m'=0}^{M-1} s_{m'}(t)$$

for $m = 0, \ldots, M - 1$ where $\{s_m(t)\}$ is a family of orthogonal pulses. Recall that

$$\int_{-\infty}^{\infty} q_m(t) q_{m'}(t) dt = \left(1 - \frac{1}{M}\right) E_p \qquad m = m'$$

$$= -\frac{E_p}{M} \qquad m \neq m'.$$

To demodulate, pass the received signal through a bank of M filters matched to the M pulses $s_m(t)$ for $m = 0, \ldots, M - 1$. The output of the bank of filters is the same as it was in the case of the original orthogonal pulses except for the additional term

$$-\frac{1}{M} \int_{-\infty}^{\infty} s_m(t) \sum_{m'=0}^{M-1} s_{m'}(t) dt = -\frac{E_p}{M},$$

which occurs in every filter output. This term is independent of m and so cannot affect the determination of the largest of the filter outputs. Hence the probability of error for simplex signaling is the same as the probability of error for orthogonal signaling. The energy of each waveform, however, is $\left(1 - \frac{1}{M}\right) E_p$. ∎

The performance of simplex waveforms, based on Corollary 3.7.2, is shown in Figure 3.14.

The probability of bit error, here denoted p_{eb}, can be computed from the probability of symbol error p_e for orthogonal or simplex signaling by noting that, when a symbol error occurs, all $M - 1$ incorrect symbols are equally probable. Errors occur with a probability that is independent of the particular symbol that is transmitted. Therefore we can fix on any symbol that is easy to analyze, in particular the symbol corresponding to the all-zero binary word. Suppose that $M = 2^k$ and that $m = 0$ is the transmitted symbol corresponding to an all-zero k-bit binary word. The expected number of ones in a nonzero symbol can be obtained by summing the total number of ones in all possible incorrect symbols, then dividing by $M - 1$. Because the correct symbol contains only zeros, we can calculate the total number of ones in all symbols, then divide by $M - 1$. But the total number of ones in the set of all k-bit binary words is easily seen to be $\frac{1}{2} k 2^k$. Therefore, on average, there are $k 2^{k-1} / (M - 1)$ incorrect bits when a symbol is incorrect. Whenever the demodulated symbol is incorrect, the conditional probability of a particular bit being incorrect is $2^{k-1} / (2^k - 1)$. Therefore the formula

$$p_{eb} = \frac{2^{k-1}}{2^k - 1} p_e$$

provides the conversion from the probability of symbol error to the probability of bit error. This is the bit error rate for orthogonal signaling or simplex signaling.

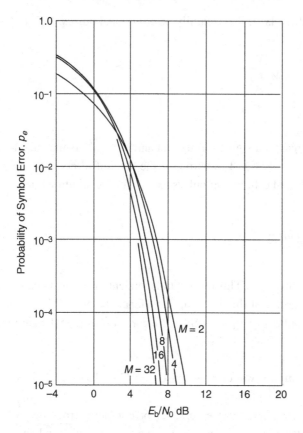

Figure 3.14. Performance of simplex signaling.

As a practical matter, orthogonal (or simplex) signaling is useful only for small or moderate values of M, at most $M = 64$ or 128. Nevertheless, it is informative to examine the performance of M-ary orthogonal signaling for very large values of M. At fixed values of E_b/N_0, the behavior of p_e as M becomes large is surprisingly simple to describe. The next theorem describes this behavior for orthogonal waveforms. The statement of the theorem also applies to simplex waveforms because, for large M, the performance of simplex signaling approaches the performance of orthogonal signaling.

Theorem 3.7.3 *As M goes to infinity, the probability of symbol error of an optimally demodulated orthogonal family of waveforms used on an additive gaussian-noise channel behaves as one of the two cases:*

$$p_e \to 0 \text{ if } E_b/N_0 > \log_e 2$$
$$p_e \to 1 \text{ if } E_b/N_0 < \log_e 2.$$

Proof   Let $\gamma = E_b/N_0$. From Theorem 3.7.1, we have

$$p_e = 1 - \int_{-\infty}^{\infty} \frac{1}{\sqrt{2\pi}} e^{-x^2/2} \left[\int_{-\infty}^{x+\sqrt{2\gamma \log_2 M}} \frac{1}{\sqrt{2\pi}} e^{-y^2/2} dy \right]^{M-1} dx$$

$$= 1 - \int_{-\infty}^{\infty} \frac{1}{\sqrt{2\pi}} e^{-x^2/2} e^{B(x,M)} dx$$

where

$$B(x,M) = \frac{\log \int_{-\infty}^{x+\sqrt{2\gamma \log_2 M}} \frac{1}{\sqrt{2\pi}} e^{-y^2/2} dy}{\frac{1}{M-1}}.$$

We need to determine the behavior of $B(x, M)$ as M goes to infinity. But as M goes to infinity, both the numerator and denominator go to zero. To find the limit of $B(x, M)$ as M goes to infinity, let $s = 1/M$ and use L'Hopital's rule[2] to find the limit as s goes to zero. Thus

$$\lim_{M \to \infty} B(x,M) = \lim_{s \to 0} \frac{\log \int_{-\infty}^{x+\sqrt{-2\gamma \log_2 s}} \frac{1}{\sqrt{2\pi}} e^{-y^2/2} dy}{\frac{s}{1-s}}.$$

The derivative of the denominator with respect to s is

$$\frac{d}{ds} \left[\frac{s}{1-s} \right] = \frac{1}{1-s} - \frac{s}{(1-s)^2}$$

which equals one in the limit as s goes to zero. The derivative of the numerator with respect to s is

$$\frac{-(\gamma/s) \frac{1}{\sqrt{2\pi}} e^{-(x+\sqrt{-2\gamma \log_2 s})^2/2}}{\sqrt{-2\gamma \log_2 s} \int_{-\infty}^{x+\sqrt{-2\gamma \log_2 s}} \frac{1}{\sqrt{2\pi}} e^{-y^2/2} dy}.$$

---

[2] L'Hopital's rule, a basic theorem of elementary calculus, states that

$$\lim_{x \to 0} \frac{A(x)}{B(x)} = \frac{A'(x)}{B'(x)}$$

whenever $A(x)/B(x)$ is indeterminate at $x = 0$, where $A'(x)$ and $B'(x)$ denote the derivatives of $A(x)$ and $B(x)$, respectively.

To find the limit of this complicated term as s goes to zero, replace s by $1/M$ and take the limit as M goes to infinity. The integral in the denominator goes to one and need not be considered further. Therefore

$$\lim_{M \to \infty} B(x,M) = \lim_{M \to \infty} \left[-\frac{\gamma M}{\sqrt{2\pi}} \frac{1}{\sqrt{2\gamma \log_2 M}} e^{-(x+\sqrt{2\gamma \log_2 M})^2/2} \right]$$

$$= \lim_{M \to \infty} \left[\frac{-\gamma}{\sqrt{2\pi}} e^{-(x^2+2x\sqrt{2\gamma \log_2 M}+2\gamma \log_2 M -2\log_e M +\log_e(2\gamma \log_2 M))/2} \right].$$

The behavior for large M is dominated by the terms that are linear in $\log M$

$$\lim_{M \to \infty} B(x,M) = \lim_{M \to \infty} [-e^{-2(\gamma -\log_e 2)\log_2 M}]f(x,M)$$

where $f(x,M)$ contains terms that are exponential in x and in $\sqrt{\gamma \log_2 M}$. For large M, the exponential in the bracket goes either to zero or to infinity depending on the sign in the exponent. Multiplying by $f(x,M)$ does not change this limit. Consequently,

$$\lim_{M \to \infty} B(x,M) = \begin{cases} 0 & \text{if } \gamma > \log_e 2 \\ -\infty & \text{if } \gamma < \log_e 2. \end{cases}$$

Hence for all x,

$$e^{B(x,M)} \to 1 \quad \text{if } \gamma > \log_e 2$$

$$e^{B(x,M)} \to 0 \quad \text{if } \gamma < \log_e 2.$$

The proof of the theorem is completed by substituting these limits for the bracketed term in the expression for p_e. ■

Theorem 3.7.3 tells us that we cannot transmit reliably with M-ary orthogonal signaling if E_b/N_0 is less than $\log_e 2$. This statement is more commonly expressed in decibels: M-ary orthogonal signaling cannot be used to transmit reliably if E_b/N_0 is less than -1.6 dB. Simplex waveforms do better for each finite M, but the relative advantage becomes insignificant for large M. Asymptotically as M goes to infinity, a family of simplex waveforms is no better than a family of orthogonal waveforms.

There still remains the question of whether there is another family of waveforms with better performance than a simplex family of waveforms. We do not answer this question for finite M, though it is widely believed that a simplex family of waveforms is optimal. However, we answer the question asymptotically in gaussian noise. The methods of information theory, as discussed in Chapter 11, tell us that it is not possible to signal reliably if E_b/N_0 is less than -1.6 dB. The information-theoretic proof completely circumvents the question of waveform design by using the ingenious methods of that subject, which is not the topic of this book. We simply assert that as M goes to infinity, an M-ary simplex family or an M-ary orthogonal family is asymptotically optimal for a channel with no bandwidth constraint.

Problems for Chapter 3

3.1.   Let

$$c(t) = \sum_{\ell=1}^{n} a_\ell s(t - \ell T).$$

Show that a matched filter for the entire waveform $c(t)$ can be constructed as the cascade (in either order) of a matched filter for the pulse $s(t)$ and a matched filter for the "array" $(a_1, \ldots, a_n)$. More generally, if

$$c(t) = a(t) * s(t).$$

Describe how the matched filter for $c(t)$ can be constructed as the cascade of two matched filters.

3.2.   Let $g(t)$ be the matched filter for the pulse $s(t)$. Suppose that instead of $g(t)$, the filter

$$g_\epsilon(t) = g(t) + \epsilon h(t)$$

is used where $h(t)$ has finite energy and ϵ is a small number. Prove that the decrease in signal-to-noise ratio at the filter output is quadratic in ϵ, that is, a series expansion in ϵ is

$$\frac{S}{N} = \left(\frac{S}{N}\right) - \epsilon^2 A + \cdots$$

for some positive constant A. What does this imply about the care with which a matched filter must be built?

3.3.   Prove that if a sinc pulse is passed through its matched filter, the output is the same sinc pulse.

3.4.   Show that

$$Q(x) = \frac{1}{2}\mathrm{erfc}\left(\frac{x}{\sqrt{2}}\right).$$

3.5. Derive an equation for the probability of error of an M-ary orthogonal waveform alphabet used on a channel with additive *colored* gaussian noise, that is, with additive gaussian noise of power density spectrum $N(f)$.

3.6. The matched filter for a complex pulse $s(t)$ is $s^*(-t)$. Sketch the matched-filter impulse response and the noise-free matched-filter output for each of the following pulses.

a.

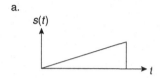

b.

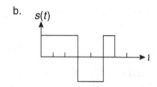

c.

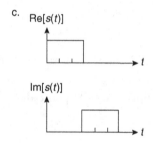

3.7. A four-ary baseband signal constellation $S = \{-3A, -A, A, 3A\}$ on the real line is to be used for a four-ary ASK waveform on an additive gaussian noise channel and demodulated by a matched filter followed by a set of decision regions.
   a. With decision regions defined by the points $-2A$, 0, $+2A$, and Gray-coded data, find an exact expression for the bit error rate.
   b. As a function of the filtered noise variance σ^2 and the energy per bit E_b, find the four points of the signal constellation and the thresholds that minimize the symbol error rate. Does this choice minimize the bit error rate?

3.8. Show how the expression

$$p_e = 1 - \int_{-\infty}^{\infty} \frac{1}{\sqrt{2\pi}} e^{-x^2/2} \left[1 - Q(x + \sqrt{2\gamma \log_2 M})\right]^{M-1} dx$$

can be used to derive the bound

$$p_e \leq (M-1)Q\left(\sqrt{\gamma \log_2 M}\right).$$

A factor of two appears in the square root of the exact equation, but is missing in the inequality. Why?

3.9. Specify the decision regions for four-ary amplitude shift keying that give the smallest bit error rate. Are the boundaries uniformly spaced? What is the bit error rate?

3.10. a. Use integration by parts to convert the formula of Theorem 3.7.1 into the expression

$$p_e = (M - 1) \int_{-\infty}^{\infty} \frac{1}{\sqrt{2\pi}} e^{-x^2/2} Q(x)^{M-2} Q\left(x - \sqrt{2\gamma \log_2 M}\right) dx.$$

  b. Derive this expression directly from the form of the demodulator.

3.11. A pulse $s(t)$ is a Nyquist pulse for period T if $s(0) = 1$ and $s(\ell T) = 0$ for every nonzero integer ℓ.

  a. Suppose that $s(t)$ has Fourier transform $S(f)$ sketched as follows.

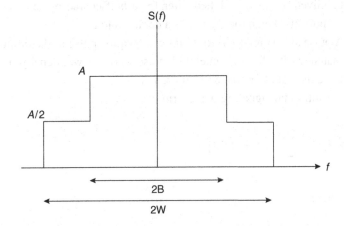

For what values of T is $s(t)$ a Nyquist pulse?

  b. If this Nyquist pulse $s(t)$ is to be the output of the matched filter in the receiver, what pulse should be transmitted?

  c. Describe a transmitted four-ary baseband ASK waveform that produces this Nyquist pulse at the output of the matched filter. What is the data rate? What signal constellation should be used?

  d. Give an approximate expression for E_b/N_0 for this ASK waveform. Sketch a demodulator.

  e. Give a bound on the largest value that the (infinitely long) transmitted signal can take at a value midway between two sampling instants.

3.12. a. Explain why, for fixed E_b/N_0, M-ary orthogonal signaling (with M larger than 2) has a smaller probability of bit error than binary signaling.

b. Show that when $M = 2$, the probability of symbol error for M-ary orthogonal signaling

$$P_e = 1 - \int_{-\infty}^{\infty} \frac{1}{\sqrt{2\pi}} e^{-x^2/2} \left[\int_{-\infty}^{x+\sqrt{(2E_b/N_0)\log_2 M}} \frac{1}{\sqrt{2\pi}} e^{-y^2/2} dy \right]^{M-1} dx$$

reduces to the probability of bit error

$$P_e = Q\left(\sqrt{\frac{E_b}{N_0}}\right)$$

for binary orthogonal signaling. (**Hint:** rotate the x, y coordinate system by 45°.)

3.13. a. Construct a four-ary orthogonal signaling waveform alphabet of duration T by using a half-cosine pulse of pulse width $T/4$ as a "chip".

b. Given a single matched filter for a half-cosine pulse as an existing component, sketch the design of a demodulator.

3.14. A pulse $s(t)$ is required to produce a Nyquist pulse at the output of a whitened matched filter for a channel with noise whose power density spectrum is $N(f)$. Give the equation that $S(f)$ must satisfy.

3.15. An antipodal signaling waveform

$$c(t) = \sum_{\ell=-\infty}^{\infty} a_\ell s(t - \ell T)$$

where

$$a_\ell = \begin{cases} A & \text{if the } \ell\text{th data bit is 1} \\ -A & \text{if the } \ell\text{th data bit is 0} \end{cases}$$

is transmitted over an additive white gaussian noise channel. At the sampled output of a matched filter, a three-level decision is used

If $x < -\frac{A}{2}$　　declare data bit is 0
If $x > \frac{A}{2}$　　declare data bit is 1
otherwise　　declare data bit is lost.

a. Give an expression for the bit error rate in terms of the Q function.
b. Give an expression for the bit lost rate in terms of the Q function.

Notes for Chapter 3

The matched filter as a filter for maximizing signal-to-noise ratio was introduced by North (1943), and was named by Van Vleck and Middleton (1946). The alternative name, *North filter*, is still used occasionally. The matched filter was studied extensively by Turin (1960). The matched filter for colored noise was introduced by Dwork (1950) and by George (1950).

4   Sequences at Baseband

Communication waveforms in which the received pulses, after filtering, are not Nyquist pulses cannot be optimally demodulated one symbol at a time. The pulses will overlap and the samples will interact. This interaction is called *intersymbol interference*. Rather than use a Nyquist pulse to prevent intersymbol interference, one may prefer to allow intersymbol interference to occur and to compensate for it in the demodulation process.

In this chapter, we shall study ways to demodulate in the presence of intersymbol interference, ways to remove intersymbol interference, and in Chapter 9, ways to precode so that the intersymbol interference seems to disappear. We will start out in this chapter thinking of the interdependence in a sequence of symbols as undesirable, but once we have developed good methods for demodulating sequences with intersymbol interference, we will be comfortable in Chapter 9 with intentionally introducing some kinds of controlled symbol interdependence in order to improve performance.

The function of modifying a channel response to obtain a required pulse shape is known as *equalization*. If the channel is not predictable, or changes slowly with time, then the equalization may be designed to slowly adjust itself by observing its own channel output; in this case, it is called *adaptive equalization*.

This chapter studies such interacting symbol sequences, both unintentional and intentional. It begins with the study of intersymbol interference and ends with the subject of adaptive equalization.

4.1   Intersymbol interference

In Chapter 3, we studied the demodulation of a single bit or a single symbol within a baseband waveform. The methods of that chapter apply whenever the individual symbols can be filtered and sampled without interaction, which requires that the signaling pulse at the filter output be a Nyquist pulse. In more general cases, the successively received symbols are overlapped or dependent. Then one speaks of demodulating the sequence of symbols rather than of demodulating one symbol at a time. The dependence between symbols is exploited in the structure of the demodulator.

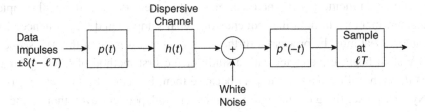

Figure 4.1. A cause of intersymbol interference.

An elementary communication waveform with intersymbol interference at the demodulator is best processed as a sequence of dependent symbols. Such dependent sequences arise in a variety of ways. The intersymbol interference may arise as a result of the pulse dispersion caused by the channel impulse response. This chapter includes ways to correct for the intersymbol interference that is created inadvertently such as by *linear dispersion* within the channel. Dispersion arises in a linear channel, which will be described as a filter, $h(t)$, as shown in Figure 4.1. The transmitted pulse and the received pulse are no longer the same. We shall denote the transmitted pulse by $p(t)$ and the received pulse by $s(t)$. For a single transmitted pulse, $p(t)$, the received pulse is $s(t) = p(t) * h(t)$. The received pulse in noise is

$$v(t) = p(t) * h(t) + n(t)$$
$$= s(t) + n(t)$$

where $n(t)$ is white noise. For a fully modulated waveform at the transmitter, the waveform at the receiver is

$$v(t) = \left[\sum_{\ell=-\infty}^{\infty} a_\ell p(t - \ell T) \right] * h(t) + n(t)$$

$$= \sum_{\ell=-\infty}^{\infty} a_\ell s(t - \ell T) + n(t).$$

How should we process $v(t)$ in the presence of the dispersive filter $h(t)$? One way is to pass the received signal through a filter matched to the pulse $p(t)$ and to sample at times ℓT. If we specify $p(t) * p(-t)$ to be the Nyquist pulse $r(t)$, then we know by Theorem 3.1.5 that the successive noise samples out of the filter will be uncorrelated, and if the noise is gaussian, independent, but the dispersion due to $h(t)$ will cause intersymbol interference in the output samples. Another way of processing $v(t)$ is to pass the received signal through a filter matched to $s(t)$ and to then sample at times ℓT. Now the output sample due to a single pulse will have maximum signal-to-noise ratio but, unless $s(t) * s(-t)$ is a Nyquist pulse, there will be intersymbol interference and the noise in the samples will be correlated. Each method results in a discrete output sequence with memory – in one case, memory in the sequence of signal samples; in the other, memory in both the sequence of signal samples and the sequence of noise samples – that must be processed further to extract an estimate of the data sequence.

We shall begin the chapter with a study of the first method of filtering – passing $v(t)$ through a filter matched to $p(t)$ because then, by requiring that $p(t) * p(-t)$ be a Nyquist pulse, the gaussian noise samples are independent even though the signal samples are dependent. In Section 7.4 of Chapter 7, we shall see how the second method of filtering – passing $v(t)$ through a filter matched to $p(t) * h(t)$ – arises as part of the maximum-likelihood demodulator. In Section 4.7 of this chapter, as an alternative to these methods, we shall study the method of linear equalization, which eliminates intersymbol interference by means of a linear filter, but which is not optimum.

With the fully modulated communication waveform stated above, the output of the channel is the waveform $v(t)$ and the output of the matched filter $p(-t)$ is

$$u(t) = v(t) * p(-t)$$

$$= \left[\sum_{\ell'=-\infty}^{\infty} a_{\ell'} p(t - \ell'T) \right] * h(t) * p(-t) + n(t) * p(-t)$$

$$= \left[\sum_{\ell'=-\infty}^{\infty} a_{\ell'} p(t - \ell'T) * p(-t) \right] * h(t) + n(t) * p(-t)$$

$$= \sum_{\ell'=-\infty}^{\infty} a_{\ell'} g(t - \ell'T) + n'(t)$$

where

$$g(t) = p(t) * p(-t) * h(t)$$
$$= r(t) * h(t).$$

Consequently

$$u_\ell = u(\ell T)$$

$$= \sum_{\ell'=-\infty}^{\infty} a_{\ell'} g_{\ell-\ell'} + n_\ell$$

$$= \sum_{\ell'=-\infty}^{\infty} g_{\ell'} a_{\ell-\ell'} + n_\ell,$$

is the sequence of samples of the waveform as seen at the output of the matched filter, where $g_\ell = g(\ell T)$, and $n_\ell = n'(\ell T)$ is the ℓth sample of the filtered noise. If

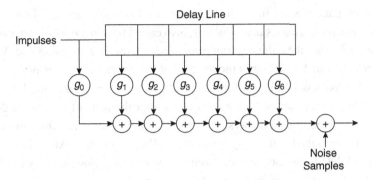

Figure 4.2. An equivalent discrete-time channel.

$p(t) * p(-t)$ is a Nyquist pulse, then $p(t)$ is orthogonal to its delayed version $p(t - \ell T)$ for all integer ℓ. Accordingly, by Theorem 3.1.5, the noise samples n_ℓ at the output of the matched filter are uncorrelated. Moreover, if $n(t)$ is gaussian, then the n_ℓ are independent, identically distributed, gaussian random variables.

The ℓth sample u_ℓ of the matched-filter output contains a superposition of responses to certain of the input pulses as well as additive noise. The sampled coefficient $g_\ell = g(\ell T)$ of the pulse $g(t)$ at the output is the ℓth coefficient of the intersymbol interference. The sequence of samples of $g(t)$ become the coefficients of a discrete-time filter describing the formation of the intersymbol interference.

Suppose that g_ℓ has only a finite number of nonzero (or nonnegligible) values, specifically that $g_\ell = 0$ if $\ell < 0$ or $\ell > v$. The statement that $g_\ell = 0$ for $\ell < 0$ holds because $g(t)$ is a causal filter, as must be the case for channel dispersion. The statement that $g_\ell = 0$ for $\ell > v$ means that the intersymbol interference persists for only $v + 1$ samples. Then

$$u_\ell = \sum_{\ell'=0}^{v} g_{\ell'} a_{\ell - \ell'} + n_\ell.$$

We call v the *constraint length* of the intersymbol interference.

Now we have arrived at the discrete-time model, shown in Figure 4.2, in which the channel is described as a finite-impulse-response filter. This model will be studied throughout the chapter.

4.2   Decision-feedback demodulation

The task of the demodulator is to recover the data sequence a_ℓ from the channel output sequence u_ℓ even when there is intersymbol interference (or interaction), as described by the coefficients g_ℓ. We shall study two demodulators that estimate the sequence of data bits a_ℓ from the sequence of filter samples u_ℓ. This section describes the *decision-feedback demodulator*, also called the *decision-feedback equalizer*. Section 4.3 describes the minimum-distance demodulator using the Viterbi algorithm, and Section 4.8 describes the method of least-mean-square error.

A decision-feedback demodulator for intersymbol interference can be used when the intersymbol interference caused by a pulse occurs after the peak of the matched-filter output, as may happen due to dispersion in the channel. Then each channel symbol creates interference only for those symbols that follow it in time. After a symbol is correctly demodulated, the interference from that symbol in subsequent symbols can be calculated and subtracted from those subsequent symbols.

A decision-feedback demodulator estimates the symbols of the datastream one by one. After each symbol is estimated, its effect on the next ν received samples is subtracted from them. In this way, each data symbol can be estimated from one received sample after the intersymbol interference is removed from that sample. With the received sample written as

$$u_\ell = \sum_{\ell'=0}^{\nu} g_{\ell'} a_{\ell-\ell'} + n_\ell,$$

we see that we can express a_ℓ as

$$a_\ell + \frac{1}{g_0} n_\ell = \frac{1}{g_0}\left[u_\ell - \sum_{\ell'=1}^{\nu} g_{\ell'} a_{\ell-\ell'}\right].$$

Because the noise is unknown, we estimate a_ℓ as the closest point of the signal constellation $\mathcal{S} = \{c_0, c_1, \ldots, c_{M-1}\}$ to the updated decision statistic at time ℓT, which is

$$u'_\ell = \frac{1}{g_0}\left[u_\ell - \sum_{\ell'=1}^{\nu} g_{\ell'} a_{\ell-\ell'}\right].$$

That is, the estimate of a_ℓ is

$$\widehat{a}_\ell = \text{argmin}_{c_m \in \mathcal{S}} d(c_m, u'_\ell).$$

This estimate is the basis of the decision-feedback demodulator. The demodulator assumes that it has demodulated all earlier symbols correctly and has formed the estimates, $\widehat{a}_\ell$, which are now used to estimate the intersymbol interference on the current received sample due to earlier symbols. The estimated value of intersymbol interference

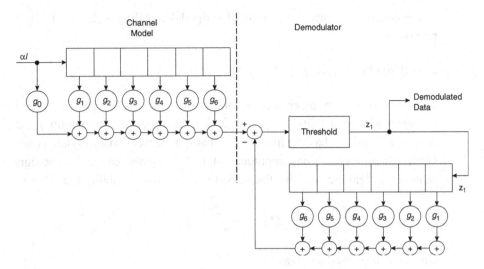

Figure 4.3. Decision-feedback demodulation.

is subtracted from the received sample, as follows:

$$\widehat{u}_\ell = \frac{1}{g_0}\left[u_\ell - \sum_{\ell'=1}^{\nu} g_{\ell'}\widehat{a}_{\ell-\ell'}\right]$$

$$= a_\ell + \frac{1}{g_0}\left[\sum_{\ell'=1}^{\nu} g_{\ell'}(a_{\ell-\ell'} - \widehat{a}_{\ell-\ell'}) + n_\ell\right].$$

If the terms under the summation are equal to zero because previous estimates are correct, the recovery of a_ℓ from $\widehat{u}_\ell$ is reduced to the task of demodulation in the absence of intersymbol interference.

Figure 4.3 shows a decision-feedback demodulator for an antipodal signaling waveform. The demodulator implements the equation

$$\widehat{a}_\ell = A\,\text{sgn}\left[u_\ell - \sum_{\ell'=\ell}^{\nu} g_{\ell'}\widehat{a}_{\ell-\ell'}\right].$$

If the demodulator has made no recent errors, the probability of error is the same as the probability of error for antipodal signaling without intersymbol interference. If the demodulator makes an error, however, it will compute the wrong feedback, and so the probability of error of subsequent bits will be larger. This phenomenon is known as *error propagation*. When the magnitude of the correction for intersymbol interference is large, error propagation will cause a significant degradation of performance. A single bit error results in the wrong feedback, which then leads to a large probability of a subsequent bit error, possibly initiating a long run of errors.

The simplest example is the case of antipodal signaling with $\nu = 1$. The decision rule is

$$\widehat{a}_\ell = A \, \text{sgn} \, [u_\ell - g_1 \widehat{a}_{\ell-1}]$$

and the probability of bit error p_e is computed as follows. The previous decision $\widehat{a}_{\ell-1}$ is correct with probability $1 - p_e$ and $\widehat{a}_{\ell-1}$ is incorrect with probability p_e. If $\widehat{a}_{\ell-1}$ is correct, $u_\ell - g_1 \widehat{a}_{\ell-1}$ has amplitude $\pm A$ just as if there were no intersymbol interference. Otherwise $u_\ell - g_1 \widehat{a}_{\ell-1}$ has amplitude $\pm A \pm 2g_1$, in which case either the signs are the same or the signs are opposite. Each case happens with probability $1/2$. This means that

$$p_e = (1 - p_e)Q\left(\frac{A}{\sigma}\right) + \frac{1}{2}p_e Q\left(\frac{A(1 + 2g_1)}{\sigma}\right) + \frac{1}{2}p_e Q\left(\frac{A(1 - 2g_1)}{\sigma}\right)$$

which can be solved for p_e as

$$p_e = \frac{Q(A/\sigma)}{1 - \frac{1}{2}Q(A(1 + 2g_1)/\sigma) - \frac{1}{2}Q(A(1 - 2g_1)/\sigma) + Q(A/\sigma)}.$$

The expression is always larger than $Q(A/\sigma)$ unless g_1 equals zero in which case it reduces to the performance of antipodal signaling. When A/σ is large and g_1 is small, the last three terms of the denominator are negligible, and $p_e \approx Q(A/\sigma)$, which is the performance of antipodal signaling without intersymbol interference.

4.3   Searching a trellis

A decision-feedback demodulator tries to cancel the effects of intersymbol interference. An alternative approach is to use the intersymbol interference as an additional source of information. Accordingly, another method of demodulating the output of a matched filter in the presence of intersymbol interference is to find the best fit to the noisy received sequence from the set of all possible noise-free received sequences. The brute-force way to do this is to compute the noise-free channel output sequence for every possible data sequence of some fixed length, and then to choose the channel output sequence that most closely resembles the received noisy sequence. The corresponding data sequence is the demodulated sequence. This is what we want to do, but it is intractable in the way that we have described it. This is because if the amplitude of the input pulse can take on q values, then there are q^n sequences of blocklength n, an exponentially large number. We will describe a better way to do this search.

The sequence demodulator that we shall describe works equally well with intersymbol interference occurring on either or both sides of the pulse maximum. It does not matter if the intersymbol-interference coefficients g_ℓ are nonzero for negative as well

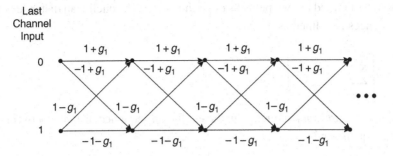

Figure 4.4. A trellis for intersymbol interference.

as positive values of ℓ. Indeed, the sequence demodulator does not even ask that one of the coefficients g_ℓ be singled out as the principal coefficient.

As a simple example, consider a simple case in which the intersymbol interference has one stage of memory with $g_0 = 1$, g_1 is nonzero, and for all other ℓ, g_ℓ is zero. The channel output symbol depends on the current and one past channel input symbol. If the channel input symbols are $+1$ and -1, corresponding to antipodal signaling, then the noise-free channel output only takes the values $\pm g_0 \pm g_1$. The channel memory is described by a two-state finite-state machine. A trellis for this problem with $g_0 = 1$ is shown in Figure 4.4. There are two states, and those states correspond to the two possible values of the previous channel bit. Every possible path through the trellis corresponds to a possible sequence of channel input sequences. Each path is labeled with the sequence of channel outputs that would occur if that path were followed and if there were no noise.

In general, a discrete-time channel is described by the finite impulse-response g_ℓ for $\ell = 0, \ldots, \nu$. The channel output depends on the current channel input symbol and the past ν channel input symbols. If the amplitude of the input pulse can take q values, then the channel can be in one of q^ν states and can be described by a trellis with q^ν states. Figure 4.4 gave an example of a trellis describing a channel with $q = 2$ and $\nu = 1$. For this channel, there are two states.

A trellis whose branches are labeled with real numbers defines an infinite collection of sequences of real numbers; each sequence is obtained by reading the sequence of labels along one of the infinite paths through the trellis. For our application, each sequence specifies one possible noise-free output sequence of the channel. So far, we have encountered only channels whose outputs form sequences of real numbers, but for full generality, we allow the branches of the trellis to be labeled with sequences of complex numbers.

We shall specify the best fit between sequences in terms of the geometrical language of *sequence distance*. Because the symbols labeling a trellis are real numbers or complex numbers, we may speak of euclidean distance between sequences. Let c and c' denote two sequences defined by two paths through the trellis. The euclidean distance between these sequences is defined as

$$d(\boldsymbol{c}, \boldsymbol{c}') = \left[\sum_{\ell=0}^{\infty} |c_\ell - c_\ell'|^2 \right]^{\frac{1}{2}} .$$

Indeed, given an arbitrary sequence v, the euclidean sequence distance between v and the trellis sequence $\boldsymbol{c}$ is

$$d(\boldsymbol{v}, \boldsymbol{c}) = \left[\sum_{\ell=0}^{\infty} |v_\ell - c_\ell|^2 \right]^{\frac{1}{2}} .$$

Every permitted sequence of labels is a possible noiseless received sequence and v is a noisy version of one of these sequences

$$v_\ell = c_\ell + n_\ell \qquad \ell = 1, 2, \ldots .$$

The task of demodulating v is the task of determining which channel sequence $\boldsymbol{c}$ gave rise to v. We can do this for a sequence of finite length by finding the trellis sequence $\boldsymbol{c}$ that is closest to v in euclidean sequence distance. When the noise is gaussian, we can justify this minimum distance criterion, in part, by the maximum-likelihood principle[1] to be given in Section 7.1. When the noise is not gaussian, the minimum-distance demodulator is still an excellent and widely used demodulator, but then it is not the maximum-likelihood demodulator.

Although each channel input symbol affects only $\nu + 1$ channel output symbols, it is not true that only $\nu + 1$ output symbols give information about any specific input symbol. We saw one manifestation of this in the form of error propagation in decision-feedback demodulation. Received symbols that are far apart become dependent because of the intermediate bits between them. The formal statement is as follows. Let $p(v_1, v_2, \ldots, v_n)$ be the probability density function on the vector random variable v. It has the structure of a Markov process. Define the *marginal distributions*

$$p(v_1, v_n) = \int_{-\infty}^{\infty} \int_{-\infty}^{\infty} \cdots \int_{-\infty}^{\infty} p(v_1, v_2, \ldots, v_n) dv_2 \cdots dv_{n-1}$$

$$p(v_1) = \int_{-\infty}^{\infty} p(v_1, v_n) dv_n$$

$$p(v_n) = \int_{-\infty}^{\infty} p(v_1, v_n) dv_1$$

---

[1] There is a subtle point here. The maximum-likelihood principle should be applied to the raw data $v(t)$ to see if it directs us to first compute the sequence v_ℓ. However, we choose to just compute the sequence v_ℓ without this justification and then apply the maximum-likelihood principle to that sequence to obtain a demodulator that is optimal, starting from the sequence v_ℓ, but might not be optimal when starting from the raw data $v(t)$.

and the *conditional distribution* $p(v_1|v_n) = p(v_1, v_n)/p(v_n)$. Then $p(v_1|v_n)$ is not equal to $p(v_1)$ even if n is much larger than the constraint length v. However, it is true that $p(v_1|v_n)$ will approach $p(v_1)$ as n grows large. This means that, even for large n, the received sample v_n can contribute something to the estimate $\hat{a}_1$, at least in theory. In practice, uncertainties in the model will eventually obscure this residual information.

Although the trellis sequences c and the received sequence v, in principle, are infinitely long, it is not meaningful to process all of the infinitely long received sequence at the same time. Rather, the demodulator will begin to estimate the data sequence after receiving only a finite number of components of the received sequence, say the first b components. Obtaining good demodulator performance will require an integer b that is larger than the constraint length v of the intersymbol interference – perhaps at least twice as large. Conceptually, the demodulator will generate the initial segment of length b of every possible trellis sequence and compare the first b samples of the senseword to each of these trellis segments. The codeword that is closest to the senseword in euclidean distance is the minimum-distance codeword for that finite segment of the datastream. The first symbol of the data sequence that produces the selected trellis sequence is chosen as the estimated first symbol of data. The channel response to this data symbol is then computed and subtracted from the senseword. The first sample of the senseword is now discarded and a new received sample is shifted into the demodulator so the observed length of the trellis is again b. The process is then repeated to find the next data symbol.

If this were implemented in this naive way, however, the minimum-distance demodulator would still be quite complex. There is a large amount of structure in the computation that can be exploited to obtain an efficient method of implementing the minimum-distance demodulator. One popular and efficient computational procedure is known as the *Viterbi algorithm.*

The Viterbi algorithm[2] is a fast algorithm for searching a labeled trellis to find the path that most closely agrees with a given noisy sequence of path labels. When applied to the trellis describing an instance of intersymbol interference, the Viterbi algorithm becomes an efficient computational procedure for implementing the minimum-distance demodulator. The Viterbi algorithm is practical for searching a trellis with a small number of states; perhaps q^v equal to 1024 or 2048 states would be reasonable.

The Viterbi algorithm operates iteratively frame by frame, tracing through the trellis in search of the correct path. At any frame of the trellis, say the bth frame, the Viterbi algorithm does not know which node the true path has reached, nor does it try to determine this node directly. Instead, it finds the best path to every node b frames into the future. Because there are q^v nodes in each frame, there are q^v such most likely paths to the q^v nodes in the frame b frames into the trellis. If all of these q^v most likely paths

---

[2] The Viterbi algorithm can be described as an application of the general techniques of the subject of dynamic programming to the task of searching a trellis.

begin the same, the demodulator has the same estimate of the initial branch of the true path regardless of which node the true sequence would reach b frames into the trellis. The demodulator also keeps a record of the euclidean distance between each of these q^ν codewords and the senseword. This distance is called the *discrepancy* of the path.

Given the first b symbols of the senseword, suppose that the demodulator has already determined the most likely path to every node $b - 1$ frames into the trellis, and the discrepancies of these paths. In its next iteration, the demodulator determines the most likely path to each of the nodes in frame b. But to get to a node in frame b, the path must pass through one of the nodes in frame $b - 1$. The candidate paths to a new node are found by extending to this new node each of the old paths that can be so extended to that node. The most likely path is found by adding the incremental discrepancy of each path extension to the discrepancy of the path to the node in frame $b - 1$. There are q such paths to any new node, and the path with the smallest total discrepancy is marked as the most likely path to the new node. The demodulator repeats this process for every node of frame b. At the end of the iteration, the demodulator knows the most likely path to every node in frame b, and the discrepancies of these paths.

Consider the set of all surviving paths to the set of nodes in the bth frame. One or more of the nodes at the first frame will be crossed by these paths. If all paths cross through the same node of the first frame, then regardless of which node the encoder visits at the bth frame, the demodulator knows the most likely node that is visited in the first frame. That is, it knows the best estimate of the first data symbol even though it has not yet made a decision for the bth data symbol.

The optimum decision delay b of the Viterbi algorithm – or of any minimum-distance trellis demodulator – is unbounded because an optimum decision cannot be made until the surviving paths to all states share a common initial subpath and there will always be rare ambiguous instances that could be resolved if just a little more data were examined. However, little degradation occurs if a sufficiently large, finite decision delay b is used.

To build a Viterbi demodulator, one must choose the decoding window width b usually validated by computer simulation of the demodulator. This sets the decision delay. At frame b, the demodulator examines all surviving paths to see that they agree in the first frame. This defines a demodulated data symbol that is passed out of the demodulator. Next, the demodulator drops the first symbol and takes in a new senseword symbol for the next iteration. If all surviving paths again pass through the same node of the now oldest surviving frame, then this data frame is demodulated. The process continues in this way, decoding frames indefinitely.

If b is chosen large enough, then a well-defined decision will almost always be made at each frame time. This decision will usually be the correct one. However, several things can go wrong. Sometimes, the sequence demodulator will reach a decision that is well-defined but incorrect. This is a *demodulation error*. When a demodulation error occurs, the demodulator will necessarily follow it with additional errors because the surviving

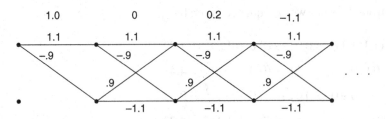

Figure 4.5. A trellis for a Viterbi demodulator.

path passes through the incorrect sequence of states. A sequence of demodulation errors is called an *error event*; an error event begins when the surviving path enters an incorrect state and ends when the surviving path returns to a correct state. Not only does one want error events to be infrequent, but one wants their duration to be short when they do occur.

Occasionally all of the surviving paths may not go through a common node in the initial frame. This is a *demodulation default*. The demodulator can be designed to put out an indication of demodulation default to the user whenever this happens, replacing each received symbol with a special symbol denoting an *erasure*. Alternatively, the demodulator can be designed to simply guess, in which case, a demodulation default becomes equivalent to a demodulation error.

An example of a Viterbi demodulator is shown in Figure 4.5. This trellis is based on a simple case of intersymbol interference in an antipodal signaling waveform with $g_0 = 1$ and $g_1 = 0.1$. Because only one bit of memory is needed, there are only two states in the trellis, corresponding to the two possible values of the previous data bit. The branches of the trellis are labeled with the values of c_ℓ, the expected values of the received sample.

Suppose the received samples at the output of the matched filter are given by the sequence

$$v = 1.0, 0.0, 0.2, -1.1, \ldots .$$

The two paths of length 1 through the trellis have squared euclidean distances from v of

$$d^2(v, (1.1)) = (1.0 - 1.1)^2 = 0.01$$
$$d^2(v, (-0.9)) = (1.0 + 0.9)^2 = 3.61.$$

The right side gives the discrepancies of the two paths to the nodes in frame 1. Each of the two paths of length one is extended in two ways to form four paths of length two, two paths to each node in frame 2. The four paths of length two have squared euclidean distances from the senseword sequence given by

$$d^2(v, (1.1, 1.1)) = 0.01 + (0.0 - 1.1)^2 = 1.22$$
$$d^2(v, (-0.9, 0.9)) = 3.61 + (0.0 - 0.9)^2 = 4.42$$
$$d^2(v, (1.1, -0.9)) = 0.01 + (0.0 + 0.9)^2 = 0.82$$
$$d^2(v, (-0.9, -1.1)) = 3.61 + (0.0 + 1.1)^2 = 4.82$$

of which the first two paths go to the top node in frame two, and the last two paths go to the bottom node in frame two. Of these, the first and third paths are selected as the two most likely paths to each of the two nodes in frame two, which implies that $a_0 = +1$ because both surviving paths begin the same.

Each of the two surviving paths of length two is extended in two ways to form four paths of length three, two such paths to each node in frame three. The four paths have squared euclidean distances from the senseword sequence, given by

$$d^2(v, (1.1, 1.1, 1.1)) = 1.22 + (0.2 - 1.1)^2 = 2.03$$
$$d^2(v, (1.1, -0.9, 0.9)) = 0.82 + (0.2 - 0.9)^2 = 1.31$$
$$d^2(v, (1.1, 1.1, -0.9)) = 1.22 + (0.2 + 0.9)^2 = 2.43$$
$$d^2(v, (1.1, 1.1, -1.1)) = 0.82 + (0.2 + 1.1)^2 = 2.51,$$

of which the first two paths go to the top node in frame three, and the last two paths go to the bottom node in frame three. Of these, the second and third paths are selected as the two most likely paths to each of the two nodes in frame three. However, in the second position, the surviving paths do not agree, so it is not yet possible to demodulate a_1.

Each of the two surviving paths of length three is extended in two ways to form four paths of length four, two paths to each node in frame four. The four paths have squared euclidean distances from the senseword sequence, given by

$$d^2(v, (1.1, -0.9, 0.9, 1.1)) = 1.31 + (-1.1 - 1.1)^2 = 6.15$$
$$d^2(v, (1.1, 1.1, -0.9, 0.9)) = 2.43 + (-1.1 - 0.9)^2 = 6.43$$
$$d^2(v, (1.1, -0.9, 0.9, -0.9)) = 1.31 + (-1.1 + 0.9)^2 = 1.35$$
$$d^2(v, (1.1, 1.1, -0.9, -1.1)) = 2.43 + (-1.1 + 1.1)^2 = 2.43,$$

of which the first two paths go to the top node in frame four, and the last two paths go to the bottom node. Of these, the first and third paths are selected as the two most likely paths to each of the two nodes in frame four. The surviving paths agree in the first three branches, which implies that $a_0 = +1$, $a_1 = -1$, and $a_2 = +1$. Notice that although a_1 could not be demodulated after the third iteration, it can be demodulated after the fourth iteration.

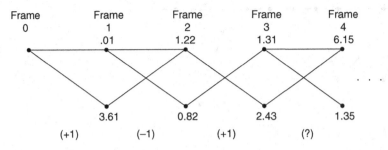

Figure 4.6. Example of a Viterbi demodulator.

The trellis in Figure 4.6 is pruned to show how the candidate paths grow and die. At each iteration, the surviving paths are extended to each of the new nodes, the cumulative squared euclidean distance is calculated, and the less likely path to each node is discarded.

4.4   Error bounds for sequence demodulation

A sequence demodulator, such as the Viterbi algorithm, works its way along a noisy received sequence, reconstructing the modulated datastream symbol by symbol. Occasionally, the demodulator makes errors. Because of the interdependence of the symbols of the sequence, we expect that errors do not occur in isolation; there will be some clustering of errors into error events.

A minimum-distance sequence demodulator finds the path through the trellis closest to the senseword in the sense of minimum euclidean distance. Our goal in this section is to determine the probability that a wrong path through the trellis is closer to the senseword than is the right path. Of course, for an infinitely long sequence, we may always expect that there will be occasional errors, and even though the error rate may be quite small, the probability of at least one error in an infinitely long sequence is one. To discuss error rates in a meaningful way, we must define a notion of local errors in a sequence.

Figure 4.7 shows two of the paths through a trellis. The first path is the upper straight line. The second path is the lower path, which re-emerges with the upper path briefly, then moves away again for a time. Suppose that the upper path depicts the correct path and the lower path depicts the incorrect path that is selected by the minimum-distance sequence demodulator because it is more similar to the senseword in the sense of euclidean distance. Twice the demodulated path through the trellis deviates from the correct path. The segment of the demodulated sequence between the time its path deviates from the correct path and the time it rejoins the correct path is an *error event*. Our example in Figure 4.7 has two error events. Within an error event, some of the

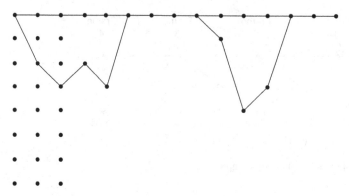

Figure 4.7. Error events in sequence demodulation.

demodulated symbols will be incorrect and some symbols will be correct. Correct symbols within an error event are rarely of value, so the probability of an error event is usually a better way to describe performance than is the probability of symbol error.

In general, we should distinguish between the probability that a symbol is in error, the probability that a symbol is contained within an error event, and the probability that a symbol is the first symbol of an error event. Each of these probabilities is difficult to compute, and we settle for bounds that can be derived conveniently, and for whatever statements that can be made.

We may design a minimum-distance demodulator so that, whenever the distance between the senseword and every trellis sequence of some fixed length is larger than some fixed value, it will refuse to demodulate that section of the sequence. Then over that section, the demodulator output is the special symbol that denotes an *erasure*. This demodulator is called a *bounded-distance demodulator*, or a demodulator with an erasure option. Such demodulators intentionally expand the role of demodulator defaults into intentional erasures and are useful when one prefers missing data to incorrect data. For such demodulators, we would also need to study the probability of erasure events. We will not further consider demodulators with an erasure option, but our methods of analysis can be extended to this case.

The exact calculation of the probability of demodulation error can be a formidable task. Fortunately, satisfactory approximations exist, so one often forgoes an exact calculation.

Definition 4.4.1 *The minimum euclidean distance, $d_{\min}$ (or free euclidean distance) of a set of real-valued sequences is the smallest euclidean distance $d(c, c')$ between any two distinct sequences c and c' in the set.*

We are interested in discrete sequences that arise in the demodulation of intersymbol interference. Such sequences can be generated by exciting a finite-impulse-response filter, starting at time zero, with sequences of values from a fixed signal constellation.

It is sufficient to compute the minimum distance only over pairs of sequences that have different values in the first symbol and differ in only a finite number of symbols thereafter. This is because, if $c_i = \Sigma_k g_{i-k} a_k$ and $c'_i = \Sigma_k g_{i-k} a'_k$, then

$$c_i - c'_i = \sum_{k=0}^{v} g_{i-k} a_k - \sum_{k=0}^{v} g_{i-k} a'_k$$

$$= \sum_{k=0}^{v} g_{i-k} (a_k - a'_k).$$

If $c_0 = c'_0$, then $a_0 = a'_0$ and the first symbol has no contribution to the distance $d(c, c')$. The distance $d(c, c')$ is equal to the distance between another pair of sequences that are both translated by one position. Furthermore, if $d(c, c')$ is finite and the a_k take values in a finite set, then the sequences c and c' must be equal except in a finite number of places.

The minimum euclidean distance in a set of sequences plays the same role in describing the probability of error in sequence demodulation as the minimum distance in a signal constellation plays in describing the probability of symbol error in quadrature amplitude demodulation in Chapter 6. Specifically, the expression

$$p_e \approx N_{d_{min}} Q \left(\frac{d_{min}}{2\sigma} \right)$$

is a widely used approximation, where $N_{d_{min}}$ is the number of *nearest neighbors*, defined as the number of sequences at distance d_{min} from a given sequence. When the signal-to-noise ratio is large enough, this approximation is justified by a satisfactory agreement with the results of many simulations of various examples, and with measured data. It is also justified in a heuristic way by the approximate upper and lower bounds we shall discuss in this section. Neither the upper bound nor the lower bound are mathematically sound. Both have derivations that are flawed. However, the bounds do aid in understanding. More importantly they are confirmed by simulation and experimentation.

To develop the upper and lower bounds, we have two tools that are available and easy to use. One tool is the union bound; the other tool is the nearest-neighbor bound. The union bound has been defined earlier. The nearest-neighbor bound deletes from consideration every erroneous sequence that is not a nearest neighbor. The union bound gives an upper bound, but it is uninformative because there are an exponentially large number of terms. The nearest-neighbor bound gives a lower bound, but it is uninformative unless the remaining terms involving the nearest neighbors are not entangled. To state an upper bound on a lower bound (or a lower bound on an upper bound) by combining the two bounds is essentially meaningless unless we can claim that the two bounds are actually good approximations. This we claim only based on experimental evidence or on simulations, and then only when the signal-to-noise ratio is sufficiently large.

We first derive a "lower bound". Let c and c' be any two sequences at distance d_{min} that have different branches in the first frame of the trellis. Suppose that whenever either of these two sequences is transmitted, a genie tells the demodulator that the transmitted sequence was one of the two, but does not say which one. The demodulator only needs to choose between the two. Such side information from the genie cannot make the probability of demodulation error worse than it would be without the side information. Hence the probability of demodulation error for the two-sequence problem can be used to form a lower bound on the probability of demodulation error whenever c is the transmitted sequence.

When given v, the minimum-distance demodulator chooses c or c' according to the distance test: choose c if $d(v, c) < d(v, c')$, and otherwise choose c'. (This is as the maximum-likelihood demodulator to be discussed in Chapter 7 whenever the noise is white gaussian noise.) This condition can be written

$$\sum_\ell |v_\ell - c'_\ell|^2 > \sum_\ell |v_\ell - c_\ell|^2.$$

An error occurs whenever

$$\sum_\ell |v_\ell - c'_\ell|^2 \le \sum_\ell |v_\ell - c_\ell|^2.$$

Because $v_\ell = c_\ell + n_\ell$, the condition for an error becomes

$$\sum_\ell |(c_\ell - c'_\ell) + n_\ell|^2 \le \sum_\ell n_\ell^2.$$

Expanding the square on the right and canceling $\sum_\ell n_\ell^2$, we can write the condition for an error as

$$\frac{\mathrm{Re}\left[\sum_\ell n_\ell (c'_\ell - c_\ell)^*\right]}{d(c, c')} \le \tfrac{1}{2} d(c, c').$$

When the noise is gaussian, the left side is a zero-mean gaussian random variable with variance σ^2, as is easily verified. Therefore, we see that this task has the same form as the task of detecting a pulse whose energy is

$$E_p = d^2(c, c')$$

in white gaussian noise of variance σ^2, as was the case in the demodulation of on–off keying. Consequently, the probability of error for the minimum-distance demodulator for distinguishing between two equally likely sequences c and c', in memoryless gaussian noise of variance σ^2, is

$$p_e(c, c') = Q\left(\frac{d_{min}}{2\sigma}\right)$$

because $d(c, c') = d_{min}$. The probability of error for the original problem, which has many codewords, can only be larger. Consequently, we have

$$p_{e|c} \geq Q\left(\frac{d_{min}}{2\sigma}\right)$$

as the lower bound on $p_{e|c}$, defined as the probability of demodulator error in the first symbol given that there are only two sequences and one of them is the correct sequence. By linearity, every c has the same pattern of neighboring sequences, so such a statement is true for every c. Therefore, the same lower bound holds for the average probability of error p_e.

This inequality is mathematically correct but rather weak because it only considers the errors that result in a particular nearest neighbor c' when c is transmitted. There are many sequences other than c' that could be the wrong code sequence. In particular, there may be many sequence nearest neighbors, the number denoted $N_{d_{min}}$, at distance d_{min} from c. Therefore, by the union bound, we may expect that

$$p_e \gtrsim N_{d_{min}} Q\left(\frac{d_{min}}{2\sigma}\right)$$

is an approximation to a lower bound where p_e denotes the probability of error in the first symbol of a sequence starting from an arbitrary trellis state. The bound is not precise because it counts as two errors the situation where there are two wrong sequences closer to the senseword sequence than is the correct sequence. This defect is not significant for high values of signal-to-noise ratio, but makes the bound questionable for low values of signal-to-noise ratio.

Next, we shall use a union-bound technique to obtain a corresponding "upper bound". Applying the union bound allows us to use the expression for the pairwise probability of error to bound the probability of error in the general problem. There is a difficulty, however: because so many codewords are in a code, the union bound can diverge. A more delicate analysis would be needed to get a rigorous upper bound.

If we apply the union bound directly, we have for the probability $p_{e|c}$ of demodulation error given that c is the correct sequence

$$p_{e|c} \leq \sum_{c' \neq c} p_e(c, c')$$

$$= \sum_{c' \neq c} Q\left(\frac{d(c, c')}{2\sigma}\right).$$

We can group all terms of the sum for which the distance $d(c, c')$ is the same; the distinct values of the distance are countable. That is, let d_i for $i = 0, 1 \ldots$ denote the distinct values that the distance $d(c, c')$ takes on. Then

$$p_{e|c} \le \sum_{i=0}^{\infty} N_{d_i} Q\left(\frac{d_i}{2\sigma}\right)$$

where N_{d_i} is the number of sequences c' at distance d_i from c and differing in the first symbol. Because of the linearity of the intersymbol interference, the set of d_i for $i = 0, 1, \ldots$, does not depend on the choice of codeword c. For values of d smaller than the minimum distance $d_{\min}$, the number of sequences at distance d is zero, and $d_0 = d_{\min}$. Consequently, averaging over all c gives

$$p_e \le \sum_{i=0}^{\infty} N_{d_i} Q\left(\frac{d_i}{2\sigma}\right)$$

$$= N_{d_{\min}} Q\left(\frac{d_{\min}}{2\sigma}\right) + \sum_{i=1}^{\infty} N_{d_i} Q\left(\frac{d_i}{2\sigma}\right).$$

. Continuing with the development, we now recall that $Q(x)$ is a steep function of its argument – at least on its tail, which is where we are interested. Consequently, we make the naive assumption that the sum on the right is dominated by the first term. Then we have the approximation to the upper bound

$$p_e \lesssim N_{d_{\min}} Q\left(\frac{d_{\min}}{2\sigma}\right),$$

which matches the approximate lower bound described earlier. The flaw in this discussion, however, is that it is not valid to discard an infinite number of terms just because each is individually small. Nevertheless this gives a nonrigorous development of a description of the probability of demodulation error which states that

$$p_e \approx N_{d_{\min}} Q\left(\frac{d_{\min}}{2\sigma}\right).$$

This agrees with the usual observations for large signal-to-noise ratio. However, because there are an infinite number of error sequences, a precise statement for p_e with a rigorous derivation has never been found.

4.5   Dominant error patterns

The nonrigorous approximations to the bounds on error probability, given in the previous section, can be quite good and the bounds are asymptotically tight for large signal-to-noise ratios, as can be verified by simulation. However, the derivations of both the upper bound and the lower bound are flawed. The derivation of the lower bound is incorrect because there are so many terms in the union bound that the right side of the union bound may very well be infinite. It is not correct to discard these many small terms in the last step under the argument that they are individually small. If desired, the derivation can be reformulated so that many of these terms are not included in the bound in the first place. This is the topic of this section.

Some sequences defined by a trellis "lie behind" other sequences when observed from a fixed sequence, c. They are, in a sense, *hidden sequences*. This means that whenever such an incorrect sequence is the most likely sequence then there is a second incorrect sequence that is also more likely than the correct sequence. That is, if c is transmitted and the hidden sequence c'' is the most likely sequence given the senseword v, then there will be another sequence c' that, given v, is also more likely than c. Whenever such a sequence is the incorrect demodulated sequence, there will be at least one other sequence closer to the senseword than the correct sequence. These hidden sequences result in double counting of certain error events, which means that these events could be omitted from the bound. But sequences are complicated and numerous, so this situation is difficult to identify.

Figure 4.8 shows three of the many paths through a trellis. The horizontal path is taken to be the correct path. The solid irregular path is taken to be the path at minimum distance from the senseword, and so is responsible for an error event. The dotted path is another error path that is closer to the senseword than is the correct path, but not as close as the chosen error path. Therefore, as we will argue, the chosen error path could be deleted from the bound on p_e without invalidating the union band. To apply the union bound more carefully to sequence demodulation, we must avoid including such extraneous terms in the sum. Accordingly, we shall introduce the *extended union bound*.

The union bound states that for any set of events,

$$\Pr\left[\bigcup_i \mathcal{E}_i\right] \leq \sum_i \Pr[\mathcal{E}_i].$$

To extend the union bound, define a *dominant set* $\mathcal{E}_i$ of the collection of sets $\{\mathcal{E}_i\}$ as a set that is not contained in any other set of $\{\mathcal{E}_i\}$, and a *hidden set* $\mathcal{E}_i \in \{\mathcal{E}_i\}$ as a set that

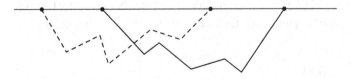

Figure 4.8. A likely and an unlikely error path.

is contained within another set of $\{\mathcal{E}_i\}$. Let $\mathcal{I}$ be the set of indices of the dominant sets. The extended union bound is

$$\Pr\left[\bigcup_i \mathcal{E}_i\right] \leq \sum_{i \in \mathcal{I}} \Pr[\mathcal{E}_i].$$

Bounds on the probability of bit error p_e for sequence demodulation will be developed as an example of the use of dominant sequences, and of the extended union bound.

An antipodal signaling waveform transmits data symbols ± 1 and, in general, the receiver sees intersymbol interference in the matched-filter samples for the antipodal signaling waveform. We shall treat the case of a finite data sequence $\{a_0, \ldots, a_{n-1}\}$ of blocklength n. The matched-filter output is the senseword

$$v_\ell = \sum_{\ell'=0}^{n-1} a_{\ell'} g_{\ell-\ell'} + n_\ell \quad \ell = 0, \pm 1, \pm 2, \ldots$$

where $a_\ell = \pm 1$ and the noise samples are independent, identically distributed, gaussian random variables. In general, the intersymbol interference can be two-sided – and even of infinite duration with little change in the discussion.

An error sequence in the estimated data sequence a of blocklength n is denoted

$$e = (e_0, e_1, \ldots, e_{n-2}, e_{n-1})$$

where, because an error always changes ± 1 to ∓ 1, e_ℓ takes only the values -2, 0, or $+2$. The weight of the vector e, denoted $\mathrm{wt}(e)$ is the number of places at which e is nonzero.

An error pattern in the estimated data sequence $\widehat{a}$ corresponds to an error in the estimated noise-free filter output sequence, which will be denoted $\delta(e)$ and defined componentwise by giving the ℓth term as

$$\delta_\ell(e) = \sum_{\ell'=0}^{n-1} (a_{\ell'} + e_{\ell'}) g_{\ell-\ell'} - \sum_{\ell'=0}^{n-1} a_{\ell'} g_{\ell-\ell'}$$

$$= \sum_{\ell'=0}^{n-1} e_{\ell'} g_{\ell-\ell'}$$

where $\delta_\ell(e)$ is the ℓth component of $\delta(e)$. Notice that $\delta_\ell(e)$ does not depend on the actual data sequence a, only on e. The euclidean distance between any two filter output sequences whose data sequences differ by the error vector e is denoted by

$$\|\delta(e)\| = \left[\sum_{\ell=-\infty}^{\infty} \delta_\ell^2(e)\right]^{\frac{1}{2}}.$$

Accordingly, the probability of confusing data vector a with data vector $a + e$ is

$$p(a + e|a) = Q\left(\frac{\|\delta(e)\|}{2\sigma}\right)$$

because this is the same binary decision problem as occurs in antipodal signaling, but in this case, the two received signals are at distance $\|\delta(e)\|$.

We shall want to average the probability of bit error over all possible data sequences and all possible error sequences. To do this, it is productive to invert our point of view by first fixing a specific error sequence, then matching it to all data sequences with which that error sequence is compatible. Because $a_\ell = \pm 1$ and the components $a_\ell + e_\ell$ must form another data sequence taking only values ± 1, not all data sequences a are compatible with a fixed error sequence e. A data sequence a is compatible with a specific e if and only if a_ℓ and e_ℓ have opposite signs for all nonzero components e_ℓ.

For a fixed error vector e, with weight $\text{wt}(e)$, the probability that a randomly and uniformly chosen binary data sequence a is compatible with e is $2^{-\text{wt}(e)}$. This is because $a + e$ is an element of $\{-1, +1\}^n$ if each nonzero component of e is opposite in sign to the corresponding component of a. Because each component of a has the appropriate sign with probability one-half, $a + e$ is an element of $\{-1, +1\}^n$ with probability $2^{-\text{wt}(e)}$.

To bound the probability of error of the kth bit, denoted $p_e(k)$, first define

$$\mathcal{E}_k = \{e \in \{-2, 0, 2\}^n : e_k \neq 0\}$$

as the set of all error sequences that have an error in the kth bit. Then, when averaged over all datawords, we have the union bound

$$p_e(k) \leq \sum_{e \in \mathcal{E}} 2^{-\text{wt}(e)} Q\left(\frac{\|\delta(e)\|}{2\sigma}\right).$$

Again, this bound is essentially useless in this form because there are so many terms in the sum that the right side may be infinite. However, we now have the problem set up in such a way that we can purge many superfluous terms from the sum. We need only replace the union bound with the extended union bound. Therefore

$$p_e(k) = \sum_{e \in \mathcal{F}_k} 2^{-\text{wt}(e)} Q\left(\frac{\|\delta(e)\|}{2\sigma}\right)$$

where $\mathcal{F}_k$ is the set of dominant error sequences with a nonzero error in the kth bit position. All that remains is to specify $\mathcal{F}_k$.

Decompose any error sequence $e \in \mathcal{E}_k$ into two nonoverlapping nonzero error sequences, as

$$e = e' + e''$$

with $e' \in \mathcal{E}_k$. The requirement that e' and e'' be nonoverlapping means that e' and e'' cannot have a common nonzero component. This means that $\text{wt}(e') + \text{wt}(e'') = \text{wt}(e)$. We can also write this property as a dot product of magnitudes: $|e'| \cdot |e''| = 0$.

We shall prove that if such a decomposition of the error pattern also satisfies

$$\text{Re} \sum_{\ell=-\infty}^{\infty} \delta_\ell(e') \delta_\ell^*(e'') \geq 0,$$

then we do not need to include the error pattern e in the sum for $p_e(k)$ because it is hidden by the error pattern e'. To anticipate this end, define the set of dominant error patterns $\mathcal{F}_k$ as the set of all error patterns that cannot be decomposed to satisfy this condition. The set of hidden error sequences is the complement of the set of dominant error patterns, and is given by

$$\mathcal{F}_k^c = \left\{ e \in \mathcal{E}_k : \ \text{Re} \sum_{\ell=-\infty}^{\infty} \delta_\ell(e') \delta_\ell^*(e'') \geq 0 \right\},$$

it being understood that $e' \in \mathcal{E}_k$, $e'' \neq 0$, $e = e' + e''$, and $|e'| \cdot |e''| = 0$. The next theorem states that only dominant error patterns need to be included in the union bound.

Theorem 4.5.1 *The probability of error in the kth bit of a sequence demodulator for antipodal signaling in memoryless white gaussian noise of variance σ^2 is bounded by*

$$p_e(k) \leq \sum_{e \in \mathcal{F}_k} 2^{-\text{wt}(e)} Q\left(\frac{\|\delta(e)\|}{2\sigma} \right)$$

where $\mathcal{F}_k$ is the set of dominant error sequences.

Proof Let $c(a)$ denote the noise-free senseword corresponding to data sequence a. Let v be the noisy received sequence $v = c(a) + n$. The minimum-distance sequence estimator, when given v, chooses a data sequence $\hat{a}$ that minimizes $d(v, c(a))$. Suppose that e is in $\mathcal{E}_k$ but not in $\mathcal{F}_k$, that a is the correct data sequence, that $\hat{a} = a + e$ is the minimum-distance data sequence, and that $e = e' + e''$ as defined previously. We shall show that if e is a hidden error pattern, then $d(v, c(a + e')) \leq d(v, c(a))$ for that v. This establishes the following containment of events:

$$\left\{ v : d(v, c(a + e)) = \min_a d(v, c(a)) \right\} \subset \left\{ v : d(v, c(a + e')) \leq d(v, c(a)) \right\}.$$

Because the set on the right is a set of dominant error patterns, the set on the left is not included in the extended union bound.

We now come to the manipulations that will establish these claims and so complete the proof of the theorem. Let $v = c(a) + n$, $\hat{a} = a + e$, and $\hat{c} = c(a) + \delta(e)$. Then, writing the components of v as $c_\ell + n_\ell$, we have

$$d(v, c(a + e)) - d(v, c(a + e')) = \sum_\ell |c_\ell + n_\ell - c_\ell - \delta_\ell(e)|^2$$

$$- \sum_\ell |c_\ell + n_\ell - c_\ell - \delta_\ell(e')|^2$$

and

$$d(v, c(a + e'')) - d(v, c(a)) = \sum_\ell |c_\ell + n_\ell - c_\ell - \delta_\ell(e'')|^2 - \sum_\ell |c_\ell + n_\ell - c_\ell|^2.$$

Now compute the difference between the right sides of these two equations, calling it Δ. Thus

$$\Delta = \sum_\ell \left[|n_\ell - \delta_\ell(e)|^2 - |n_\ell - \delta_\ell(e')|^2 - |n_\ell - \delta_\ell(e'')|^2 + |n_\ell|^2 \right].$$

Next, recall that $\delta_\ell(e) = \delta_\ell(e') + \delta_\ell(e'')$ and reduce the expression for Δ to the form

$$\Delta = \sum_\ell \left[|\delta_\ell(e') + \delta_\ell(e'')|^2 - |\delta_\ell(e')|^2 - |\delta_\ell(e'')|^2 \right]$$

$$= 2\mathrm{Re} \sum_\ell \delta_\ell(e')\delta_\ell^*(e'')$$

$$\geq 0$$

where the inequality holds as an assumption of the theorem. Consequently,

$$d(v, c(a + e)) - d(v, c(a + e')) - d(v, c(a + e'')) + d(v, c(a)) \geq 0.$$

This can be rewritten as

$$d(v, c(a + e')) - d(v, c(a)) \leq d(v, c(a + e)) - d(v, c(a + e''))$$

$$\leq 0$$

where the second inequality here holds because $d(v, c(a + e))$, by definition, is the minimum distance between v and any c. Thus

$$d(v, c(a + e')) \leq d(v, c(a)),$$

and the theorem is proved.   ■

The usefulness of Theorem 4.5.1 is helped by the fact that there are only a finite number of dominant error patterns of any given weight because if the support of e' and e'' are separated by more than the constraint length, then $\delta_\ell(e')$ and $\delta(e'')$ do not overlap, so $\sum_\ell \delta_\ell(e')\delta_\ell(e'')$ is zero.

4.6   Linear equalization

A signaling waveform that has no intersymbol interference must use a Nyquist pulse. Otherwise, the intersymbol interference should be compensated in some way, or an unnecessary performance degradation will occur. This process of compensation for undesired intersymbol interference is called *equalization*. We have already studied the decision-feedback demodulator and the Viterbi algorithm as methods of demodulating in the presence of intersymbol interference. Accordingly, these are also called *decision-feedback equalization* and *Viterbi equalization*, respectively.

Linear equalization is another method of dealing with intersymbol interference. This method consists of cascading a filter, called an *equalization filter*, with the channel so that the combination of channel and filter now has a predetermined transfer function. The cascade of the actual channel and the equalization filter has the transfer function of the desired channel.

The underlying idea of linear equalization, as shown in Figure 4.9, is simple. If we desire a channel transfer function $H(f)$, and the channel has an actual transfer function that is different from the desired channel transfer function, we can cascade an equalization filter with the channel. The frequency-domain transfer function of the equalization filter is equal to the desired transfer function divided by the actual transfer function. The cascade of the equalization filter with the actual channel then gives the desired channel response.

The linear equalizer does not rest on the theory of the matched filter or on the maximum-likelihood principle, so it is not an optimum demodulator in either of these senses. Because the actual channel transfer function $H'(f)$ appears in the denominator of the equalization filter, there will be noise amplification at frequencies where $H'(f)$ is small, and corresponding performance degradation. This is the disadvantage

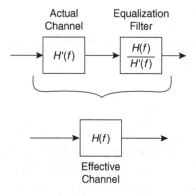

Figure 4.9. Equalization.

of the linear equalizer. The advantage of a linear equalizer is its comparative ease of implementation. Whereas a Viterbi demodulator has a complexity that depends exponentially on the constraint length of the intersymbol interference, a linear equalizer has a complexity that depends only linearly on the constraint length.

An equalization filter can be placed either in the receiver or in the transmitter. In magnetic recording, the latter is called *write equalization*. If the system were linear and noiseless, it would not matter to the theory whether the equalizer was in the transmitter or in the receiver but, because noise is always present, the two locations are not equivalent. The equalizer can even be factored, with part in the transmitter and part in the receiver. We usually think of the equalizer as residing in the receiver, as shown in Figure 4.9. This is often the more convenient place to put it, especially if the equalizer is an adaptive equalizer as described in Section 4.7. Moreover, the equalization filter may be implemented partially in continuous time before the sampler and partially in discrete time after the sampler.

When the channel input pulse is $p(t)$, the channel output is

$$v(t) = p(t) * h(t) + n(t)$$
$$= s(t) + n(t),$$

where $h(t)$ is the impulse response of the channel. The output of the equalization filter $g(t)$ is

$$u(t) = v(t) * g(t)$$
$$= p(t) * h(t) * g(t) + n(t) * g(t).$$

Suppose that $p(t)$ has been selected so that $p(t) * p(-t)$ is the Nyquist pulse $r(t)$. We can easily find the equalization filter $g(t)$ that recreates the Nyquist pulse by working in the frequency domain. The output pulse of the equalizer has Fourier transform $P(f)H(f)G(f)$, which we want to equal $R(f)$. Thus

$$P(f)H(f)G(f) = P(f)P^*(f),$$

from which we conclude that

$$G(f) = \begin{cases} P^*(f)/H(f) & P(f) \neq 0 \\ 0 & P(f) = 0. \end{cases}$$

This simple equalizer is called a *zero-forcing equalizer* because it restores the equalizer output to a Nyquist pulse, as shown in Figure 4.10. The *zero-forcing equalizer* is an elementary equalizer that gives no consideration to its effect on the noise in the received waveform. The filter may clash with the goals of a matched filter and may even amplify

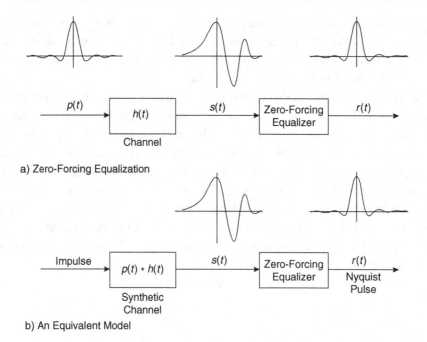

a) Zero-Forcing Equalization

b) An Equivalent Model

Figure 4.10. The zero-forcing equalizer.

the noise. A more sophisticated equalizer compromises between shaping the output pulse and rejecting channel noise.

An equalizer that consists of a linear finite-impulse-response filter of finite length cannot, in general, force all intersymbol interference coefficients to zero; there must be a compromise. This compromise can be based on defining a performance measure, then selecting the tap weights in order to optimize this performance measure. The obvious performance measure to minimize is the bit error rate at the output of the demodulator because bit error rate is a standard performance measure of a modem. However, the bit error rate is an intractable function of the tap weights so other elementary performance measures are employed for designing an equalizer, such as the *mean-square error* or the *peak interference*. Of these, the least-mean-square equalizer is noteworthy and tractable. It is developed in Section 4.8.

4.7   Adaptive equalization

A linear equalizer adjusts the actual linear channel in order to change it to the desired linear channel. To accomplish this, the equalizer must know the actual channel or an adequate model of the actual channel. This amounts to knowing the channel impulse response $h(t)$. In many applications, the channel impulse response is not completely known at the receiver; it must be inferred from the received signal itself. This is inherently a statistical process because of noise and because the data symbols may not be known. An *adaptive equalizer* is an equalizer that adjusts itself by using an observation of a sample output sequence of the actual channel. The output sequence may be the response of the channel to a prespecified channel input, known as a *training sequence*, that is known to the receiver. Alternatively, the output sequence may be the response of the channel to an ordinary input waveform carrying a random data sequence. Designing an adaptive equalizer is more difficult if the data itself is not known.

A training sequence may be appropriate for a point-to-point communication system in which messages are of finite length with a beginning and an end. The training sequence may be placed as a preamble to the overall message, and the receiver is equalized before the data-bearing portion of the message begins. In a broadcast communication system, however, there may be many receivers. An individual receiver may choose to tune in to an ongoing broadcast at its own convenience. In such a system, one might choose to adjust the equalization based on the ordinary data-carrying waveform. Even if a training sequence is used initially, it may be appropriate to continue to adjust the equalization based on the received channel signal so that any variations that occur in the channel transfer function will be neutralized by changes in the equalization filter.

Figure 4.11 shows the general plan of an adaptive equalizer. A replica of the modulator is contained in the receiver. This local modulator generates a signal that will be compared with the equalizer output to form an error signal, which is the input to an adaptive equalization filter. The adaptive equalization filter adjusts itself in such a way that the error signal is driven toward zero. The input to the local modulator may be either a training sequence or a regenerated sequence of data from the demodulator. However, there are two reasons why a training sequence is to be preferred. First, the demodulator will make occasional errors, especially when the adaptive equalizer is far

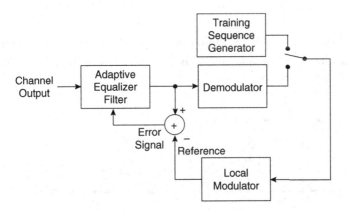

Figure 4.11. Schema for adaptive equalization.

from its correct setting. This will cause errors in the output of the local modulator, which will tend to drive the adaptive equalizer incorrectly. If the equalizer begins when it is severely out of adjustment, it may not converge. Second, the datastream coming from the demodulator might not be sufficiently random. This can fail to expose some of the errors in the equalizer setting. If the actual datastream results in a waveform without enough energy in some frequency region, then the equalizer will not be able to adapt properly in this region, and the demodulator may be vulnerable to future data errors when this frequency region is used. For example, an antipodal signaling datastream with a long run of alternating zeros and ones will have little energy near zero frequency so the equalization may be poor near zero frequency.

4.8 Least-mean-square equalization

A variety of criteria can be used for designing a linear equalizer, whether it is adaptive or nonadaptive. We shall describe the popular least-mean-square-error criterion. The goal of this equalizer is to minimize the mean-square error resulting from the combination of noise and intersymbol interference.

We will consider a real baseband channel. In this case the time samples are real. The same techniques can also be used with a complex baseband channel, in which case the time samples would be complex.

A common structure for a discrete-time linear equalizer is the transversal filter, which consists of a delay line tapped at uniformly spaced intervals. When time is discrete, a transversal filter is called a *finite-impulse-response filter*. The output of a typical finite-impulse-response filter, as shown in Figure 4.12, is a linear convolution of the input sequence and the sequence described by the filter tap weights

$$r_\ell = \sum_{k=0}^{K-1} g_k s_{\ell-k} \quad \ell = 0, \dots$$

where g_k for $k = 0, \dots, K-1$ is the kth tap weight of the filter having a total of K taps, s_ℓ is the ℓth filter input sample, and r_ℓ is the ℓth filter output sample. The task

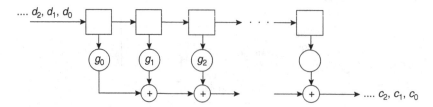

Figure 4.12. A finite-impulse-response filter.

of designing a finite-impulse-response equalizer consists of choosing K and specifying g_k for $k = 0, \ldots, K - 1$. One way of designing this FIR filter of a given length K is to approximate the ideal linear equalizer with a finite-impulse-response filter. Another way is to restate the criteria for the equalizer design and to rederive the best linear equalizer within this class of finite-impulse-response filters of a given length K.

The only linear equalization filters that need be used for our usual waveform of quadrature-amplitude modulation (QAM) symbols are those that have the form of a whitened matched filter before the sampler, followed by a discrete-time filter after the sampler. Later, we shall see in Section 7.3, in the conclusion of Theorem 7.3.1, that there is no advantage in having another filter prior to the matched filter. Even when intersymbol interference is present, so long as only the second-order properties of the noise are considered, the matched filter, followed by a sampler, preserves all relevant information in the received signal that can be recovered by a linear filter.

The least-mean-square equalizer minimizes the mean-square value of the error signal at the output of the equalizer, an error signal that is due to both the residual intersymbol interference and the unfiltered channel noise. Minimization of mean-square error is an ad hoc compromise that cannot be motivated by any deeper theory. Minimizing mean-square-error at the equalizer output does not necessarily minimize the probability of demodulated symbol error because intersymbol interference does not have the same probability density function as gaussian noise. Further, dependence will remain in the sequence of samples; hence following the equalizer by an isolated-symbol demodulator will ignore available residual information. Nevertheless, the least-mean-square-error equalizer followed by an isolated-symbol demodulator can be entirely adequate in many applications.

For the equalization filter, we choose a finite-impulse-response filter of length K so that the equalized output samples are

$$a'_\ell = \sum_{k=0}^{K-1} g_k v_{\ell-k}.$$

The mean-square error criterion requires that the ℓth sample a'_ℓ out of the filter should be equal to the ℓth transmitted pulse amplitude a_ℓ with minimum mean-square error due to the combination of intersymbol interference and noise. The tap weights g_k will be chosen to make the combined result of these two sources of error as small as possible.

The mean-square error of the vector of n output samples is defined as

$$\epsilon = \frac{1}{n} \sum_{\ell=0}^{n-1} e_\ell^2$$

where e_ℓ is the error in the ℓth sample

$$e_\ell = a'_\ell - a_\ell$$

$$= \sum_{k=0}^{K-1} g_k v_{\ell-k} - a_\ell.$$

The mean-square error is the expectation of a quadratic function of the K tap weights as follows

$$\epsilon = \mathrm{E} \frac{1}{n} \sum_{\ell=0}^{n-1} \left[\sum_{k=0}^{K-1} g_k v_{\ell-k} - a_\ell \right] \left[\sum_{k'=0}^{K-1} g_{k'} v_{\ell-k'} - a_\ell \right]$$

$$= \frac{1}{n} \sum_{\ell=0}^{n-1} \left[\sum_{k=0}^{K-1} \sum_{k'=0}^{K-1} \mathrm{E} \left[g_k g_{k'} v_{\ell-k} v_{\ell-k'} \right] - 2a_\ell \mathrm{E} \sum_{k=0}^{K-1} \left[g_k v_{\ell-k} \right] + a_\ell^2 \right].$$

Now recall that, for white noise of variance σ^2,

$$\mathrm{E}[v_{\ell-k} v_{\ell-k'}] = s_{\ell-k} s_{\ell-k'} + \sigma^2 \delta_{kk'}$$

where the discrete impulse function $\delta_{kk'}$ is equal to one if k is equal to k', and otherwise is equal to zero. Now we can write

$$\epsilon = \frac{1}{n} \sum_{\ell=0}^{n-1} \left[\sum_{k=0}^{K-1} \sum_{k'=0}^{K-1} g_k g_{k'} s_{\ell-k} s_{\ell-k'} + \sum_{k=0}^{K-1} g_k^2 \sigma^2 - 2a_\ell \sum_{k=0}^{K-1} g_k s_{\ell-k} + a_\ell^2 \right].$$

We can visualize this quadratic function in the g_k as a bowl in K-dimensional space. The goal of the adaptive equalizer is to find the minimum of this bowl. A necessary and sufficient condition on the optimal setting of the tap weights is that the partial derivatives $\partial \epsilon / \partial g_k$ are equal to zero for $k = 0, \ldots, K-1$. The partials are evaluated as

$$\frac{\partial \epsilon}{\partial g_k} = \frac{2}{n} \left[\sum_{\ell=0}^{n-1} \sum_{j=0}^{K-1} g_j s_{\ell-j} s_{\ell-k} + n\sigma^2 g_j \delta_{jk} - \sum_{\ell=0}^{n-1} a_\ell s_{\ell-k} \right] = 0.$$

Therefore with the notation

$$R_{jk} = \frac{1}{n} \sum_{\ell=0}^{n-1} s_{\ell-j} s_{\ell-k} + n\sigma^2 \delta_{jk}$$

and

$$R'_k = \frac{1}{n} \sum_{\ell=0}^{n-1} a_\ell s_{\ell-k},$$

the tap vector g is the solution of the equation

$$\sum_{j=0}^{K-1} g_j R_{jk} = R'_k.$$

An explicit solution is

$$g_j = \sum_{k=0}^{K-1} R_{jk}^{-1} R'_k.$$

The explicit solution given above is usually considered unsatisfactory, especially for adaptive equalization, not only because it may be computationally excessive but because it weighs all data equally, even the oldest data in use. The preferred approaches are iterative, updating the tap weights in a way that drives the partial derivatives toward zero. Iterative computational procedures, such as the *stochastic gradient* adaptive equalizer, are in wide use, and form part of a more complete treatment of the subject of adaptive filtering.

In theory, it is sufficient to use a *symbol-spaced equalizer*, that is, one with tap spacing equal to the symbol spacing of T seconds. In practice, a *fractionally-spaced equalizer*, such as one with a tap spacing of $T/2$ is more robust and its performance is more tolerant of compromises in the complexity of the implementation. We shall only discuss the simpler case of a symbol-spaced equalizer.

An alternative linear equalizer is the *zero-forcing equalizer*. The zero-forcing equalizer must equalize the channel so that when the received pulse is $s(t)$, the sampled output of the equalized channel has no intersymbol interference. Specifically, we require $r(t)$ to have samples

$$r_\ell = r(\ell T) = \begin{cases} 0 & \ell = -N, \ldots, -1 \\ 1 & \ell = 0 \\ 0 & \ell = 1, \ldots, N \end{cases}$$

where we have imposed constraints on only $2N+1$ samples because of the finite length of the FIR filter. Because the constraints are specified for negative values of ℓ, we must rewrite the filter response in the symmetric form as

$$r_\ell = \sum_{k=-N}^{N} g_k s_{\ell-k}.$$

Writing this out in matrix form gives

$$
\begin{bmatrix}
s_0 & s_{-1} & \cdots & s_{-2N+1} & s_{-2N} \\
s_1 & s_0 & \cdots & s_{-2N+2} & s_{-2N+1} \\
\vdots & & & & \\
s_N & s_{N-1} & \cdots & s_{-N+1} & s_{-N} \\
\vdots & & & & \\
s_{2N-1} & s_{2N-2} & \cdots & s_0 & s_{-1} \\
s_{2N} & s_{2N-1} & \cdots & s_1 & s_0
\end{bmatrix}
\begin{bmatrix}
g_{-N} \\
g_{-N+1} \\
\vdots \\
g_0 \\
\vdots \\
g_{N-1} \\
g_N
\end{bmatrix}
=
\begin{bmatrix}
0 \\
0 \\
\vdots \\
1 \\
\vdots \\
0 \\
0
\end{bmatrix}.
$$

In compact form, this is the matrix-vector equation

$$Sg = r.$$

The vector of tap weights is then given as the solution of this equation

$$g = S^{-1}r.$$

Because N is finite, the zero-forcing equalizer does not prevent the impulse response from being nonzero at sampling instants far from the origin, so intersymbol interference is not completely eliminated. Therefore the zero-forcing equalizer forms only an approximation to a Nyquist pulse at the channel output. However, if the number of filter taps is large in comparison to the number of sampling intervals in which the channel output is significant, then we may expect the residual effects to be negligible, although the theory does not guarantee that this is so.

Problems for Chapter 4

4.1. An antipodal signaling waveform is used on a channel that has intersymbol interference described by the discrete sequence

$$g = (1,\ 0.25,\ 0,\ 0,\ 0,\dots).$$

Assuming that the binary data symbols are equiprobable, and that each data symbol is demodulated by a simple threshold without consideration for the intersymbol interference, find an expression for the probability of error as a function of E_b/N_0. On the same graph, plot p_e versus E_b/N_0 for antipodal signaling without intersymbol interference, and with this model of uncompensated intersymbol interference. At $p_e = 10^{-5}$, what is the difference in E_b/N_0?

4.2. The pulse

$$s(t) = \frac{\sin \pi t/T}{\pi t/T} \frac{\cos \pi t/T}{1 - 4t^2/T^2}$$

is a Nyquist pulse for signaling interval T.

a. Describe the zeros of $s(t)$.

b. Describe the intersymbol interference for the situation where the pulse is used with antipodal signaling, but the channel data rate is doubled. That is, $T_b = T/2$ and

$$c(t) = \sum_{\ell=-\infty}^{\infty} a_\ell s(t - \ell T_b).$$

c. Sketch and label a trellis for demodulation.

4.3. A discrete-time channel has intersymbol interference described by the polynomial

$$g(x) = g_{-1}x^{-1} + g_0 + g_1 x$$
$$= 0.1x^{-1} + 1 + 0.1x.$$

By combining the techniques of a decision-feedback demodulator and a Viterbi demodulator, design a demodulator in which the Viterbi algorithm searches a trellis that has only two states.

4.4. a. Prove that for $x \geq 0$ and $y \geq 0$,

$$Q(x + y) \leq Q(x)e^{-y^2/2}.$$

b. Show that the inequality

$$P_e \leq \sum_i Q\left(\frac{d_i}{2\sigma}\right)$$

can be weakened to give

$$P_e \leq Q\left(\frac{d_{\min}}{2\sigma}\right) \sum_i e^{-(d_i - d_{\min})^2/8\sigma^2}.$$

Does the inequality apply to the demodulation of sequences as well as to the demodulation of amplitude shift keying?

4.5. With the channel output matched-filtered to the data pulse $s(t)$, because of channel dispersion the samples of a binary antipodal signaling waveform display intersymbol interference described by $g_0 = 1$, $g_1 = 0.8$, and $g_i = 0$ otherwise.

    a. Given that the previous bit was demodulated correctly, what is the probability of bit error as a function of E_b/N_0? What is the value of p_e if $E_b/N_0 = 10\,\mathrm{dB}$?

    b. Given that the previous bit was demodulated in error, what is the probability of bit error as a function of E_b/N_0? What is the value of p_e if $E_b/N_0 = 10\,\mathrm{dB}$?

    c. What is the bit error rate if $E_b/N_0 = 10\,\mathrm{dB}$?

4.6. (Echo cancellation.) An echo channel has the form

$$g(t) = h(t) + ah(t - \Delta)$$

where Δ is large in comparison with the signaling symbol duration T of the waveform

$$c(t) = \sum_{\ell=0}^{\infty} a_\ell s(t - \ell T).$$

Describe a decision-feedback demodulator to remove the intersymbol interference.

4.7. a. Given a channel with intersymbol interference and two sequences c and c' at distance $d_{\min}$, one of which is known to have been transmitted, a demodulator selects sequence c or c' if $d(v, c) < d_{\min}/6$, or $d(v, c') < d_{\min}/6$, and otherwise declares a sequence erasure (undemodulatable data). Give an approximate expression for the probability of message error and the probability of message erasure.

    b. Give approximate expressions for these probabilities for a demodulator for a fully modulated waveform.

4.8. a. Show that the equalization filter

$$G(f) = \frac{E_c \dfrac{S^*(f)}{N(f)}}{1 + \dfrac{E_c}{T} \displaystyle\sum_{k=-\infty}^{\infty} \frac{\left| S\left(f + \frac{k}{T}\right)\right|^2}{N\left(f + \frac{k}{T}\right)}}$$

reduces to a matched filter when the signal-to-noise ratio is small.

    b. Show that the output of the equalization filter reduces to a Nyquist pulse as the signal-to-noise ratio becomes large.

    c. Show that when the noise is white and $s(t) * s^*(-t)$ is a Nyquist pulse, the equalization filter reduces to a matched filter for all values of the noise power.

    d. How should the condition of part c be restated when the noise is not white?

4.9. a. A linear channel is known to have a transfer function $H(f)$ and additive white noise. How should the transmitted pulse shape be designed for binary antipodal signaling so that the channel output can be filtered to have a maximum signal-to-noise ratio, and so that the samples of the filter output have no intersymbol interference and uncorrelated noise components?

b. Suppose now that the noise has power density spectrum $N(f)$. How does the answer to part a change?

4.10. The following example is contrived to show that the samples out of a filter matched to the channel input pulse $p(t)$ can be inferior to the samples out of a filter matched to the channel output pulse $s(t)$. Let

$$p(t) = \text{rect}\left(\frac{t}{T}\right)$$

and

$$c(t) = \sum_{\ell=-\infty}^{\infty} a_\ell \text{rect}\left(\frac{t - \ell T_b}{T}\right)$$

where $T_b = 8T$. Sketch the signal out of the matched filter $p(-t)$. Is the pulse $p(t) * p(-t)$ a Nyquist pulse? Given the channel impulse response

$$h(t) = \text{rect}\left(\frac{t - 6T}{T}\right),$$

describe the samples when the channel output $c(t) * h(t)$ is sampled at the symbol signaling instants ℓT_b. How does this compare with the output samples of a filter matched to $p(t) * h(t)$?

4.11. A binary antipodal signaling waveform is used on a channel that has intersymbol interference described by the discrete sequence

$$g = (1, 0.25, 0, 0, 0, \ldots).$$

Assuming that the binary data symbols are equiprobable and the noise is white and gaussian, and that each data symbol is demodulated by a decision-feedback demodulator, find an expression for the probability of error as a function of E_b/N_0. On the same graph, plot p_e versus E_b/N_0 for binary antipodal signaling without intersymbol interference, and with this model of intersymbol interference. At $p_e = 10^{-5}$, what is the difference in E_b/N_0?

4.12. A binary antipodal signaling waveform is used on a channel that has intersymbol interference described by the discrete sequence

$$g = (1, 0.25, 0, 0, 0, \ldots).$$

Given a received sequence $v = (v_0, v_1, v_2, \ldots) = (1.2, -0, 8, -0.4, -0.5, 0.8, 1.4, \ldots)$, find the first six iterations of the Viterbi demodulator.

Notes for Chapter 4

Linear equalization, which is perhaps the simplest way of dealing with intersymbol interference, goes back at least as far as Nyquist (1928). The more difficult topic of *adaptive equalization* has its own extensive literature. The alternative method of decision-feedback demodulation was introduced by Austin (1967) and advanced by Monsen (1970), Mueller and Salz (1981), and others. However, the essentials of the idea had been applied earlier by Mathes (1919) for telegraphy over submarine cables. Salz (1973) and Proakis (1975) discussed the superior performance of a decision-feedback demodulator relative to linear equalization with regard to noise, and Qureshi (1982) showed the superior performance with regard to clock-synchronization errors. Tufts (1965) and Aaron and Tufts (1966) discussed the difficult problem of designing a linear equalizer to minimize the probability of error, and Ericson (1971) formulated the decomposition of an equalizer into a matched filter followed by a discrete-time filter after the sampler. The merits of the fractionally-spaced linear equalizer were recognized independently by many modem designers. Some early published papers on this subject were by Lucky (1969), Guidoux (1975), Ungerboeck (1976), and Qureshi and Forney (1977). Adaptive equalization began with the work of Lucky (1965, 1966).

The trellis was so named by Forney (1973) as a graphical description of the behavior of any finite state machine. The Viterbi (1967) algorithm was originally devised as a pedagogical aid for understanding the decoding of convolutional codes. It was quickly realized that the algorithm is quite practical for convolutional codes of modest blocklength. Kobayashi (1971) and Omura (1970) established that the Viterbi algorithm could also be used as a demodulator in the presence of intersymbol interference. Forney (1973) promoted the Viterbi algorithm as a general method of searching a trellis and as useful for a broad variety of problems, which include demodulation in the presence of intersymbol interference. The use of the union bound to bound the probability of error is standard. The upper bound, and a corresponding lower bound appeared in Viterbi's 1967 paper for the decoding of convolutional codes.

5 Passband Modulation

A *waveform channel* is a channel whose inputs are continuous functions of time. A *passband channel* is a waveform channel suitable for an input waveform that has a spectrum confined to an appropriately narrow interval of frequencies centered about a nonzero reference frequency, f_0. A *complex baseband channel* is a waveform channel whose input waveform is a complex function of time that has a spectrum confined to an interval of frequencies containing the zero frequency. We shall see that every passband channel can be converted to or from a complex baseband channel by using standard techniques in the modulator and demodulator.

The function of a digital modulator for a passband channel is to convert a digital datastream into a waveform representation of the data that can be accepted by the passband channel. The waveform from the modulator is designed to accommodate the spectral characteristics of the channel, to obtain high rates of data transmission, to minimize transmitted power, and to keep the bit error rate small.

A passband modulation waveform cannot be judged independently of the performance of the demodulator. To understand how a modem works, it is necessary to study both the passband modulation techniques of this chapter and the passband demodulation techniques of Chapter 6. The final test of a modem is in the ability of the demodulator to recover the input datastream from the signal received by the demodulator in the presence of noise, interference, distortion, and other impairments.

5.1 Passband waveforms

A *passband waveform*, denoted $\tilde{v}(t)$[1], is a waveform of the form

$$\tilde{v}(t) = v_R(t) \cos 2\pi f_0 t - v_I(t) \sin 2\pi f_0 t,$$

where $v_R(t)$ and $v_I(t)$ are baseband waveforms and the reference frequency f_0 is large in comparison to the spectral components in the Fourier transforms $V_R(f)$ and $V_I(f)$.

---

[1] Passband waveforms will be indicated by the tilde over the letter that denotes the waveform.

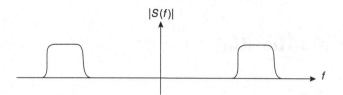

Figure 5.1. Magnitude spectrum of a passband signal.

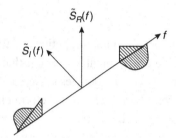

Figure 5.2. Complex spectrum of a passband signal.

A passband waveform $\widetilde{v}(t)$ has a Fourier transform, or spectrum, $\widetilde{V}(f)$, whose nonzero values lie in an interval of the frequency axis that does not include the origin.

A *passband pulse*, denoted $\widetilde{s}(t)$, is a passband waveform with finite energy of the form

$$\widetilde{s}(t) = s_R(t) \cos 2\pi f_0 t - s_I(t) \sin 2\pi f_0 t.$$

We usually use the term "passband pulse" for a relatively simple function and the term "passband waveform" for a relatively complicated function, although this distinction is not precisely defined. Figure 5.1 gives an illustration of the magnitude of the spectrum of a passband pulse. The transform $\widetilde{S}(f)$ is complex in general, so it has both a real part, $\text{Re}[\widetilde{S}(f)]$, and an imaginary part, $\text{Im}[\widetilde{S}(f)]$. These are *not* the translated copies of the Fourier transforms of $s_R(t)$ and $s_I(t)$. Figure 5.2 gives an illustration of a transform $\widetilde{S}(f)$, including both the real part and the imaginary part in a single graph, where $\widetilde{S}(f) = \text{Re}[\widetilde{S}(f)] + j\text{Im}[\widetilde{S}(f)]$. Because the pulse $\widetilde{s}(t)$ is real, the transform satisfies $\widetilde{S}^*(f) = \widetilde{S}(-f)$, so the real and imaginary parts satisfy

$$\text{Re}[\widetilde{S}(f)] = \text{Re}[\widetilde{S}(-f)]$$
$$\text{Im}[\widetilde{S}(f)] = -\text{Im}[\widetilde{S}(-f)].$$

This relationship can be seen in Figure 5.2.

A passband pulse has a Fourier transform that is related to the Fourier transforms of the modulation components $s_R(t)$ and $s_I(t)$ by the modulation theorem

$$\widetilde{S}(f) = \frac{1}{2}[S_R(f - f_0) + S_R(f + f_0)] - \frac{1}{2}j[S_I(f - f_0) - S_I(f + f_0)]$$

where $S_R(f)$ and $S_I(f)$ may each themselves be complex. This is why $\text{Re}[\widetilde{S}(f)]$ and $\text{Im}[\widetilde{S}(f)]$ are not, in general, equal to frequency shifted copies of $S_R(f)$ and $S_I(f)$.

It is always possible to convert a passband waveform $\widetilde{v}(t)$ into a complex baseband waveform $v(t)$. Just as the two baseband waveforms $v_R(t)$ and $v_I(t)$ can be assembled into the passband waveform $\widetilde{v}(t)$, so the passband waveform $\widetilde{v}(t)$ can be decomposed to form two real baseband waveforms $v_R(t)$ and $v_I(t)$. Specifically, given any arbitrary frequency f_0 in (or near) the passband, the arbitrary passband waveform $\widetilde{v}(t)$ can be decomposed into a representation consisting of two baseband waveforms $v_R(t)$ and $v_I(t)$ multiplying sine and cosine carriers in the standard form

$$\widetilde{v}(t) = v_R(t) \cos 2\pi f_0 t - v_I(t) \sin 2\pi f_0 t.$$

To verify that any arbitrary passband waveform can be expressed in this form, one can employ an argument in the frequency domain defining the baseband spectra $V_R(f)$ and $V_I(f)$ in terms of $\widetilde{V}(f)$, or an argument in the time domain defining the baseband signals $v_R(t)$ and $v_I(t)$ by the mixing operations defined in the next paragraph.[2] Figure 5.3 illustrates this decomposition of a passband signal. The frequency f_0 is called the *carrier frequency* in many applications, though for an arbitrary passband signal it is better called the *reference frequency* because then f_0 is only an arbitrary point of reference on the frequency axis. It need not have any other special role in $\widetilde{v}(t)$. But if there were some frequency that played a special role in $\widetilde{v}(t)$, as when there is an explicit sinusoid, we would likely choose that frequency as the reference frequency f_0. The frequency f_0 only needs to have the property that

$$\widetilde{V}(f) = 0 \quad \text{if } |(|f| - f_0)| > \frac{B}{2}$$

for some constant B smaller than $2f_0$. In practice, the condition that $\widetilde{V}(f)$ is zero is replaced by the somewhat vague condition that $\widetilde{V}(f)$ is negligible.

---

[2] If a more formal procedure is desired, define

$v_R(t) = \text{Re}[\bar{v}(t)e^{-j2\pi f_0 t}]$

$v_I(t) = \text{Im}[\bar{v}(t)e^{-j2\pi f_0 t}]$

where $\bar{v}(t)$ is the *analytic signal* associated with $v(t)$ and defined by

$\bar{v}(t) = v(t) + j\widehat{v}(t),$

and $\widehat{v}(t)$ is the *Hilbert transform* of $v(t)$ defined by

$\widehat{v}(t) = \int_{-\infty}^{\infty} \frac{1}{\pi} \frac{v(\xi)}{t - \xi} d\xi.$

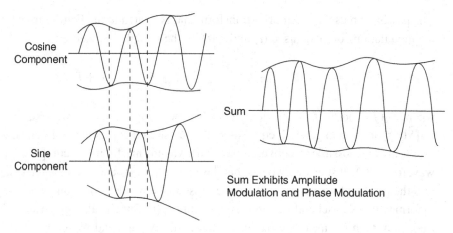

Figure 5.3. A passband signal.

To identify $v_R(t)$ and $v_I(t)$ directly from $\tilde{v}(t)$, use the standard trigonometric identities

$$2\cos^2 A = 1 + \cos 2A$$

$$2\sin^2 A = 1 - \cos 2A$$

$$2\sin A \cos A = \sin 2A$$

to observe that

$$2\tilde{v}(t)\cos 2\pi f_0 t = 2v_R(t)\cos^2 2\pi f_0 t - 2v_I(t)\sin 2\pi f_0 t \cos 2\pi f_0 t$$

$$= v_R(t) + v_R(t)\cos 4\pi f_0 t - v_I(t)\sin 4\pi f_0 t$$

$$-2\tilde{v}(t)\sin 2\pi f_0 t = -2v_R(t)\cos 2\pi f_0 t \sin 2\pi f_0 t + 2v_I(t)\sin^2 2\pi f_0 t$$

$$= v_I(t) - v_R(t)\sin 4\pi f_0 t - v_I(t)\cos 4\pi f_0 t.$$

In each case, the sinusoids are at frequency $2f_0$. An ideal lowpass filter will reject these sinusoids, thereby resulting in $v_R(t)$ and $v_I(t)$.

The choice of reference frequency is arbitrary; any frequency in the vicinity of the passband will do. The choice of a different reference frequency changes the modulation components but does not change the form of the passband representation. Thus we can change the reference frequency to f_0' as follows:

$$\tilde{v}(t) = v_R(t)\cos 2\pi f_0 t - v_I(t)\sin 2\pi f_0 t$$

$$= v_R(t)\cos[2\pi f_0' t + 2\pi(f_0 - f_0')t] - v_I(t)\sin[2\pi f_0' t + 2\pi(f_0 - f_0')t]$$

$$= [v_R(t)\cos 2\pi(f_0 - f_0')t - v_I(t)\sin 2\pi(f_0 - f_0')t]\cos 2\pi f_0' t$$

$$- [v_R(t)\sin 2\pi(f_0 - f_0')t + v_I(t)\cos 2\pi(f_0 - f_0')t]\sin 2\pi f_0' t.$$

Therefore, with new reference frequency f_0',

$$\tilde{v}(t) = v_R'(t) \cos 2\pi f_0' t - v_I'(t) \sin 2\pi f_0' t$$

now with modulation components

$$v_R'(t) = v_R(t) \cos 2\pi (f_0 - f_0')t - v_I(t) \sin 2\pi (f_0 - f_0')t$$
$$v_I'(t) = v_R(t) \sin 2\pi (f_0 - f_0')t + v_I(t) \cos 2\pi (f_0 - f_0')t.$$

The receiver may even use a reference frequency that is a little different from the transmitter, say f_0' in place of f_0. The receiver still sees a passband signal, but the modulation components will be different when referenced to a different reference frequency. When it is possible, we should choose the reference frequency f_0 according to convenience, that is, so as to make the modulation components simple to describe or to process.

A passband signal at carrier frequency f_0 can be converted to a passband signal at carrier frequency f_1 without changing the modulation components by an operation known as *mixing*, provided f_1 and f_0 are sufficiently far apart. This involves first multiplying the passband waveform by $2 \cos 2\pi (f_0 - f_1)t$ as follows:

$$\tilde{v}'(t) = \tilde{v}'(t) 2 \cos 2\pi (f_0 - f_1)t$$
$$= [v_R(t) \cos 2\pi f_0 t - v_I(t) \sin 2\pi f_0 t] 2 \cos 2\pi (f_0 - f_1)t.$$

The product of a passband signal and a sinusoid can be expanded by using the standard trigonometric identities:

$$2 \cos A \cos B = \cos(A + B) + \cos(A - B)$$
$$2 \sin A \cos B = \sin(A + B) + \sin(A - B).$$

Therefore

$$\tilde{v}'(t) = [v_R(t) \cos 2\pi f_1 t - v_I(t) \sin 2\pi f_1 t]$$
$$+ [v_R(t) \cos 2\pi (2f_0 - f_1)t - v_I(t) \sin 2\pi (2f_0 - f_1)t].$$

The two brackets are called the *upper and lower sidebands* of the mixing operation. The frequency f_1 is chosen so that the spectra of the two sidebands do not overlap. The undesired sideband is called the *image* of the desired sideband. Figure 5.4 shows the case where f_1 is chosen smaller than f_0. The upper sideband can be rejected by a filtering operation to produce the passband signal, which now we also call $\tilde{v}(t)$, but at the new carrier frequency f_1,

$$\tilde{v}(t) = v_R(t) \cos 2\pi f_1 t - v_I(t) \sin 2\pi f_1 t.$$

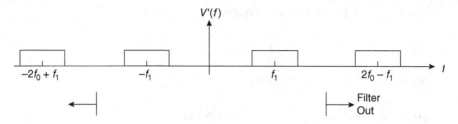

Figure 5.4. Upper and lower mixing sidebands.

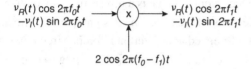

Figure 5.5. Symbolic representation of a mixer.

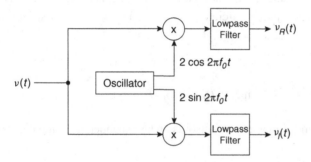

Figure 5.6. Forming the complex representation of a passband signal.

The function of a mixer is shown symbolically in Figure 5.5. The image suppression filter is not shown in the figure, but it is understood to be present.

Figure 5.6 shows how the carrier sinusoids of a passband signal can be completely stripped by the operation of mixing – this time using a reference sinusoid at frequency f_0. Multiplication by $2 \cos 2\pi f_0 t$ produces the signal

$$v'(t) = \tilde{v}(t) 2 \cos 2\pi f_0 t$$
$$= [v_R(t) \cos 2\pi f_0 t - v_I(t) \sin 2\pi f_0 t] 2 \cos 2\pi f_0 t$$
$$= v_R(t) + v_R(t) \cos 2\pi (2f_0)t - v_I(t) \sin 2\pi (2f_0)t$$

where the second line follows from the standard trigonometric identities. A lowpass filter rejects the terms at frequency $2f_0$, thereby producing a signal equal to $v_R(t)$, the in-phase modulation component. Similarly, multiplication by $2 \sin 2\pi f_0 t$ produces a signal at baseband equal to the quadrature modulation component $v_I(t)$. In this way,

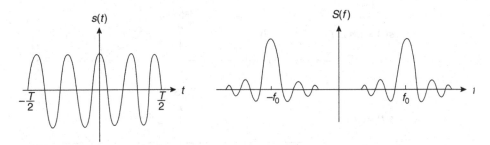

Figure 5.7. An example of a passband pulse.

every passband waveform $\widetilde{v}(t)$ with reference frequency f_0 maps into a unique pair of baseband waveforms $(v_R(t), v_I(t))$ from which the passband waveform $\widetilde{v}(t)$ can be recovered by the modulation process.

Figure 5.7 shows a simple example: the modulation of a square pulse by a cosine wave together with the Fourier transform of this passband pulse. In the general case, a passband pulse $\widetilde{s}(t)$ will have both an in-phase and a quadrature modulation component, and each of these will have a Fourier transform with a real part and an imaginary part. To see how these components combine in the process of modulation, it is instructive to look at the examples of the Fourier transforms of simple passband waveforms, shown in Figure 5.8. These examples are intended to show the interplay between the real and imaginary components of the spectrum.

The passband pulse

$$\widetilde{s}(t) = s_R(t) \cos 2\pi f_0 t - s_I(t) \sin 2\pi f_0 t$$

has energy

$$E_p = \int_{-\infty}^{\infty} \widetilde{s}^2(t) dt.$$

This can be expressed in terms of the modulation components. To show this in the frequency domain, first write

$$E_p = \int_{-\infty}^{\infty} |\widetilde{S}(f)|^2 df$$

$$= \frac{1}{4} \int_{-\infty}^{\infty} |S_R(f - f_0) - jS_I(f - f_0) + S_R(f + f_0) + jS_I(f + f_0)|^2 df.$$

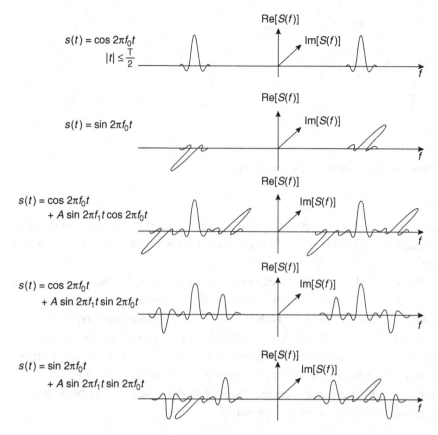

Figure 5.8. Complex spectra of some simple passband waveforms.

When the square is expanded, any product of a term centered at $-f_0$, and a term centered at $+f_0$, will equal zero by the definition of a passband waveform. Therefore

$$E_p = \frac{1}{4}\int_{-\infty}^{\infty} |S_R(f - f_0) - jS_I(f - f_0)|^2 df + \frac{1}{4}\int_{-\infty}^{\infty} |S_R(f + f_0) + jS_I(f + f_0)|^2 df$$

$$= \frac{1}{4}\int_{-\infty}^{\infty} |S_R(f) - jS_I(f)|^2 df + \frac{1}{4}\int_{-\infty}^{\infty} |S_R(f) + jS_I(f)|^2 df.$$

When the squares are expanded, the cross terms cancel, leaving the expression

$$E_p = \frac{1}{2}\int_{-\infty}^{\infty} |S_R(f)|^2 df + \frac{1}{2}\int_{-\infty}^{\infty} |S_I(f)|^2 df$$

$$= \frac{1}{2}\int_{-\infty}^{\infty} [s_R^2(t) + s_I^2(t)]dt.$$

The energy of the passband pulse is one-half of the energy in the complex baseband representation developed in the next section. This is easy to see in the time domain.

Multiplying a baseband pulse by a sinusoid of sufficiently high frequency reduces its energy by one-half.

5.2 Complex baseband waveforms

The most modern and compact way – and ultimately the simplest way – to think about a passband waveform $v(t)$ for digital communication is to think of it as a *complex baseband waveform* defined as

$$v(t) = v_R(t) + jv_I(t).$$

The complex baseband waveform $v(t)$ is a representation of the passband waveform $v(t)$, and is called the *complex baseband representation* of $\tilde{v}(t)$. The complex baseband representation of a waveform is usually emphasized throughout the book in preference to a passband representation. Not only is the complex baseband representation notationally convenient, but it allows one to draw on one's geometrical intuition when dealing with the complex plane. Treating the complex representation as the more fundamental version of the waveform may also be justified by the fact that the waveforms usually exist as complex baseband waveforms as they pass through major portions of the transmitter or receiver, and this is where most of the design details are found.

We shall use the complex representation in two ways. Sometimes it is just a convenient shorthand for a signal that is physically a passband signal. We have already encountered some fairly complicated equations that result when writing out passband equations explicitly. The complex notation is often used instead because it is much easier to work with. The complex notation might also represent a signal that has been physically converted into the form of a pair of baseband signals by stripping off the carrier sinusoids. Then the complex signal $v(t)$ represents the pair of baseband signals $(v_R(t), v_I(t))$. Figure 5.9 shows a passband pulse and its complex representation. A

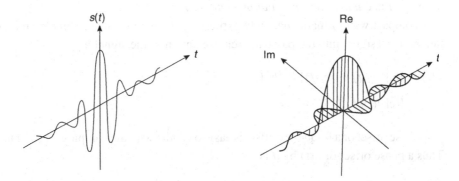

Figure 5.9. A passband pulse and its complex representation.

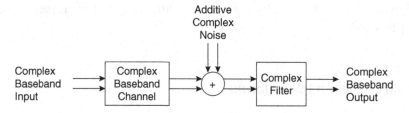

Figure 5.10. Convention for depicting complex signals.

complex signal might be denoted in a functional block diagram, as in Figure 5.10, by a double line, one line depicting the real part and one depicting the imaginary part. However, a single line will usually suffice with the understanding that it represents a complex signal.

For passage through a passband channel, the complex baseband waveform is temporarily converted to the passband representation

$$\tilde{v}(t) = v_R(t) \cos 2\pi f_0 t - v_I(t) \sin 2\pi f_0 t.$$

The passband output of the passband channel may be reconverted back to the complex baseband representation for the computational purpose of demodulation. Because we do not study propagation, the complex representation is of primary interest in most of this book.

Moreover, any complex baseband signal, $v(t) = v_R(t) + jv_I(t)$, can be used to define the passband signal $\tilde{v}(t)$ by

$$\tilde{v}(t) = v_R(t) \cos 2\pi f_0 t - v_I(t) \sin 2\pi f_0 t$$

where the carrier frequency f_0 is a fixed reference frequency. The carrier frequency must be sufficiently large to make $v(t)$ a proper passband waveform, and the map from $v_R(t)$ and $v_I(t)$ into the passband representation $\tilde{v}(t)$ is called *passband modulation*. The terms $v_R(t)$ and $v_I(t)$ are called the *in-phase* and *quadrature modulation components*, or the *real* and *imaginary modulation components*.

A compact way of mathematically expressing the map from the complex representation of the signal into the passband representation of the signal is

$$\tilde{v}(t) = \text{Re}[[v_R(t) + jv_I(t)]e^{j2\pi f_0 t}]$$
$$= \text{Re}[v(t)e^{j2\pi f_0 t}].$$

A phase offset or a frequency offset is easy to handle with the complex representation. Thus a phase offset of $v(t)$ by θ is

$$v(t) = [v_R(t) + jv_I(t)]e^{j\theta},$$

corresponding to the passband representation

$$\tilde{v}(t) = v_R(t)\cos(2\pi f_0 + \theta) - v_I(t)\sin(2\pi f_0 t + \theta).$$

A frequency offset of $v(t)$ by ν is

$$v(t) = [v_R(t) + jv_I(t)]e^{j2\pi \nu t}$$
$$= [v_R(t)\cos 2\pi \nu t - v_I(t)\sin 2\pi \nu t]$$
$$+ j[v_R(t)\sin 2\pi \nu t + v_I(t)\cos 2\pi \nu t]$$

corresponding to the passband representation

$$\tilde{v}(t) = v_R(t)\cos 2\pi (f_0 + \nu)t - v_I(t)\sin 2\pi (f_0 + \nu)t.$$

A comparison with the manipulations of the passband waveforms, which we worked through earlier, should make it evident that great simplicity results when using the complex representation. This is one reason we usually prefer the complex representation.

The energy in the complex baseband pulse

$$s(t) = s_R(t) + js_I(t)$$

is defined as

$$E_p = \int_{-\infty}^{\infty} |s(t)|^2 dt$$
$$= \int_{-\infty}^{\infty} [s_R^2(t) + s_I^2(t)]dt.$$

This is exactly twice the energy[3] of the passband version of $s(t)$. Because we will usually be interested in the ratio of signal energy to noise density, this factor of two will cancel in the ratio. Consequently, this factor of two will rarely be a consideration affecting the interests of this book.

Because our complex baseband representation and passband representation differ in energy (or power) by a factor of two, one may prefer an alternative definition of the passband waveform given by

$$\tilde{v}(t) = v_R(t)\sqrt{2}\cos 2\pi f_0 t - v_I(t)\sqrt{2}\sin 2\pi f_0 t$$

so that the complex representation and the passband representation have the same energy. For notational simplicity, however, the usual convention is to suppress the $\sqrt{2}$ when studying waveforms.

---

[3] To avoid this behavior, one could define the passband pulse to include an amplitude of $\sqrt{2}$. We prefer to avoid complicating the formulas in this manner.

5.3   Passband filtering

A passband filter is a filter whose impulse response is a passband signal. The impulse response of the passband filter $\widetilde{h}(t)$ is

$$\widetilde{h}(t) = h_R(t) \cos 2\pi f_0 t - h_I(t) \sin 2\pi f_0 t.$$

The complex representation of the impulse response is

$$h(t) = h_R(t) + jh_I(t).$$

One reason why the complex notation is useful is that the operation of passband filtering of a passband waveform behaves the same as the operation of complex filtering of a complex waveform. In Figure 5.11, we illustrate this principle by showing how to implement a passband filter by using the baseband filters $h_R(t)$ and $h_I(t)$ following down-conversion and followed by up-conversion. The complex baseband filter is sometimes a convenient implementation. It is also a useful way of thinking. The following theorem verifies that passband filtering of a passband waveform exactly corresponds to the corresponding complex baseband filtering of a complex baseband waveform.

Theorem 5.3.1   *If $\widetilde{u}(t)$ is given by*

$$\widetilde{u}(t) = \int_{-\infty}^{\infty} \widetilde{v}(\xi)\widetilde{h}(t - \xi)d\xi$$

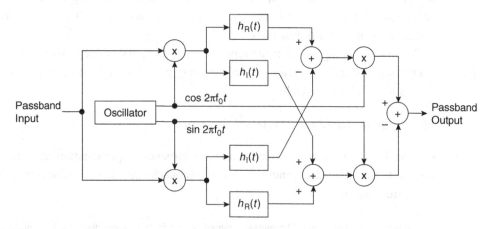

Figure 5.11. Implementing a passband filter at baseband.

where $\widetilde{h}(t)$ and $\widetilde{v}(t)$ are both passband signals at carrier frequency f_0, then $\widetilde{u}(t)$ is also a passband signal at carrier frequency f_0, and can be expressed as

$$\widetilde{u}(t) = \left[\frac{1}{2}\int_{-\infty}^{\infty}(v_R(\xi)h_R(t-\xi) - v_I(\xi)h_I(t-\xi))d\xi\right]\cos 2\pi f_0 t$$

$$-\left[\frac{1}{2}\int_{-\infty}^{\infty}(v_R(\xi)h_I(t-\xi) + v_I(\xi)h_R(t-\xi))d\xi\right]\sin 2\pi f_0 t$$

or, more concisely,

$$\widetilde{u}(t) = \frac{1}{2}[v_R(t)*h_R(t) - v_I(t)*h_I(t)]\cos 2\pi f_0 t$$

$$-\frac{1}{2}[v_R(t)*h_I(t) + v_I(t)*h_R(t)]\sin 2\pi f_0 t.$$

Proof   The term $\widetilde{v}(\xi)\widetilde{h}(t-\xi)$ when stated in terms of the passband representation and expanded gives four terms

$$\widetilde{v}(\xi)\widetilde{h}(t-\xi) = v_R(\xi)h_R(t-\xi)\cos 2\pi f_0\xi \cos 2\pi f_0(t-\xi)$$

$$+ v_I(\xi)h_I(t-\xi)\sin 2\pi f_0\xi \sin 2\pi f_0(t-\xi)$$

$$- v_R(\xi)h_I(t-\xi)\cos 2\pi f_0\xi \sin 2\pi f_0(t-\xi)$$

$$- v_I(\xi)h_R(t-\xi)\sin 2\pi f_0\xi \cos 2\pi f_0(t-\xi).$$

Each of these terms expands into two terms by using standard trigonometric identities. We write these eight terms as follows:

$$\widetilde{v}(\xi)\widetilde{h}(t-\xi) = \frac{1}{2}v_R(\xi)h_R(t-\xi)\cos 2\pi f_0 t - \frac{1}{2}v_R(\xi)h_R(t-\xi)\cos 2\pi 2f_0(\xi - t/2)$$

$$- \frac{1}{2}v_I(\xi)h_I(t-\xi)\cos 2\pi f_0 t + \frac{1}{2}v_I(\xi)h_I(t-\xi)\cos 2\pi 2f_0(\xi - t/2)$$

$$- \frac{1}{2}v_R(\xi)h_I(t-\xi)\sin 2\pi f_0 t + \frac{1}{2}v_R(\xi)h_I(t-\xi)\sin 2\pi 2f_0(\xi - t/2)$$

$$- \frac{1}{2}v_I(\xi)h_R(t-\xi)\sin 2\pi f_0 t - \frac{1}{2}v_I(\xi)h_R(t-\xi)\sin 2\pi 2f_0(\xi - t/2).$$

Now integrate with respect to ξ. The theorem will be established if we can show that each of the terms involving $\cos 4\pi f_0(\xi - t/2)$ integrates to zero. We show this for the first of such terms by using Parseval's formula. The other three terms can be treated in the same way. Let

$$a(\xi) = h_R(t-\xi)$$

$$b(\xi) = v_R(\xi)\cos 2\pi 2f_0(\xi - t/2).$$

By Parseval's formula,

$$\int_{-\infty}^{\infty} a(\xi)b^*(\xi)d\xi = \int_{-\infty}^{\infty} A(f)B^*(f)df.$$

By the properties of the Fourier transform,

$$A(f) = H_R^*(f)e^{-j2\pi\xi f}$$

where $H_R(f)$ is the Fourier transform of $h_R(t)$. By definition of $h(t)$ as a passband filter, $H_R^*(f)$ equals zero for $|f| \geq f_0$. On the other hand, by the definition of $v(t)$, the Fourier transform of $v_R(\xi)$ also equals zero for $|f| \geq f_0$. Consequently, by the modulation theorem, $B(f)$ must be zero for $|f| \leq f_0$. Therefore $A(f)B^*(f)$ is zero everywhere; so by Parseval's formula,

$$\int_{-\infty}^{\infty} a(\xi)b^*(\xi)d\xi = 0.$$

This completes the proof of the theorem.   ∎

5.4   Passband waveforms for binary signaling

A complex baseband waveform can be viewed as two real waveforms; the real component forms one waveform and the imaginary component forms the other. Similarly, a passband waveform can be viewed as two waveforms, the in-phase component and the quadrature component. Each component of the passband waveform can be used independently to convey a baseband waveform. It is not even necessary that the same kind of baseband waveform be used on each component.

The simplest passband waveform for digital communication uses binary antipodal signaling on one component, and no modulation on the other. This passband version of antipodal signaling is called *binary phase-shift keying* (BPSK). The simplest digital passband waveform that uses both modulation components uses binary antipodal signaling on each component. This is called *quadrature phase-shift keying* (QPSK) provided the time reference is the same on both components. The in-phase and quadrature components of the passband channel can be viewed as independent channels, modulating two binary datastreams into two baseband waveforms that are modulated onto the two components of the carrier. A single datastream can be modulated into QPSK, two bits at a time, with alternate bits going onto the in-phase and quadrature carrier components. With QPSK, two bits are transmitted in a time interval of duration T, and the average time per bit T_b is given by $T_b = T/2$.

Any passband waveform can be presented as a complex baseband waveform, so we can view a QPSK waveform as a single complex baseband waveform. Then the modulation is

$$c(t) = \sum_{\ell=-\infty}^{\infty} a_\ell s(t - \ell T)$$

where

$$a_\ell = \begin{cases} (1+j)A & \text{if the } \ell\text{th bit pair is } 11 \\ (1-j)A & \text{if the } \ell\text{th bit pair is } 10 \\ (-1+j)A & \text{if the } \ell\text{th bit pair is } 01 \\ (-1-j)A & \text{if the } \ell\text{th bit pair is } 00. \end{cases}$$

This may also be expressed by using the notation of the impulse function as

$$c(t) = \left[\sum_{\ell=-\infty}^{\infty} a_\ell \delta(t - \ell T) \right] * s(t).$$

This factorization makes evident the role of the complex amplitudes a_ℓ in representing the data and the role of the pulse $s(t)$ in creating a continuous-time waveform suitable for the channel.

Each pair of data bits defines a point in the complex plane. These four complex points, shown in Figure 5.12, form a simple example of what is known as a *complex signal constellation*. This constellation is called the four-ary PSK signal constellation, or the QPSK signal constellation. The QPSK waveform can be visualized in terms of the functional modulator shown in Figure 5.13. Bit pairs are mapped into a stream of complex impulses, with complex amplitudes $(\pm 1 \pm j)A$. The stream of complex impulses is then passed through a filter with an impulse response $s(t)$ to form the complex waveform. The waveform itself is a complex function of time and is somewhat difficult to visualize. It consists of a stream of pulses; each pulse can be at one of four possible phases. The pulses could be Nyquist pulses, such as those shown in

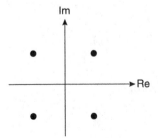

Figure 5.12. The QPSK signal constellation.

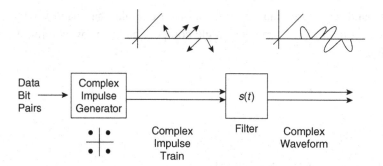

Figure 5.13. Functional description of a QPSK modulator.

Figure 2.11. Then the complex waveforms would have a complicated phase history. The real and imaginary components of $c(t)$ will each be as described by the eye diagram of Figure 2.12.

In QPSK using rectangular pulses, the phase angle can change by multiples of 90° between one time interval and the next. Because rectangular pulses, in practice, do not have perfectly sharp edges, the complex envelope of the waveform will pass through zero amplitude when the phase changes by 180°. An alternative method of using rectangular pulses, called *offset* QPSK (OQPSK), delays the quadrature waveform by half a pulse width so that the phase angle only changes by 90° at any given time. The waveform is then written

$$c(t) = \sum_{\ell=-\infty}^{\infty} a_{2\ell}s(t - \ell T) + j \sum_{\ell=-\infty}^{\infty} a_{2\ell+1}s(t - \ell T - T/2)$$

where

$$a_\ell = \begin{cases} A & \text{if the } \ell\text{th data bit is a one} \\ -A & \text{if the } \ell\text{th data bit is a zero.} \end{cases}$$

The time T is the duration of a pair of bits. Expressed in terms of bit time, $T_b = T/2$, the equation becomes

$$c(t) = \sum_{\ell=-\infty}^{\infty} a_{2\ell}s(t - 2\ell T_b) + j \sum_{\ell=-\infty}^{\infty} a_{2\ell+1}s(t - (2\ell + 1)T_b).$$

An alternative way of writing the OQPSK waveform is

$$c(t) = \sum_{\ell=-\infty}^{\infty} b_\ell s(t - \ell T_b)$$

State

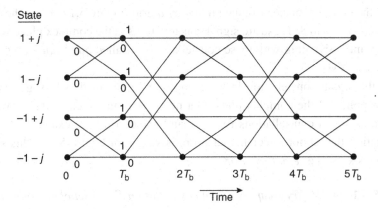

Figure 5.14. Trellis for OQPSK modulation.

where now the b_ℓ are purely real numbers equal to a_ℓ when ℓ is even, and are purely imaginary numbers equal to ja_ℓ when ℓ is odd. If $s(t)$ is a rectangular pulse of duration $2T_b$, the waveform $c(t)$ will take the values $(\pm A \pm jA)$ because $s(t)$ has duration $2T_b$.

Because of the time offset, OQPSK cannot be described simply in terms of choosing a point of a signal constellation. However, if a rule is imposed to constrain the choice of the constellation point as a function of the previous point, the OQPSK waveform can be viewed in terms of a signal constellation. This means that the encoder is a finite-state machine. When taking this point of view, it is simpler to think of the data as modulated into the waveform in a different way that can be represented by a trellis.

In Figure 5.14, we show the trellis for OQPSK. There are four nodes in each column, corresponding to the four points in the QPSK signal constellation. At even bit times, only the imaginary component can change, and this is reflected in the destination of the path leaving each node. Similarly, at odd bit times, only the real component can change. The sequence of states through which a path passes defines a sequence of complex numbers that specifies the OQPSK waveform.

5.5   Multilevel signaling at passband

Now we turn attention to methods that transmit multiple bits at one time within a narrow bandwidth by modulating the complex amplitude of a single pulse $s(t)$. Any of the signal constellations on the real line that were shown in Chapter 2 (in Figure 2.6) can be used to define a set of signal amplitudes for the in-phase modulation and also a set of signal amplitudes for the quadrature modulation. If the signaling alphabets are binary on each component, and the sample times are synchronized on the two axes, then this is QPSK. We have seen that QPSK can be represented as four points in the complex plane. More generally, a q-ary signal constellation on each modulation component becomes a set of

q^2 points in the complex plane, arranged in a q by q grid. Figure 5.15 shows the cases where q equals 2, 4, and 8. These are signal constellations in the complex plane with 4, 16, and 64 points. Signaling with these constellations is called *quadrature amplitude modulation* or QAM.

Viewing the signal constellation in the complex plane immediately suggests that other sets of points of the complex plane, not necessarily in a rectangular array, can also be used as signal constellations for passband waveforms. An example of a signal constellation that is not a rectangular array is shown in Figure 5.16. This signal constellation is known as *eight-ary PSK*.

Definition 5.5.1   *An M-ary complex signal constellation S is a set of M points in the complex plane.*

$$c_m = c_{mR} + jc_{mI} \quad m = 0, \ldots, M - 1.$$

Usually $M = 2^k$ for some integer k.

Figure 5.17 shows several examples of complex (or two-dimensional) signal constellations.

The *average energy per symbol* of the signal constellation is

$$E_c = \frac{1}{M} \sum_m [c_{mR}^2 + c_{mI}^2].$$

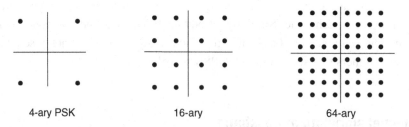

4-ary PSK   16-ary   64-ary

Figure 5.15. Signal constellations for quadrature-amplitude modulation.

8-ary PSK

Figure 5.16. A signal constellation for phase-shift keying.

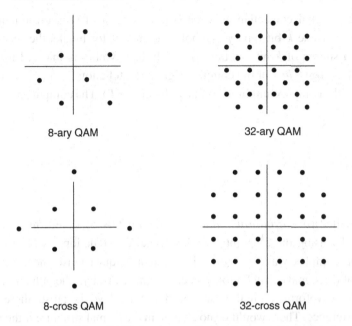

8-ary QAM   32-ary QAM

8-cross QAM   32-cross QAM

Figure 5.17. More QAM signal constellations with amplitude and phase.

The *average energy per bit* of the signal constellation is

$$E_b = E_c/\log_2 M$$

$$= \frac{1}{k}E_c.$$

The *minimum distance* of the signal constellation is defined as the smallest euclidean distance between any two distinct points in the signal constellation. That is,

$$d_{\min} = \min_{m \neq m'} [(c_{mR} - c_{m'R})^2 + (c_{mI} - c_{m'I})^2]^{\frac{1}{2}}.$$

The size of a signal constellation will interest us, and also the relative arrangement of the points. Therefore we may also judge a signal constellation by the figure of merit ρ_{min} given by $\rho_{min}^2 = d_{min}^2/E_b$ and called the *normalized minimum distance* of the signal constellation. The normalized minimum distance is defined such that BPSK and QPSK have a normalized minimum distance of 2. Other signal constellations must have a normalized minimum distance of less than 2. In the next chapter, we shall see that the probability of demodulation error in gaussian noise depends on the structure of the signal constellation primarily through the minimum distance. With E_b held constant, the probability of demodulation error depends on the normalized minimum distance. Therefore two signal constellations with the same number of points are compared, in part, by a comparison of their normalized minimum distances.

To use a 2^k-ary signal constellation, each complex point of the constellation is assigned to represent one k-bit binary symbol. The modulator breaks the incoming datastream into a stream of k-bit symbols. The ℓth data symbol is mapped into the assigned point $a_\ell = a_{\ell R} + ja_{\ell I}$ of the complex signal constellation $\{c_0, c_1, \ldots, c_{M-1}\}$. Then a_ℓ is used as the complex amplitude of the pulse $s(t - \ell T)$. The complex baseband waveform is

$$c(t) = \sum_{\ell=-\infty}^{\infty} a_\ell s(t - \ell T).$$

This waveform, called a *quadrature amplitude modulation* (QAM) waveform, transmits k bits in time T by using the 2^k-point complex signal constellation, as is shown in Figure 5.18. The complex waveform $c(t)$, if $s(t)$ is a Nyquist pulse, neatly passes through a point of the signal constellation at each sampling instant. The pulse $s(t)$ used to form the QAM waveform is a real pulse, and if $s(t)$ is a Nyquist pulse, there is no intersymbol interference. There would be no change in the formal structure if the pulse $s(t)$ were complex, but there seems to be no reason to use a complex pulse.

The real and imaginary parts of $c(t)$ form the in-phase and quadrature modulation components. The transmitted passband waveform is

$$\widetilde{c}(t) = \left[\sum_{\ell=-\infty}^{\infty} \mathrm{Re}[a_\ell s(t - \ell T)] \right] \cos 2\pi f_0 t - \left[\sum_{\ell=-\infty}^{\infty} \mathrm{Im}[a_\ell s(t - \ell T)] \right] \sin 2\pi f_0 t.$$

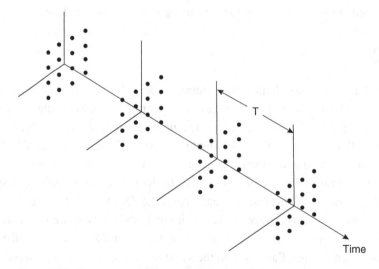

Figure 5.18. Sixteen-ary quadrature amplitude modulation.

Because the pulse $s(t)$ is a real pulse, this passband waveform can be rewritten as

$$\widetilde{c}(t) = \sum_{\ell=-\infty}^{\infty} [a_{\ell R} \cos 2\pi f_0 t - a_{\ell I} \sin 2\pi f_0 t] s(t - \ell T).$$

The reorganization of the equation suggests that the pulse shape $s(t)$ can be imposed after the passband structure is formed. Thus, for example, $s(t)$ might even be formed as the convolution of two pulses, one pulse introduced at complex baseband and the second introduced by a passband filter so that the composition of the two pulses is the actual baseband pulse $s(t)$.

5.6   Power density spectra of passband waveforms

The power density spectrum of a stationary passband random process will be studied by reference to the power density spectrum of its complex baseband representation. The power density spectrum of a complex data waveform such as QPSK is determined primarily by the spectrum $|s(t)|^2$ of the pulse $s(t)$, as was shown for real baseband waveforms in Section 2.9. Just as for real baseband waveforms, $c(t)$ is not stationary so it does not have a power density spectrum. To make the waveform stationary, we replace the expression for $c(t)$ by

$$c(t) = \sum_{\ell=-\infty}^{\infty} a_{\ell} s(t - \ell T - \alpha)$$

where α is a random variable uniformly distributed on $[0, T]$, and a_{ℓ} for $\ell = 0, \pm 1, \pm 2, \ldots$ is a series of independent, identically distributed, complex random variables taking values in some complex signal constellation. Now the waveform is a stationary random process because the ensemble of waveforms is invariant under a translation of the time origin.

Theorem 5.6.1 *Let*

$$c(t) = \sum_{\ell=-\infty}^{\infty} a_{\ell} s(t - \ell T - \alpha)$$

where the a_{ℓ} are independent, identically distributed, complex random variables of zero mean and variance 2, and α is a random variable uniformly distributed over $[0, T]$ and independent of the a_{ℓ}. Then $c(t)$ is stationary with power density spectrum

$$\Phi_c(f) = \frac{2}{T} |S(f)|^2.$$

Proof   The random process is stationary because the ensemble is independent of the time origin. The autocorrelation function of the complex random waveform is

$$R(\tau) = E[c(t)c^*(t+\tau)]$$

$$= E\left[\sum_{\ell=-\infty}^{\infty} \sum_{\ell'=-\infty}^{\infty} a_\ell a_{\ell'}^* s(t - \ell T - \alpha) s^*(t + \tau - \ell' T - \alpha) \right].$$

Move the expectation inside the sum and recall that the random variables a_ℓ and α are independent to obtain

$$R(\tau) = 2 \sum_{\ell=-\infty}^{\infty} \sum_{\ell'=-\infty}^{\infty} E[a_\ell a_{\ell'}^*] E[s(t - \ell T - \alpha) s^*(t + \tau - \ell' T - \alpha)]$$

$$= \sum_{\ell=-\infty}^{\infty} E[s(t - \ell T - \alpha) s^*(t + \tau - \ell T - \alpha)]$$

because $E[a_\ell a_{\ell'}^*] = \delta_{\ell\ell'} E[(\pm 1 \pm j)(\pm 1 \mp j)] = 2\delta_{\ell\ell'}$. Moreover, because the delay is uniformly distributed, we can write

$$E[s(t - \ell T - \alpha) s^*(t + \tau - \ell T - \alpha)] = \frac{1}{T} \int_0^T s(t - \ell T) s^*(t + \tau - \ell T) dt.$$

This brings us to the final string of equalities

$$R(\tau) = \frac{2}{T} \sum_{\ell=-\infty}^{\infty} \int_0^T s(t - \ell T) s^*(t + \tau - \ell T) dt$$

$$= \frac{2}{T} \sum_{\ell=-\infty}^{\infty} \int_{\ell T}^{(\ell+1)T} s(t) s^*(t + \tau) dt$$

$$= \frac{2}{T} \int_{-\infty}^{\infty} s(t) s^*(t + \tau) dt,$$

which is independent of t. Therefore $c(t)$ is indeed stationary. The Fourier transform of $R(\tau)$ is

$$\Phi_c(f) = \frac{2}{T} |S(f)|^2,$$

the power density spectrum.   ∎

5.7   Minimum-shift keying

An often preferred variation of OQPSK is obtained by taking $s(t)$ to be a half-cosine pulse

$$s(t) = \begin{cases} \cos \pi t/T & |t| \leq T/2 \\ 0 & \text{otherwise.} \end{cases}$$

This version of OQPSK is called *minimum-shift keying* (MSK). A representation of a minimum-shift keying signaling waveform as a complex function of time is shown in Figure 5.19. The construction of an MSK waveform as a passband waveform is shown in Figure 5.20.

An MSK waveform

$$c(t) = \sum_{\ell=-\infty}^{\infty} a_{2\ell}s(t - 2\ell T_b) + j \sum_{\ell=-\infty}^{\infty} a_{2\ell+1}s(t - (2\ell + 1)T_b)$$

has a constant amplitude $|c(t)| = A$. To see this, notice that when offset by $T/2$, the half-cosine pulse becomes a half-sine pulse

$$s(t - T/2) = \sin \pi t/T \quad 0 \leq t \leq T.$$

Choosing $A = 1$, each a_ℓ is either $+1$ or -1. Therefore we can write $c(t)$ as

$$c(t) = \pm \cos \pi t/T \pm j \sin \pi t/T$$

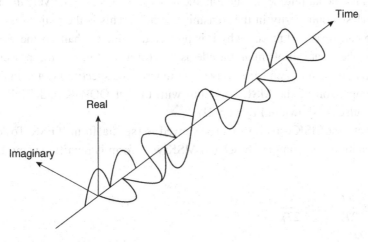

Figure 5.19. The MSK waveform at complex baseband.

Carrier

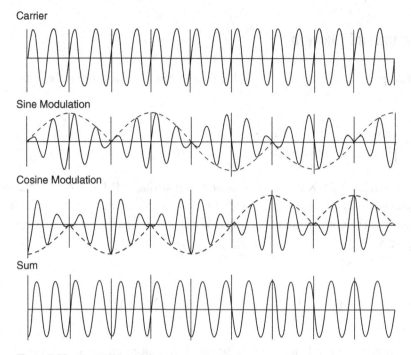

Sine Modulation

Cosine Modulation

Sum

Figure 5.20. The MSK waveform at passband.

where the signs keep changing according to the a_ℓ. Therefore

$$|c(t)|^2 = (\pm\cos\pi t/T)^2 + (\pm\sin\pi t/T)^2$$
$$= 1.$$

Because it has no amplitude fluctuation, the MSK waveform is relatively insensitive to a memoryless nonlinearity in the transmit amplifier. This is the major reason that MSK is popular. A second reason why it is preferred is that the half-cosine pulse has spectral sidelobes that fall off in amplitude as f^{-2} (or as f^{-4} in spectral power). The power density spectrum of an MSK waveform inherits the spectrum of the individual pulse. A comparison of the MSK spectrum with that of OQPSK and BPSK using rectangular pulses is shown in Figure 5.21.

Alternatively, the MSK waveform can be viewed as a special form of FSK. To develop this representation, the complex baseband MSK waveform is rewritten using Euler's formula as

$$c(t) = \pm\cos\frac{\pi t}{2T_b} \pm \sin\frac{\pi t}{2T_b}$$
$$= \pm e^{\pm j2\pi(t/4T_b)}$$

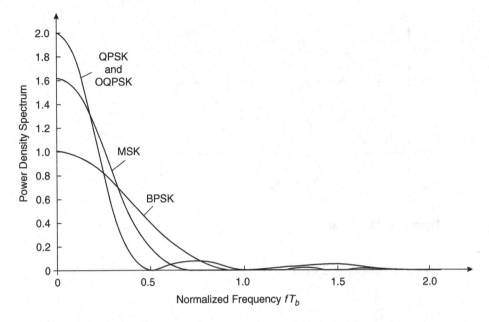

Figure 5.21. Power spectra for binary signaling.

where the signs remain constant during the interval $\ell T_b \leq t < (\ell + 1)T_b$, and the signs change at the start of the next such interval as determined by the next data bit. Thus within each interval, we see a frequency of $\pm(1/4T_b)$. Because the phase changes linearly with time, it is possible to rewrite $c(t)$ as

$$c(t) = e^{j\theta(t)}$$

where

$$\theta(t) = \theta_\ell \pm 2\pi \frac{t - \ell T_b}{4T_b} \qquad \ell T_b \leq t < (\ell + 1)T_b$$

as can be seen from Figure 5.22. Each phase sample $\theta(\ell T_b) = \theta_\ell$, expressed in degrees, equals either $0°$ or $180°$ whenever ℓ is even, and equals either $90°$ or $270°$ whenever ℓ is odd. The actual values depend on the datastream being modulated. The plus or minus sign in the second term of $\theta(t)$ is chosen according to how θ_ℓ must change to become $\theta_{\ell+1}$. Figure 5.22 shows the set of all such phase functions $\theta(t)$ superimposed on a common graph, which should be interpreted modulo $360°$.

As a passband waveform, $c(t)$ is

$$c(t) = \cos(2\pi(f_0 t \pm t/4T_b \mp 1/4) + \theta_\ell)$$
$$= \cos(2\pi(f_i t \mp 1/4) + \theta_\ell) \qquad \ell T_b < t \leq (\ell - 1)T_b$$

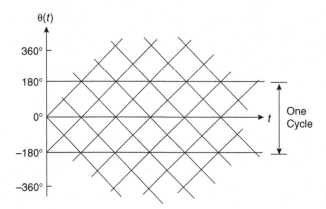

Figure 5.22. MSK phase trajectories.

for i equal to 1 or to 2 where

$$f_1 = f_0 + \frac{1}{4T_b}$$

$$f_2 = f_0 - \frac{1}{4T_b}.$$

Any path through the graph in Figure 5.22 corresponds to a possible phase history; the actual path is determined by the data. If we desire, we can map the data directly into the phase history by using the alternative modulation rule at each node that a data bit zero maps into an increasing phase corresponding to frequency f_1, and a data bit one maps into a decreasing phase, corresponding to frequency f_2. This is the FSK modulation rule. Mapping the data into the waveform by the FSK modulation rule will give an MSK waveform, but not the same MSK waveform as when the data is mapped into the waveform by the I and Q modulation rule in which the data bits directly define the in-phase and quadrature components. Under the FSK modulation rule, the data bits appear in the waveform in a different way than when mapped into the in-phase and quadrature components. As long as the demodulator reads bits out of the waveform in the same way that the modulator reads bits into the waveform, then it does not matter which modulation rule is used. If the demodulator reads bits out of the waveform in a different way than was used by the modulator, then the demodulated data will not be the same as the original data. Conversion between the two forms by means of a precoder or postcoder is necessary.

Of course, the demodulator could be designed to read the data out of the I and Q samples in the same way that the data was read into the I and Q samples, but we want to point out that there is no fundamental reason to avoid the combination of an FSK demodulator with an I and Q modulator. Working through the relationship is an excellent way of exploring the structure of an MSK waveform.

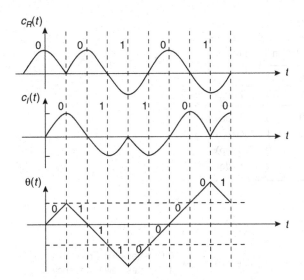

Figure 5.23. Relationship between the I and Q format and the FSK format.

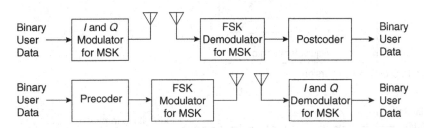

Figure 5.24. The function of a precoder or postcoder.

Figure 5.23 gives an example of an MSK waveform produced by an I and Q modulator. The datastream specifies the polarities of the in-phase and quadrature pulses. Also shown in Figure 5.23 is the phase history of that same MSK waveform annotated with a data record that describes that phase history. Clearly, if the demodulator reads the phase history, it will not produce the original data. However, the original data can be recovered by a postcoder, as shown in Figure 5.24. Alternatively, if the modulator anticipates the demodulation method, the data can be modified by a precoder prior to modulation, as shown in Figure 5.24, so that it will be correct at the demodulator output.

To design the postcoder, we will first represent the combined modulator/demodulator as an equivalent logical transformation of the datastream. It should be clear that the output of the demodulator depends only on the latest bit into the modulator and on the immediately previous bit. Older bits will have no effect. In addition, the functional dependence may be different for odd and even bit times. This motivates us to form Table 5.1.

Table 5.1. *Bit time indices*

Old Time Index			Even Time Index		
Old Bit	New Bit	Frequency	Old Bit	New Bit	Frequency
0	0	0 (up)	0	0	1 (down)
0	1	1 (down)	0	1	0 (up)
1	0	1 (down)	1	0	0 (up)
1	1	0 (up)	1	1	1 (down)

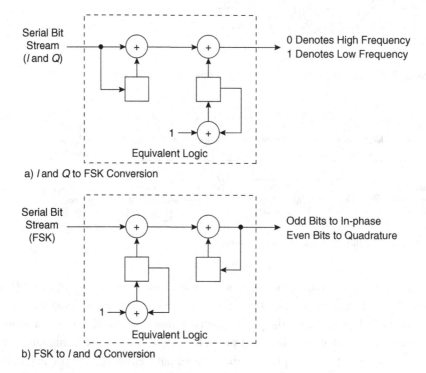

a) *I* and *Q* to FSK Conversion

b) FSK to *I* and *Q* Conversion

Figure 5.25. MSK data format conversion.

Each entry in the frequency column of the table is found by seeking an example of that combination of bits described by the in-phase and quadrature bits in Figure 5.23. An inspection of the table reveals that the relationship between the I and Q format and the FSK format is simply an exclusive-or operation on the last two data bits, followed by complementation at even time indices. An equivalent logic diagram is shown in Figure 5.25, with the exclusive-or operation implemented as a modulo-two addition.

The complementation is effected in the diagram by modulo-two adding the output of a modulo-two counter to the data. In Figure 5.25b, binary logic is shown that will invert the effect of the logic in part a of the figure. To verify this, observe that when the circuits of part a and part b are cascaded, the effect of the modulo-two adders will cancel. The remaining memory cells always contain the same value, and the modulo-two addition of this value twice will cancel and have no effect. Consequently, the circuit in part b is the postcoder that will return the data to its original form.

5.8 *M*-ary orthogonal signaling at passband

Now we turn attention to methods of passband signaling that do not satisfy a bandwidth constraint. Just as for baseband waveforms, as discussed in Section 2.8, one can design passband signaling waveforms that modulate multiple bits at a time using low power. In this section, we shall study passband signaling techniques that make no attempt to confine bandwidth but do reduce the energy per bit.

Definition 5.8.1 *A complex M-ary orthogonal pulse alphabet is a set of M complex functions $s_m(t)$, for $m = 0, \ldots, M - 1$, having the properties of unit energy and orthogonality, as described by*

$$\int_{-\infty}^{\infty} s_m(t)s_n^*(t)dt = \begin{cases} 1 & m = n \\ 0 & m \neq n \end{cases}$$

and

$$\int_{-\infty}^{\infty} s_m(t)s_n^*(t - \ell T)dt = 0 \qquad \ell \neq 0.$$

Any M-ary alphabet, or family, of orthogonal baseband pulses or simplex baseband pulses, as introduced in Section 2.8, can be modulated onto a carrier to produce an M-ary alphabet of orthogonal passband pulses or an M-ary simplex pulse alphabet at passband. Suppose that the set $\{s_m(t) : m = 0, \ldots, M - 1\}$ forms an orthogonal alphabet of real baseband pulses. Then the set $\{s_m(t) \cos 2\pi f_0 t : m = 0, \ldots, M - 1\}$ forms an M-ary orthogonal alphabet of passband signaling pulses, provided that f_0 is large compared to any frequency contained in any $s_m(t)$. These orthogonal passband pulses, formed in this way from baseband pulses, use only the in-phase modulation component, but do not use the quadrature modulation component. More general ways of constructing passband pulses for M-ary orthogonal signaling use the quadrature modulation component as well. In the general case, one can use passband pulses of the form

$$\tilde{s}_m(t) = s_{mR}(t) \cos 2\pi f_0 t - s_{mI}(t) \sin 2\pi f_0 t$$

where f_0 is larger than the largest spectral frequency of the modulation components. A set of M orthogonal pulses of this form, possibly with either $s_{mR}(t)$ or $s_{mI}(t)$ equal to zero, is called an M-ary orthogonal passband pulse alphabet. These waveforms can be expressed in the complex baseband form as

$$s_m(t) = s_{mR}(t) + js_{mI}(t).$$

The passband pulses $\tilde{s}_m(t)$ and $\tilde{s}_{m'}(t)$ are orthogonal

$$\int_{-\infty}^{\infty} \tilde{s}_m(t)\tilde{s}_{m'}(t)dt = 0$$

if and only if their complex baseband pulses $s_m(t)$ and $s_{m'}(t)$ are orthogonal

$$\int_{-\infty}^{\infty} s_m(t)s_{m'}^*(t)dt = 0.$$

The *autocorrelation function* of the pulse $s_m(t)$ is defined as

$$r_{mm}(\tau) = \int_{-\infty}^{\infty} s_m(t)s_m^*(t+\tau)dt,$$

and the *cross-correlation function* of $s_m(t)$ with $s_{m'}(t)$ is defined as

$$r_{mm'}(\tau) = \int_{-\infty}^{\infty} s_m(t)s_{m'}^*(t+\tau)dt.$$

Figure 5.26 illustrates how some of these correlation functions might look for a typical family of orthogonal pulses. Specifically, Figure 5.26 shows the autocorrelation function of $s_m(t)$, which equals E_p at $t = 0$; the autocorrelation function of $s_{m'}(t)$, which

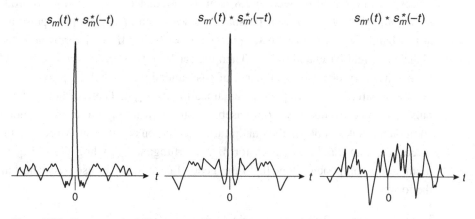

$s_m(t) \star s_m^*(-t)$ $s_{m'}(t) \star s_{m'}^*(-t)$ $s_{m'}(t) \star s_m^*(-t)$

Figure 5.26. Some matched-filter outputs for an orthogonal family.

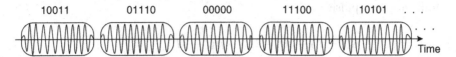

Figure 5.27. A 32-ary orthogonal waveform using MSK pulses.

also equals E_p at $t = 0$; and the cross-correlation function of $s_m(t)$ and $s_{m'}(t)$, which equals zero at $t = 0$.

Let M be a power of 2; $M = 2^k$. To transmit a binary datastream using an M-ary orthogonal alphabet, the datastream is broken into k-bit symbols, the ℓth such symbol represents k data bits by the number m_ℓ, which is mapped into the pulse $s_{m_\ell}(t)$. The transmitted waveform is

$$c(t) = \sum_{\ell=-\infty}^{\infty} s_{m_\ell}(t - \ell T).$$

In this way, the particular pulse contained in the waveform $c(t)$ at the time ℓT conveys the k bits of data that were transmitted at that time.

Figure 5.27 illustrates one example of an orthogonal signaling waveform at passband. The figure suggests that all pulses of that waveform have the same amplitude profile, and the uniqueness resides in the phase structure of the pulses. Each of the 32 pulses in such an alphabet represents one of the 32 five-bit numbers. The pulses superficially look the same, but the phase history within each pulse makes the 32 pulses actually quite different.

We will design a set of M-ary pulses as an example of an orthogonal waveform family. The following four sequences

$$a_0 = (1, j, 1, j)$$
$$a_1 = (1, j, -1, -j)$$
$$a_2 = (1, -j, -1, j)$$
$$a_3 = (1, -j, 1, -j)$$

are orthogonal sequences

$$\sum_{\ell=0}^{3} a_{m\ell} a_{m'\ell}^* = 0.$$

Now choose a pulse $s(t)$, which in this context is called a *chip* or a *pulselet*. Using the real pulse $s(t)$, and the four orthogonal sequences given earlier, we can form the following four pulses:

$$s_0(t) = s(t) + js(t - T) + s(t - 2T) + js(t - 3T)$$
$$s_1(t) = s(t) + js(t - T) - s(t - 2T) - js(t - 3T)$$
$$s_2(t) = s(t) - js(t - T) - s(t - 2T) + js(t - 3T)$$
$$s_3(t) = s(t) - js(t - T) + s(t - 2T) - js(t - 3T).$$

The four new pulses will be orthogonal if the pulselet $s(t)$ satisfies

$$\int_{-\infty}^{\infty} s(t)s(t - \ell T)dt = 0$$

for all nonzero ℓ. Any $s(t)$ will do as the pulselet, provided $s(t) * s(-t)$ is a Nyquist pulse.

We can modify this example, recalling the structure of an MSK waveform, to get another attractive set of M-ary pulses that are not orthogonal, but are partially so. For the pulselet $s(t)$, choose the half-cosine pulse

$$s(t) = \begin{cases} \cos \pi t/T & |t| < T/2 \\ 0 & |t| \geq T/2 \end{cases}$$

which we will space by $T/2$. This is not a Nyquist pulse at spacing $T/2$. Define the four pulses

$$s_0(t) = s(t) + js(t - T/2) + s(t - T) + js(t - 3T/2)$$
$$s_1(t) = s(t) + js(t - T/2) - s(t - T) - js(t - 3T/2)$$
$$s_2(t) = s(t) - js(t - T/2) - s(t - T) + js(t - 3T/2)$$
$$s_3(t) = s(t) - js(t - T/2) + s(t - T) - js(t - 3T/2).$$

Each of these four pulses has the same amplitude profile, given by

$$|s_m(t)| = \begin{cases} 0 & t \leq -T/2 \\ \cos \pi t/T & -T/2 \leq t \leq 0 \\ 1 & 0 \leq t \leq 3T/2 \\ \cos \pi \left(t + \frac{T}{2}\right)/T & 3T/2 \leq t < 2T \\ 0 & 2T < t. \end{cases}$$

Each pulse has sinusoidal-shaped rising and trailing edges and is constant in the middle. Consequently, only the energy in the edges of the pulse is affected by a nonlinearity in the transmitter.

Even though the underlying sequences of amplitudes are orthogonal as sequences, these four pulses might not form an orthogonal family of pulses. This is because the pulselets are of width T and the spacing between pulselets is $T/2$, so in computing the correlation

$$\int_{-\infty}^{\infty} s_m(t)s_{m'}^*(t)dt$$

one pulselet of the pulse $s_m(t)$ can overlap as many as three pulselets of the pulse $s_{m'}(t)$. For example, for the correlation between $s_0(t)$ and $s_3(t)$, we have

$$\int_{-\infty}^{\infty} s_0(t)s_3^*(t)dt = \int_{-\infty}^{\infty} \left[s^2(t) - s^2(t-T/2) + s^2(t-T) - s^2(t-3T/2) \right] dt$$

$$+ j \int_{-\infty}^{\infty} [s(t)s(t-T/2) + s(t-T/2)s(t-T) + s(t-T)s(t-3T/2)] \, dt$$

$$+ j \int_{-\infty}^{\infty} [s(t-T/2)s(t) + s(t-T)s(t-T/2) + s(t-3T/2)s(t-T)] \, dt.$$

All other terms are equal to zero. The reason that other terms, such as the integral of $s(t)s(t-T)$, are not written in the equation is that these pulselets do not overlap, so their product is zero.

In the first integral on the right, two of the four terms are equal to the pulselet energy E_s, and two are equal to the negative of E_s, so that integral is zero. In the second and third integrals, each term is $j \int_{-\infty}^{\infty} s(t)s(t-T/2)dt$. Therefore we conclude that

$$\int_{-\infty}^{\infty} s_0(t)s_3^*(t)dt = 6j \int_0^{T/2} \cos\frac{\pi t}{T} \sin\frac{\pi t}{T} dt$$

$$= j3\frac{T}{\pi} = j\frac{6}{\pi}E_s$$

$$= j\frac{3}{2\pi}E_p$$

where E_s is the energy in each pulselet and $E_p = 4E_s$ is the energy in a pulse. The cross-correlation between $s_0(t)$ and $s_3(t)$ is purely imaginary and has a magnitude equal to 48 percent of the magnitude of the autocorrelation E_p.

For M equal to any power of two, an M-ary family of pulses – MSK pulses – can be designed in this way. These families of MSK pulses are generally not orthogonal, but they do satisfy the weaker condition

$$\text{Re}\left[\int_{-\infty}^{\infty} s_m(t)s_{m'}^*(t)dt\right] = 0.$$

Even though they lack orthogonality, such families are in wide use – say for $M = 32$ or 64 – because the constant amplitude property of MSK pulses is desirable and, for large M, such a family may be close to orthogonal, and the correlation can be made small by the choice of code sequences.

5.9   Signal space

The task of designing a communication waveform can be formulated from an abstract point of view by using the language of geometry. The geometric language develops intuition because it suggests modes of visualization. Moreover the abstract setting can often streamline the discussion of many topics. The geometric approach consists of defining an abstract space as a set of points and also defining a distance between the points of this space. The points of the space are the waveforms of finite energy that satisfy the timewidth and bandwidth constraints. This space is called *signal space*, or *function space*. In this setting, we can describe the task of designing a set of modulation waveforms as the task of choosing a set of points within signal space that are sufficiently far apart. A modulator then is a rule for mapping data sequences into this set of points of signal space.

The energy of any waveform $w(t)$ is given by

$$E_w = \int_{-\infty}^{\infty} |w(t)|^2 dt.$$

For a waveform of infinite duration, this energy would usually be infinite. We shall usually study only signaling waveforms of finite energy. For example, consider a BPSK waveform of finite duration

$$c(t) = \sum_{\ell=0}^{n-1} a_\ell s(t - \ell T),$$

using the pulse $s(t)$ such that the filtered pulse $r(t) = s(t) * s(-t)$ is a real Nyquist pulse. Then the energy in the waveform is

$$E_w = \int_{-\infty}^{\infty} c^2(t) dt$$

$$= \sum_{\ell=0}^{n-1} \sum_{\ell'=0}^{n-1} a_\ell a_{\ell'} \int_{-\infty}^{\infty} s(t - \ell T) s(t - \ell' T) dt$$

$$= \sum_{\ell=0}^{n-1} \sum_{\ell'=0}^{n-1} a_\ell a_{\ell'} \delta_{\ell \ell'}$$

where $\delta_{\ell\ell'} = 1$ if $\ell = \ell'$, and otherwise $\delta_{\ell\ell'} = 0$. Consequently,

$$E_w = \sum_{\ell=0}^{n-1} a_\ell^2.$$

The energy per bit is defined as $E_b = E_w/n$.

In the set of all real square-integrable functions, the *euclidean distance* between two functions $c(t)$ and $c'(t)$ is defined as

$$d(c(t), c'(t)) = \sqrt{\int_{-\infty}^{\infty} |c(t) - c'(t)|^2 dt}.$$

The euclidean distance between $c(t)$ and the zero signal is the *norm* of $c(t)$, denoted $\|c(t)\|$. The square of the norm $\|c(t)\|$ is equal to the energy in $c(t)$. The distance in signal space satisfies both the Schwarz inequality and the triangle inequality.

For an example of distance, within the set of all BPSK waveforms of blocklength n, consider the two waveforms

$$c(t) = \sum_{\ell=0}^{n-1} a_\ell s(t - \ell T)$$

and

$$c'(t) = \sum_{\ell=0}^{n-1} a'_\ell s(t - \ell T)$$

where $a_\ell = \pm A$ and $a'_\ell = \pm A$. Suppose, again, that $s(t) * s^*(-t)$ is a Nyquist pulse. The squared euclidean distance between $c(t)$ and $c'(t)$ is given by

$$d^2(c(t), c'(t)) = \int_{-\infty}^{\infty} |c(t) - c'(t)|^2 dt$$

$$= \sum_{\ell=0}^{n-1} \sum_{\ell'=0}^{n-1} (a_\ell - a'_\ell)(a_{\ell'} - a'_{\ell'}) \int_{-\infty}^{\infty} s(t - \ell T) s^*(t - \ell' T) dt$$

$$= \sum_{\ell=0}^{n-1} \sum_{\ell'=0}^{n-1} (a_\ell - a'_\ell)(a_{\ell'} - a'_{\ell'}) \delta_{\ell\ell'}$$

$$= \sum_{\ell=0}^{n-1} (a_\ell - a'_\ell)^2 = d^2(a, a').$$

For such a case, we can speak interchangeably of the euclidean distance between two waveforms in signal space, or of the euclidean distance between two sequences of discrete symbols from the signal constellation.

What is the smallest euclidean distance between any two such BPSK waveforms? There will be pairs of data sequences that differ in only a single bit position. For such sequences,

$$d^2(c(t), c'(t)) = (A + A)^2$$

or

$$d(c(t), c'(t)) = 2A.$$

We say that the minimum distance of this set of BPSK waveforms is $2A$. This means that there is at least one pair of waveforms separated by a euclidean distance of $2A$, and there is no pair of distinct waveforms spaced more closely than $2A$. In fact, there are a great many pairs of waveforms separated by distance $2A$; these are the pairs differing only in a single bit position.

In general, the minimum distance, denoted d_{min}, of a set of waveforms is the smallest euclidean distance between any pair of distinct waveforms in the set,

$$d_{min} = \min_{c(t) \neq c'(t)} d(c(t), c'(t)).$$

The normalized minimum distance is $\rho_{min} = d_{min} / \sqrt{E_b}$. The minimum distance between any two BPSK waveforms is $2A$, and the normalized minimum distance is 2. We shall see eventually that the performance of the set of communication waveforms is, in large part, an immediate consequence of the minimum distance. We should seek sets of waveforms that have large minimum distances. Every pair of waveforms should be far apart in the sense of euclidean distance.

Let $v(t)$ and $v'(t)$ be two real waveforms, each with energy E_w. How far apart can we make them?

$$d^2(v(t), v'(t)) = \int_{-\infty}^{\infty} [v(t) - v'(t)]^2 dt$$

$$= E_w - 2 \int_{-\infty}^{\infty} v(t)v'(t) dt + E_w.$$

This implies that we should make the correlation negative and as large as possible to make the distance large. The best we can do is to choose $v'(t) = -v(t)$ in which case

$$d^2(v(t), v'(t)) = 4E_w.$$

However, if there are a great many such waveforms to be chosen, we will want every pair of them to be separated by a large distance. We know that we can make every pairwise correlation negative by using a simplex pulse alphabet, but for large M, the negative correlation is not very large. For a simplex family of waveforms,

$$d_{min}^2 = 2 \left(1 + \frac{1}{M} \right) E_w.$$

When M is large, we can do nearly as well if the pairwise correlation is equal to zero. Then the family of waveforms is an orthogonal pulse alphabet, and the squared euclidean distance between any two waveforms is $2E_w$.

There are two elementary examples of such families of orthogonal pulses: the families of orthogonal sinusoids, and the families of orthogonal sinc pulses. Given a frequency band $[-W_0, W_0]$, we have the sinc pulses as Nyquist pulses within that band

$$s(t - \ell T) = \frac{\sin 2\pi W_0(t - \ell T)}{2\pi W_0(t - \ell T)} .$$

where $2W_0 T = 1$. These pulses are pairwise orthogonal, all with the same energy. Within a long duration T_0, there are about $2T_0 W_0$ of these pulses, though to get this many, we must allow the sidelobes of the sinc pulse to extend outside of the specified interval. Thus, although the bandwidth constraint is precisely satisfied, the timewidth of this orthogonal alphabet is only loosely constrained.

Alternatively, given a long interval of time, say from $-T_0/2$ to $T_0/2$, we can specify the orthogonal sinusoids

$$s_1(t) = \sin 2\pi \frac{\ell}{T_0} t \qquad -T_0/2 \leq 0 \leq T_0/2$$

$$s_1'(t) = \cos 2\pi \frac{\ell}{T_0} t \qquad -T_0/2 \leq 0 \leq T_0/2$$

where $1/T_0 \leq W_0$ so that the bandwidth constraint is satisfied. Again, there are $2T_0 W_0$ waveforms in this alphabet of orthogonal waveforms, and they can be normalized so that each has energy E_w. Because the time duration is restricted, these sinusoids have spectra that are sinc functions in the frequency domain. Again, although the timewidth constraint is precisely satisfied, the bandwidth of this family is only loosely constrained.

As we have seen, there are practical limits on the use of an orthogonal (or simplex) family because the required bandwidth would be exponential in the number of bits transmitted. How many orthogonal waveforms with energy E_w can we find that simultaneously fit within the time interval $[-T_0/2, T_0/2]$ and the frequency interval $[-W_0, W_0]$? The properties of the Fourier transform tell us that there are no such waveforms; a waveform cannot be simultaneously time-limited and band-limited. However, if the product $T_0 W_0$ is large and we are allowed to fudge a little in exactly fitting the time and frequency intervals, then there are approximately $2T_0 W_0$ waveforms in any set of orthogonal waveforms that fit the time and bandwidth constraints. We can choose these waveforms so that each has energy E_w.

We can now see that, though an orthogonal pulse alphabet can be used as a set of building blocks to construct a family of waveforms, the communication waveforms themselves cannot be orthogonal, even approximately, on an arbitrarily long interval if the data rate is to be maintained. This is because the number of orthogonal waveforms can increase only linearly with time, whereas to transmit at a rate of R bits per second, 2^{RT_0} waveforms are needed in time T_0.

In practice, because bandwidth is limited, we can choose an M-ary orthogonal alphabet with M fixed. The number of bits in a message is allowed to grow without limit by concatenating a sequence of M-ary symbols to form the waveform

$$c(t) = \sum_{\ell=-\infty}^{\infty} s_{m_\ell}(t - \ell T).$$

At each time ℓT, any of the M symbols may be chosen. The minimum distance of this waveform family occurs for two waveforms $c(t)$ and $c'(t)$ that differ in only a single symbol, say at $\ell = 0$. Then

$$d_{min}^2 = \int_{-\infty}^{\infty} |s_{m_0}(t) - s_{m_0'}(t)|^2 dt$$

$$= 2E_p.$$

Thus, the minimum distance for orthogonal signaling is established by the energy in a single symbol, not by the energy in the entire waveform.

Problems for Chapter 5

5.1. Show that the amplitude and phase of a passband waveform are the same as the amplitude and phase of the complex representation of that passband waveform. If the phase of the passband waveform is offset by ϕ, how does the phase of the complex representation change? If the reference frequency of the passband waveform is changed by Δf, what happens to the complex representation?

5.2. The legacy telephone channel can be characterized as an ideal passband channel from 300 Hz to 2700 Hz. Choose a signal constellation and a symbol rate to obtain a 9600-bits/second telephone line modem. (See Problem 2.7.) This time use a 32-ary signal constellation and 1920 symbols per second. Use a pulse shape $s(t) = \text{sinc}(at)\text{sinc}(bt)$ where one sinc function is related to the 1920 symbol rate and the other is adjusted to fill out the channel bandwidth. Explain what the use of $s(t)$ does to the sidelobes. Estimate at what width the pulse can be truncated if a 1 percent interference is allowed between one pulse and another at the output of the ideal channel.

5.3. Show that, when using an MSK waveform with an I and Q modulator and an FSK demodulator, one can use either a precoder or a postcoder to obtain consistency between the data sent and the data received, but the use of a post-coder leaves an ambiguity in the datastream if the demodulation does not begin with the very first bit. Show that, when using an FSK modulator and an I and Q demodulator, the use of a precoder causes an ambiguity. In each case that uses a postcoder, describe what happens at the output of the postcoder if the demodulator makes a single bit error.

5.4. Given the passband waveform $v(t)$ such that $V(f) = 0$ if $|(|f| - f_0)| \geq W/2$, define $V_R(f)$ and $V_I(f)$ so that

$$v(t) = v_R(t) \cos 2\pi f_0 t - v_I(t) \sin 2\pi f_0 t.$$

5.5. An MSK waveform transmits $1/T_b$ bits per second at carrier frequency f_0. Another communication system operates in a narrow band of W hertz (W small compared to $1/T_b$) at carrier frequency $f_0 - 10/T_b$ (see illustration).

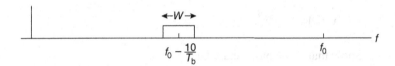

A cosine pulse of width $T = 2T_b$ has Fourier transform

$$S(f) = \frac{2T \cos \pi T f}{\pi (4T^2 f^2 - 1)}.$$

a. Give a rough bound on the percentage of power of the MSK waveform that will show up as interference with the second communication system.
b. To satisfy settlement of a lawsuit brought by the owner of the second communication system because of interference, an ideal passband filter is inserted between the MSK modulator and the transmitting antenna. What do you expect this to do to the MSK waveform? Qualitatively discuss (and sketch) what happens to an individual bit pulse and the interference between bit pulses.

5.6. An I and Q modulator for minimum-shift keying requires that the two baseband modulators be precisely matched in time delay and carrier phase. At very high data rates, perhaps above 100 Mbps, this can be difficult. An alternative, and somewhat subtle, MSK modulator is the *serial* MSK modulator. It consists of a BPSK waveform (with rectangular pulses of bit duration T_b) modulated onto an offset carrier frequency $f_0 - 1/4T_b$, which is then passed through a passband filter with impulse response

$$g(t) = \begin{cases} \sin 2\pi (f_0 + 1/4T_b)t & 0 \le t \le T_b \\ 0 & \text{otherwise.} \end{cases}$$

(Notice that the apparent frequency of the passband filter and the frequency of the BPSK waveform are offset in opposite directions.)

a. Sketch a functional block diagram of the serial MSK modulator.

b. Prove that the serial MSK modulator does indeed give an MSK waveform. (**Hint:** Work with only the first bit, and show in the Fourier transform domain that the right Fourier transform is obtained, then extend this to the other bits.)

c. What is the relationship between the bit stream that is modulated into the BPSK waveform and the apparent datastream in the I and Q components of the final MSK waveform?

5.7. Design a serial demodulator for an MSK waveform by reasoning that one must reverse the modulator structure given in Problem 5.6.

5.8. A cubic nonlinearity is given by

$$y = x + Ax^2 + Bx^3.$$

Show that if the passband signal

$$x(t) = \cos[2\pi f_0 t + \theta(t)]$$

is passed through the nonlinearity, the output signal $y(t)$ in the vicinity of f_0 is a scaled version of $x(t)$, while if the passband signal

$$x(t) = a(t) \cos[2\pi f_0 t + \theta(t)]$$

is passed through the nonlinearity, the output signal in the vicinity of f_0 is not a scaled version of $x(t)$ unless $a(t)$ is a constant. Specifically, show that the error term near f_0, when expressed in the frequency domain, is proportional to $A(f) * A(f) * X(f)$. Sketch this if $\theta(t) = 0$ and $A(f)$ is a rectangle.

5.9. A passband MSK waveform, whose carrier frequency f_0 is large, is amplified in a high-power transmitter tube prior to transmission. Because of a desire to press the transmitted power to its maximum, the tube is operated in a saturating mode with output

$$d(t) = \text{sgn } [c(t)]$$

when $c(t)$ is the passband input, where sgn $(x) = \pm 1$ according to the sign of x. This form of saturation is called a (passband) *hardlimiter*.

a. Use the Fourier series expansion of a square wave to write $d(t)$ in the form

$$d(t) = \frac{4}{\pi} \sum_{k=0}^{\infty} \frac{(-1)^k \cos(2k+1)(2\pi f_0 t + \theta(t))}{2k+1}.$$

b. Show that, except for an apparent loss in E_b, an MSK demodulator that receives only the component of $d(t)$ in the vicinity of f_0 will be unaware of the hardlimiter in the modulator.

c. What is the loss in E_b (expressed in decibels)?

d. Is the received energy more or less than would be received if a linear transmitter tube were used with gain adjusted so that the signal has the same peak amplitude as above?

e. The nonlinearity acting on the carrier does "splatter" energy into the harmonics of the carrier frequency. How much energy is splattered into each harmonic? Why is it important to keep this splattered energy small? Comment on the possible role of the transmit antenna in determining the radiated splatter energy.

5.10. a. Show that the "in-phase" modulated waveform

$$c(t) = c_R(t) \cos 2\pi f_0 t$$

is completely determined by its spectrum $C(f)$ at those frequencies f satisfying $|f| \geq f_0$. That is, if

$$C'(f) = \begin{cases} C(f) & \text{if } |f| \geq f_0 \\ 0 & \text{if } |f| < f_0, \end{cases}$$

then $c(t)$, in principle, can be exactly recovered from the "single-sideband" waveform $c'(t)$.

b. Show that this is not true for the fully modulated waveform

$$c(t) = c_R(t) \cos 2\pi f_0 t - c_I(t) \sin 2\pi f_0 t.$$

(This shows why single-sideband modulation is not important for modern digital communication systems. To halve the bandwidth, it is easier to divide the data between $c_R(t)$ and $c_I(t)$ and use a fully modulated waveform than it is to put all the data in $c_R(t)$ and then use a single-sideband waveform. Double-sideband, suppressed-carrier, quadrature-amplitude modulation is more convenient to use than is single-sideband amplitude modulation, especially for digital communications.)

5.11. Prove that if two passband waveforms, $\tilde{s}_m(t)$ and $\tilde{s}_{m'}(t)$ are orthogonal, then their complex baseband representatives $s_m(t)$ and $s_{m'}(t)$ are orthogonal also. Explain the circumstances under which the converse fails to be true. If two passband waveforms $\widetilde{s}_m(t)$ and $\widetilde{s}_{m'}(t)$ are orthogonal, will they remain orthogonal if the carrier of $\widetilde{s}_{m'}(t)$ is shifted by θ?

5.12. A sixteen-ary complex signal constellation is sketched below.

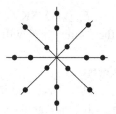

By using the perpendicular bisectors of the lines between points of the constellation, sketch the decision regions for this constellation. Prove that an arbitrary point v can be assigned to its proper region by finding the i that minimizes the set of distances $d(v, c_i) = \sqrt{(v_R - c_{Ri})^2 + (v_I - c_{Ii})^2}$.

5.13. What is the appropriate definition of a *complex* Nyquist pulse? Does this generalization of a Nyquist pulse depend on the notions of coherent and noncoherent that are given in Chapter 6? Is the notion of a complex Nyquist pulse ever useful?

5.14. Which of the following two sixteen-ary signal constellations has the largest minimum distance if the axes are scaled so that both have the same average energy per bit?

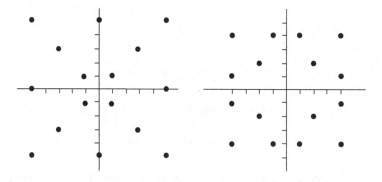

5.15. A signal constellation for sixteen-ary quadrature-amplitude modulation (QAM) is as follows:

Q

a. If the information rate is 100 Mbps, what is the channel symbol rate?

b. What amplitudes and phases will be seen in the waveform?

c. Assign four data bits to each point of the signal constellation such that adjacent points (horizontally or vertically) differ in only one bit.

d. Prove that the in-phase and quadrature channels can be processed independently if the noise is bandlimited gaussian noise.

e. Partition the plane with the decision regions for which, if the complex number v in the mth decision region is received, the demodulator decides that the mth point of the signal constellations was transmitted.

5.16. A signal constellation for a sixteen-ary PSK is as follows

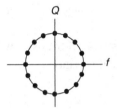

a. Sketch the decision regions.

b. Label the points of the signal constellation with four data bits so as to minimize error probability.

c. Would you prefer QAM or PSK on a channel with amplitude nonlinearities?

d. Would you prefer QAM or PSK on a channel with phase errors (phase noise)?

e. Would you prefer QAM or PSK on a channel with additive white gaussian noise?

5.17. Two four-ary complex signal constellations in the complex plane are shown in the following illustration:

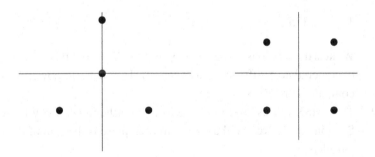

In the first signal constellation, three of the points are equispaced on a circle of radius r_1 about the origin. In the second, four points are equispaced on a circle of radius r_2 about the origin. These two complex signal constellations are used to signal in additive white gaussian noise.

    a. Sketch a demodulator for each signal constellation.

    b. For the two constellations, find (union bound) expressions for the probability of bit error p_e as a function of E_b/N_0.

    c. What is the (approximate) ratio of r_1 to r_2 if p_e is to be the same for both constellations?

    d. Now suppose that the gaussian-noise power is negligible, and the carrier phase reference is in error by an error angle of θ_e where θ_e is a random variable that is uniform on the interval $\left[-\frac{\pi}{3}, \frac{\pi}{3}\right]$. Give an expression for p_e.

    e. What are the advantages and disadvantages of the two signal constellations?

    f. Suppose that the power of the gaussian noise is not negligible, and the carrier phase reference is in error as described in part d. Using part b and part d to form asymptotes, sketch graphs of p_e versus E_b/N_0 for the two signal constellations.

5.18. An eight-ary modulation scheme known as $\pi/4$-QPSK uses two QPSK signal constellations offset by $45°$ as shown in the illustration

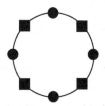

Data is differentially encoded, two bits at a time as

$00 \rightarrow +45°$

$01 \rightarrow +135°$

$10 \rightarrow -45°$

$11 \rightarrow -135°.$

What is the relationship between p_e and E_b/N_0? Does this depend on the choice of demodulator (differential or coherent)? How does the power density spectrum compare with QPSK?

5.19. The "MSK pulses" were defined as a nonorthogonal M-ary pulse alphabet in Section 5.8. Prove that the energy in each pulse is the sum of the energy in the pulselets.

5.20. Prove or disprove the following: two complex baseband pulses $s_0(t)$ and $s_1(t)$ are orthogonal if and only if their passband representations are orthogonal.

5.21. a. Prove that the *triangle inequality*

$$d(c(t), c'(t)) + d(c(t), c''(t)) \geq d(c'(t), c''(t))$$

is equivalent to

$$\|c'(t)\| + \|c''(t)\| \geq \|c'(t) - c''(t)\|.$$

b. Prove that the triangle inequality holds in signal space.

Notes for Chapter 5

The complex representation of a passband signal was introduced by Gabor (1946). Complex envelopes and pre-envelopes of passband signals have been developed by Arens (1957) and Dungundji (1958), and surveyed by Rice (1982).

The MSK waveform was introduced by Doelz and Heald (1961) and has been described in many ways. Amoroso (1976) studied the relationships among the many ways of viewing MSK, and Amoroso and Kivett (1977) showed how the waveform can be generated by filtering binary PSK. A popular variant of MSK was introduced by Murota and Hirada (1981). The design of signal constellations was discussed in papers by Cahn (1959, 1960), Hancock and Lucky (1960), Lucky and Hancock (1962), Campopiano and Glazer (1962), and others. Four-dimensional signal constellations were studied by Welti and Lee (1974). They are related to the permutation codes contributed by Slepian (1965).

6 Passband Demodulation

The demodulation of a passband waveform or of a complex baseband waveform uses methods similar to those used to demodulate baseband signals. However, there are many new details that emerge in the larger setting of passband or complex baseband demodulation. This is because a complex baseband function (or a passband function) can be expressed either in terms of real and imaginary components or in terms of amplitude and phase. It is obvious that phase is meaningful only if there is an absolute phase reference. A new set of topics arises when the modulator and demodulator do not share a common phase reference. This is the distinction between coherent and noncoherent demodulation. When the phase reference is known to the demodulator, the demodulator is called a *coherent demodulator*. When the phase reference is not known to the demodulator, that demodulator is called a *noncoherent demodulator*.

We begin the chapter with a study of the matched filter at passband. Then we use the matched filter as a central component in the development of a variety of demodulators, both coherent and noncoherent, for the passband waveforms that were introduced in Chapter 5.

The methods for the demodulation of baseband sequences that were described in Chapter 4 can be restated in the setting of passband waveforms. We shall prefer, however, the equivalent formulation in terms of complex baseband waveforms. It becomes obvious immediately how to generalize methods of demodulation from sequences of real numbers to sequences of complex numbers, so the chapter starts out with a straightforward reformulation of the topic of demodulation. Soon, however, new details for the complex case enter the discussion and the topic becomes richer and more subtle.

6.1 The matched filter at passband

We are interested in passband pulses

$$\tilde{s}(t) = s_R(t) \cos 2\pi f_0 t - s_I(t) \sin 2\pi f_0 t$$

received in passband noise

$$\tilde{n}(t) = n_R(t) \cos 2\pi f_0 t - n_I(t) \sin 2\pi f_0 t.$$

The signal received in noise, given by

$$\tilde{v}(t) = \tilde{s}(t) + \tilde{n}(t)$$
$$= v_R(t) \cos 2\pi f_0 t - v_I(t) \sin 2\pi f_0 t$$

is a passband waveform as well.

A passband pulse $\tilde{s}(t)$ is a real pulse and has finite energy. Thus, we are free to ignore the passband property and to treat $\tilde{s}(t)$ as any other pulse. The optimality of the matched filter holds for any pulse of finite energy, so it holds equally well for any passband pulse. We can simply treat the passband pulse (including the carrier), directly by Theorem 3.1.2 and Corollary 3.1.3. Thus, the filter matched to $\tilde{s}(t)$ is

$$\tilde{g}(t) = \tilde{s}(-t)$$
$$= s_R(-t) \cos(-2\pi f_0 t) - s_I(-t) \sin(-2\pi f_0 t)$$
$$= s_R(-t) \cos 2\pi f_0 t + s_I(-t) \sin 2\pi f_0 t$$

which is a passband filter matched to the entire passband pulse. This filter is illustrated in Figure 6.1 for the case in which $s_I(t) = 0$. In Figure 6.2, we show how this can be alternatively implemented at complex baseband.

The passband pulse $\tilde{s}(t)$ and the matched filter $\tilde{g}(t)$ can be represented as complex baseband pulses. Thus

$$s(t) = s_R(t) + js_I(t)$$
$$g(t) = g_R(t) + jg_I(t).$$

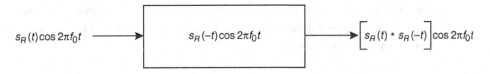

Figure 6.1. A passband matched filter for a passband pulse.

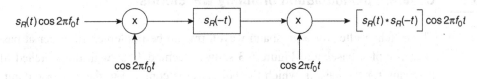

Figure 6.2. A baseband matched filter for a passband pulse.

The complex-baseband matched filter is easily seen to be

$$g(t) = s_R(-t) - js_I(-t)$$
$$= s^*(-t)$$

by inspection of the passband matched filter.

A concise expression for the complex-baseband signal output $r(t)$ of the complex-baseband matched filter $g(t)$ uses the abbreviation of the convolution, given by

$$r(t) = s(t) * s^*(-t).$$

In detail, this is

$$r_R(t) + jr_I(t) = [s_R(t) * s_R(-t) + s_I(t) * s_I(-t)] + j[s_I(t) * s_R(-t) - s_R(t) * s_I(-t)].$$

Notice that a purely real pulse has a purely real matched-filter output, and a purely imaginary pulse also has a purely real matched-filter output. The imaginary component at the output of the matched filter is produced by cross-convolution terms between the real and imaginary components at the input. Furthermore, the imaginary component of the output is always zero at $t = 0$. The imaginary component need not be computed if only the value at $t = 0$ is of interest.

At the sampling instant $t = 0$, the output

$$r(0) = \int_{-\infty}^{\infty} [s_R^2(\xi) + s_I^2(\xi)] d\xi$$

equals the pulse energy, which is always real.

The complex signals with which we deal are contaminated by complex noise; this is noise with both a real part and an imaginary part. Rather than deal with the total noise power, we prefer to deal with the noise power per component of the complex noise. With this convention, signal-to-noise analyses of real waveforms and of complex waveforms share the same formulas.

6.2   Coherent demodulation of binary waveforms

The matched filter for a passband waveform can be implemented either at passband or at complex baseband. Figure 6.3 shows a demodulator with the matched filter at passband for the case in which the pulse $s(t)$ is confined to the in-phase component only. The passband demodulator must sample the cosine carrier of the matched-filter

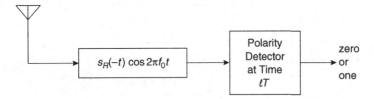

Figure 6.3. A BPSK demodulator at passband.

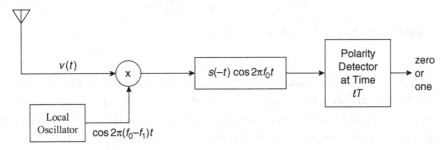

Figure 6.4. A BPSK demodulator at an intermediate frequency.

output

$$u(t) = \left[\sum_{\ell=-\infty}^{\infty} a_\ell r(t - \ell T) \right] \cos 2\pi f_0 t + n_R'(t) \cos 2\pi f_0 t - n_I'(t) \sin 2\pi f_0 t,$$

where $r(t) = s(t) * s^*(-t)$ is a Nyquist pulse. Set $t = \ell T$. Then

$$u(\ell T) = a_\ell \cos 2\pi f_0 \ell T + n_R'(\ell T) \cos 2\pi f_0 \ell T - n_I'(\ell T) \sin 2\pi f_0 \ell T.$$

If $f_0 T = 1$, then $\cos 2\pi f_0 \ell T = 1$ and $\sin 2\pi f_0 \ell T = 0$, so this becomes

$$u(\ell T) = a_\ell + n_R'(\ell T).$$

Thus we see a need for a relationship between the sampling interval T and the carrier frequency f_0. This requires that time be known to an accuracy that is small compared to $1/f_0$. Otherwise, if the sampling instant is in error by δt, some of the signal will be missed because $\cos 2\pi f_0(\ell T + \delta t) \neq 1$. In the extreme case, when the sampling is incorrectly performed at $t = lT + 1/4 f_0$, the entire signal is lost because $\cos 2\pi f_0 t = 0$. A small timing error can be very destructive if f_0 is large.

The problem here is that the time reference for sampling the output must be aligned with the time reference of the carrier to within a small fraction of a carrier period. For this reason, when f_0 is very large, the passband demodulator might be performed, as in Figure 6.4, at an *intermediate frequency* f_1 that is much smaller than f_0. To change the carrier of the passband signal $\tilde{v}(t)$ from f_0 to f_1, a mixer is used to multiply $\tilde{v}(t)$ by

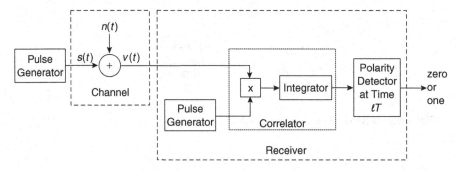

Figure 6.5. Local-replica demodulation of BPSK.

$\cos 2\pi (f_0 - f_1)t$ and to accept only the sideband at frequency f_1. This process is called "down-conversion". The cosine wave entering the mixer is formed for this purpose by a device called a *local oscillator*, where the term "local" refers to the fact that this device is within the receiver itself; it is not a global reference. This alternative method – moving the passband signal to an "intermediate" frequency – has not eliminated the need for precise synchronization; it has only moved the need from the sampler where it appears as a time synchronization, to the local oscillator, where it appears as a phase synchronization and may be easier to deal with.

Sometimes a correlator is used, as shown in Figure 6.5, in place of a matched filter, to compute the value

$$r(0) = \int_{-\infty}^{\infty} v(t)s^*(t)dt.$$

This correlator is sometimes called a *local replica correlator* to underscore the fact that $s(t)$ is generated locally in the receiver as a copy, or replica, of the pulse $s(t)$ sent by the transmitter. This terminology evolved as a reminder that the two copies of $s(t)$ might be generated imperfectly, one in the transmitter and one in the receiver, and with some mismatch. The replica pulse must be synchronized in time with the received pulse that is contained in $v(t)$. The local replica correlator may be more convenient to use in some applications and less convenient in others. It is not convenient when the pulses have overlapping tails, as do many of the Nyquist pulses we deal with, because then it is not possible to time-share a single integrator.

Now we turn to the demodulation of a binary passband orthogonal signaling waveform, shown in Figure 6.6. This is similar to the demodulation of a binary baseband orthogonal signaling waveform. The received waveform

$$v(t) = \sum_{\ell'=-\infty}^{\infty} [a_{\ell'}\tilde{s}_0(t - \ell'T) + \bar{a}_{\ell'}\tilde{s}_1(t - \ell'T)] + \tilde{n}(t)$$

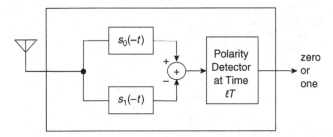

Figure 6.6. Demodulation of binary FSK at passband.

is the input to two passband filters matched to the two passband orthogonal pulses, given by

$$\tilde{s}_0(t) = s_{0R}(t) \cos 2\pi f_0 t - s_{0I}(t) \sin 2\pi f_0 t$$
$$\tilde{s}_1(t) = s_{1R}(t) \cos 2\pi f_0 t - s_{1I}(t) \sin 2\pi f_0 t,$$

and the output of each filter is sampled at time ℓT.

The output of filter $\tilde{s}_0^*(-t)$ is

$$\tilde{u}_0(t) = \sum_{\ell'=-\infty}^{\infty} a_{\ell'} \tilde{s}_0(t - \ell'T) * \tilde{s}_0^*(-t) + \bar{a}_{\ell'} \tilde{s}_1(t - \ell'T) * \tilde{s}_0^*(-t) + \tilde{n}_0'(t)$$

where

$$\tilde{n}_0'(t) = \tilde{n}(t) * \tilde{s}_0^*(-t).$$

If $\tilde{s}_0(t) * \tilde{s}(-t)$ is a Nyquist pulse and $\tilde{s}_0(t)$ is orthogonal to $\tilde{s}_1(t - \ell T)$ for all ℓ, then the samples are

$$\tilde{u}_0(\ell T) = a_\ell + n_{0\ell}'$$

where $n_{0\ell}' = \tilde{n}_0'(\ell T)$. Similarly, the samples of the second filter output are

$$\tilde{u}_1(\ell T) = \bar{a}_\ell + n_{1\ell}'.$$

All output noise samples, $n_{0\ell}'$ and $n_{1\ell}'$ for $\ell = \ldots, -1, 0, 1, \ldots$, are uncorrelated and are independent, identically distributed random variables if $n(t)$ is gaussian, as a consequence of Theorem 3.1.5. Either a_ℓ or $\bar{a}_\ell$ is equal to zero. The decision at the ℓth bit time is based on which of $u_0(\ell T)$ or $u_1(\ell T)$ is larger. One way to test which of the two filter outputs is larger is to test whether the difference in the two filter outputs is positive or negative, as shown in Figure 6.6.

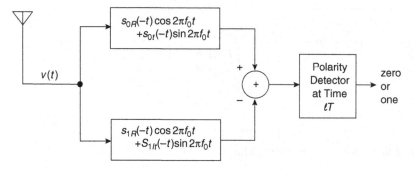

Figure 6.7. A coherent FSK demodulator at passband.

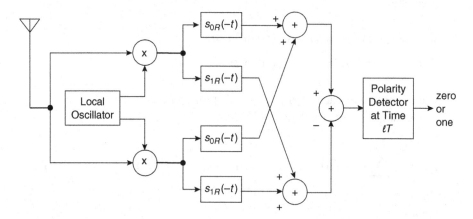

Figure 6.8. A coherent FSK demodulator at baseband.

Figure 6.7 shows the coherent demodulator for binary orthogonal signaling – this time with the matched filters written out at passband.

The development above can be immediately transferred to complex baseband simply by replacing $\tilde{s}_0(t)$ and $\tilde{s}_1(t)$ by their complex baseband representation, $s_0(t)$ and $s_1(t)$, given by

$$s_0(t) = s_{0R}(t) + js_{0I}(t)$$
$$s_1(t) = s_{1R}(t) + js_{1I}(t).$$

Figure 6.8 shows the coherent demodulator for binary orthogonal signaling with the matched filters at complex baseband. Only the real part of the matched-filter output is formed because only the real part contains a nonzero signal component at the sampling instant.

6.3   Noncoherent demodulation of binary waveforms

A passband waveform with an unknown phase is a waveform of the form

$$v(t) = v_R(t) \cos(2\pi f_0 t + \theta) - v_I(t) \sin(2\pi f_0 t + \theta).$$

The complex baseband representation of this waveform is

$$v(t) = [v_R(t) + j v_I(t)] e^{j\theta}.$$

The phase angle θ is unknown to the demodulator. It may be due to phase shifts in the atmosphere, in the transmitter power amplifier, in the antennas, or in the front end of the receiver. When f_0 is very large, it is generally not possible to calibrate all of these parts of the system. The receiver must either demodulate in the presence of an unknown θ, which is called a *noncoherent demodulator*, or try to estimate θ from the datastream itself and then use a coherent demodulator. In this section, we shall study receiver strategies that demodulate a symbol without knowing the value of θ.

If the complex input to a complex filter is multiplied by $e^{j\theta}$, then, by linearity, the output of the filter is also multiplied by $e^{j\theta}$ and is otherwise unchanged. The corresponding statement for a passband pulse must also be true, and is given by the following theorem. The point of the theorem is that an undesired phase can be dealt with either before the filter or after, as is convenient.

Theorem 6.3.1 *If the phase of the carrier of a passband signal at the input to a passband filter is shifted by θ, then the passband signal at the output of that passband filter has the same modulation components as before but now modulated onto a carrier that is also shifted by θ.*

Proof A passband signal whose passband carrier has a phase offset by θ can be written as

$$\widetilde{s}(t) = s_R(t) \cos(2\pi f_0 t + \theta) - s_I(t) \sin(2\pi f_0 t + \theta).$$

This signal is passed through the passband filter

$$\widetilde{g}(t) = g_R(t) \cos 2\pi f_0 t - g_I(t) \sin 2\pi f_0 t.$$

The filter output can be computed with the aid of Theorem 5.3.1 by representing the passband signal as a complex baseband signal. In the complex representation, the phase angle appears in a convenient way,

$$s'(t) = [s_R(t) + js_I(t)]e^{j\theta} = s(t)e^{j\theta}$$

$$g(t) = [g_R(t) + jg_I(t)].$$

Then

$$s'(t) * g(t) = [s(t) * g(t)]e^{j\theta}$$

which means that the passband signal is

$$\widetilde{s}'(t) * \widetilde{g}(t) = [\widetilde{s}(t) * \widetilde{g}(t)]_R \cos(2\pi f_0 t + \theta) - [\widetilde{s}(t) * \widetilde{g}(t)]_I \sin(2\pi f_0 t + \theta),$$

as in the statement of the theorem. ∎

The complex baseband signal multiplied by $e^{j\theta}$ can be expanded as

$$v(t) = [v_R(t)\cos\theta - v_I(t)\sin\theta] + j[v_R(t)\sin\theta + v_I(t)\cos\theta],$$

and θ is unknown. In a simple waveform such as on–off keying, the quadrature modulation is zero, and $v_R(t)$ is composed of a stream of pulses of the form $As(t)$ in additive noise. In the absence of noise, a single data bit at complex baseband leads to

$$v(t) = As(t)\cos\theta + jAs(t)\sin\theta$$

if a one is transmitted, and

$$v(t) = 0$$

if a zero is transmitted. The noncoherent receiver, shown in Figure 6.9, passes the complex signal $v(t)$ through a complex matched filter with real part and imaginary part each matched to $s(t)$ and then takes the square root of the sum of the squares of the real part and the imaginary part to suppress the unknown phase angle.

In general, a fully modulated OOK waveform in the absence of noise is received as

$$v(t) = \left[\sum_{\ell'=-\infty}^{\infty} a_{\ell'}s(t - \ell'T)\right]\cos\theta + j\left[\sum_{\ell'=-\infty}^{\infty} a_{\ell'}s(t - \ell'T)\right]\sin\theta.$$

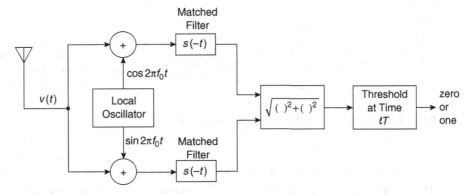

Figure 6.9. A noncoherent OOK demodulator at baseband.

When this waveform is passed through the matched filter, the filter output is

$$u(t) = \left[\sum_{\ell'=-\infty}^{\infty} a_{\ell'} r(t - \ell'T) \right] \cos\theta + j \left[\sum_{\ell'=-\infty}^{\infty} a_{\ell'} r(t - \ell'T) \right] \sin\theta.$$

Whenever $r(t)$ is a Nyquist pulse, the sample of $u(t)$ at time ℓT, in the absence of noise, is

$$u(\ell T) = a_\ell \cos\theta + ja_\ell \sin\theta$$
$$= u_R(\ell T) + ju_I(\ell T).$$

In the absence of noise,

$$a_\ell = \sqrt{(u_R(\ell T))^2 + (u_I(\ell T))^2}$$

where a_ℓ is either zero or A depending on the value of the databit. In the presence of noise, the demodulation decision is

$$\widehat{a}_\ell = \begin{cases} A & \text{if } \sqrt{(u_R(\ell T))^2 + (u_I(\ell T))^2} \geq \Theta \\ 0 & \text{if } \sqrt{(u_R(\ell T))^2 + (u_I(\ell T))^2} < \Theta. \end{cases}$$

The calculation of the magnitude rather than the real part of the matched-filter output is the feature of the noncoherent demodulator that distinguishes it from the coherent demodulator. Later, we shall analyze the effect of noise in the received signal on this operation. Then we shall see that the penalty for having a noncoherent demodulator is that there is more noise sensitivity because both the real part and the imaginary part of the complex noise play a role.

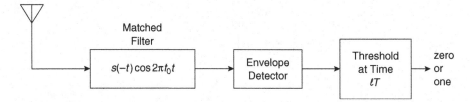

Figure 6.10. A noncoherent OOK demodulator at passband.

Figure 6.10 shows a noncoherent demodulator of OOK implemented at passband. The envelope detector of the passband waveform in Figure 6.10 plays the same role that the square root of the sum of the squares of the complex waveform does in Figure 6.9.

The noncoherent demodulation of FSK is based on similar reasoning. Suppose that $s_0(t)$ and $s_1(t)$ are real orthogonal pulses such that $r_0(t)$ and $r_1(t)$ are Nyquist pulses. The received noisy signal at complex baseband is

$$v(t) = \sum_{\ell'=-\infty}^{\infty} [a_{\ell'} s_0(t - \ell'T) + \bar{a}_{\ell'} s_1(t - \ell'T)]e^{j\theta} + n_R(t) + jn_I(t),$$

and the sampled outputs of the two filters matched to $s_0(t)$ and $s_1(t)$ are

$$u_0(\ell T) = a_\ell \cos\theta + ja_\ell \sin\theta + n'_{0\ell,R} + jn'_{0\ell,I}$$
$$u_1(\ell T) = \bar{a}_\ell \cos\theta + j\bar{a}_\ell \sin\theta + n'_{1\ell,R} + jn'_{1\ell,I}.$$

If there were no noise, the pulse amplitudes a_ℓ and $\bar{a}_\ell$ could be recovered by taking the square root of the sum of the squares of the components $u_0(\ell T)$ or $u_1(\ell T)$. When there is noise, the magnitudes, given by

$$|u_0(\ell T)| = \sqrt{([u_{0R}(\ell T)])^2 + ([u_{0I}(\ell T)])^2}$$
$$|u_1(\ell T)| = \sqrt{([u_{1R}(\ell T)])^2 + ([u_{1I}(\ell T)])^2}$$

are compared to choose the largest. This test is often expressed as

$$|u_0(\ell T)| - |u_1(\ell T)| \gtrless \Theta.$$

A noncoherent demodulator at complex baseband for FSK, using real pulses, is shown in Figure 6.11. An alternative noncoherent demodulator for FSK – this one at passband – is shown in Figure 6.12.

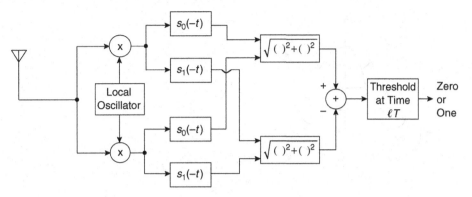

Figure 6.11. A noncoherent FSK demodulator at baseband.

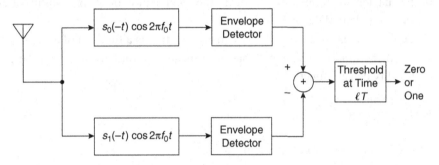

Figure 6.12. A noncoherent FSK demodulator at passband.

6.4   Rayleigh and ricean probability distributions

In problems of noncoherent demodulation at complex baseband, each sample of the output of a matched filter is complex, consisting of a real part and an imaginary part. The decision statistic is the square root of the sum of the squares of the real part and the imaginary part. This can be viewed as a transformation of the complex baseband signal from rectangular coordinates to polar coordinates, keeping only the magnitude. It corresponds to a transformation of the passband signal from the in-phase and quadrature representation to an amplitude and phase representation. When the input is a signal in additive gaussian noise, the real and imaginary outputs of the complex baseband matched filter are each an independent gaussian random variable and the magnitude is the square root of the sum of the squares. To analyze the probability of error of a noncoherent demodulator, we must study what happens to two independent gaussian random variables under the computation of the square root of the sum of the squares. We shall study this transformation of random variables in this section. In the remainder

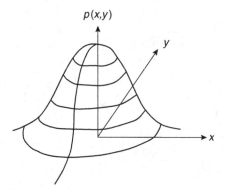

$p(x,y)$

Figure 6.13. Two-dimensional gaussian probability density function.

of the chapter, we apply these results to the computation of the probability of error of noncoherent demodulators.

A univariate gaussian probability density function is a function of a single variable, say x. If the random variable X has zero mean and variance σ^2, then the gaussian probability density function is

$$p(x) = \frac{1}{\sqrt{2\pi}\sigma} e^{-x^2/2\sigma^2}$$

or, if there is a nonzero mean $(\bar{x}, \bar{y})$,

$$p(x) = \frac{1}{\sqrt{2\pi}\sigma} e^{-(x-\bar{x})^2/2\sigma^2}.$$

A bivariate gaussian probability density function, shown in Figure 6.13, is defined in terms of two variables, say x and y. If the random variables X and Y are independent, zero mean, and have equal variance, then the bivariate gaussian probability density function is

$$p(x,y) = \frac{1}{2\pi\sigma^2} e^{-(x^2+y^2)/2\sigma^2}$$

or, if there is a nonzero mean,

$$p(x,y) = \frac{1}{2\pi\sigma^2} e^{-[(x-\bar{x})^2+(y-\bar{y})^2]/2\sigma^2}.$$

The transformation from rectangular coordinates (x, y) to polar coordinates (r, ϕ) is given by

$$r = \sqrt{x^2 + y^2}$$

$$\phi = \tan^{-1}\frac{x}{y}.$$

We need expressions for the probability density functions of the amplitude r and the phase ϕ. Although we are mostly interested in r, the calculations for ϕ are necessary as an intermediate step.

The probability of any region $\mathcal{A}$ must be the same whether the region is expressed in rectangular coordinates or in polar coordinates. That is, for any region $\mathcal{A}$,

$$\int_{\mathcal{A}} p(r,\phi)drd\phi = \int_{\mathcal{A}} \frac{1}{2\pi\sigma^2} e^{-(x^2+y^2)/2\sigma^2} dxdy$$

where $p(r,\phi)$ is the probability density function in polar coordinates. On the right side, substitute

$$x^2 + y^2 = r^2$$
$$dxdy = rd\phi dr.$$

This gives

$$\int_{\mathcal{A}} p(r,\phi)drd\phi = \int_{\mathcal{A}} \frac{1}{2\pi\sigma^2} e^{-r^2/2\sigma^2} rdrd\phi$$

from which it follows that

$$p(r,\phi) = \frac{r}{2\pi\sigma^2} e^{-r^2/2\sigma^2}$$

where $0 \le \phi < 2\pi$ and $r \ge 0$. Clearly, the probability density function in ϕ is uniform, given by $p(\phi) = \frac{1}{2\pi}$. Therefore, integrating $p(r,\phi)$ over ϕ from 0 to 2π must give

$$p(r) = \frac{r}{\sigma^2} e^{-r^2/2\sigma^2} \quad r \ge 0.$$

This probability density function is known as a *rayleigh probability density function*. By direct calculation we can see that the rayleigh density function has a mean $\sigma\sqrt{\pi/2}$ and a variance $(2 - \pi/2)\sigma^2$. The rayleigh density function is the probability density function for the amplitude of unbiased complex gaussian noise. It is also the probability density function for the envelope of unbiased passband gaussian noise at any instant.

To write the rayleigh density function in a standardized parameter-free form, make the change in variables

$$z = \frac{r}{\sigma}$$

and define

$$p_{Ra}(z) = ze^{-z^2/2} \quad z \ge 0$$

so that

$$p(r) = \frac{1}{\sigma} p_{Ra} \left(\frac{r}{\sigma} \right).$$

Next, to compute the probability density function of the amplitude of a nonzero signal in complex gaussian noise, we must convert a bivariate gaussian density function with a mean $(\bar{x}, \bar{y})$ to polar coordinates. Write the mean in the form

$$\bar{x} = A \cos \theta$$
$$\bar{y} = A \sin \theta$$

where θ is an unknown phase angle. Then

$$p_R(x) = \frac{1}{\sqrt{2\pi}\sigma} e^{-(x-A\cos\theta)^2/2\sigma^2}$$

$$p_I(y) = \frac{1}{\sqrt{2\pi}\sigma} e^{-(y-A\sin\theta)^2/2\sigma^2}.$$

Carrying through the transformation of variables, as before, gives

$$p(r, \phi) = \frac{r}{2\pi\sigma^2} e^{-(r^2-2Ar\cos(\phi-\theta)+A^2)/2\sigma^2}.$$

Integrating over ϕ gives

$$p(r) = \int_0^{2\pi} \frac{r}{2\pi\sigma^2} e^{-(r^2+A^2)/2\sigma^2} e^{-Ar\cos(\phi-\theta)/\sigma^2} d\phi.$$

The integral is clearly independent of θ because the integral is periodic and the integral extends over one period for any value of θ. The integral can be expressed in terms of a standard function known as the *modified Bessel function* of the first kind and order zero, and defined by the integral

$$I_0(x) = \frac{1}{2\pi} \int_0^{2\pi} e^{x\cos\phi} d\phi.$$

The function $I_0(x)$ is shown in Figure 6.14. Hence we can write

$$p(r) = \frac{r}{\sigma^2} e^{-(r^2+A^2)/2\sigma^2} I_0 \left(\frac{Ar}{\sigma^2} \right) \quad r \geq 0.$$

This probability density function is known as a *ricean probability density function*.

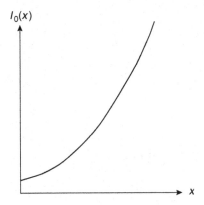

$I_0(x)$

Figure 6.14. The function $I_0(x)$.

To write the ricean density function in a standardized form, make the change in variables

$$z = \frac{r}{\sigma}$$

$$\lambda = \frac{A}{\sigma}$$

and define

$$p_{Ri}(z, \lambda) = z e^{-(z^2 + \lambda^2)/2} I_0(\lambda z)$$

so that

$$p(r) = \frac{1}{\sigma} p_{Ri}\left(\frac{r}{\sigma}, \frac{A}{\sigma}\right).$$

The probability density function $p_{Ri}(z, \lambda)$ is known as a ricean probability density function with parameter λ. It is a family of probability densities, one for each value of λ, as shown in Figure 6.15. The ricean probability density function reduces to a rayleigh probability density function when λ is equal to zero. When λ is large, the ricean probability density resembles a gaussian probability density.

The problem of detecting a passband pulse noncoherently in gaussian noise can be solved by passing the pulse through a passband matched filter, taking the magnitude of the output and applying it to a threshold. The probability distributions for the two hypotheses of pulse absent and pulse present are

$$p_0(r) = \frac{r}{\sigma^2} e^{-r^2/2\sigma^2} \quad r \geq 0$$

$$p_1(r) = \frac{r}{\sigma^2} e^{-(r^2 + A^2)/2\sigma^2} I_0\left(\frac{Ar}{\sigma^2}\right) \quad r \geq 0.$$

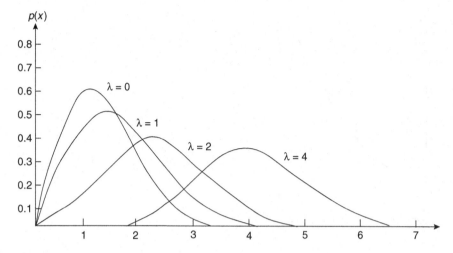

Figure 6.15. Some ricean probability density functions.

6.5   Error rates for noncoherent signaling

We now derive expressions for the bit error rate for noncoherent binary signaling over an additive, white gaussian-noise channel, starting with noncoherent binary OOK. The OOK demodulator passes the received complex signal through a filter matched to the pulse $s(t)$. If the pulse is present, the output of the filter at the sampling instant ℓT will be the complex value

$$x + jy = Ar(\ell T)e^{j\theta} + n'_R + jn'_I$$
$$= Ae^{j\theta} + n'_R + jn'_I,$$

provided $r(t)$ is a Nyquist pulse. If the pulse is absent, the output will be the complex value

$$x + jy = n'_R + jn'_I.$$

The complex output sample of the matched filter is described by the two real gaussian random variables (x, y), and from these the decision is made. The two-dimensional random variable is characterized by one two-dimensional gaussian probability density function when the pulse is absent and by another when the pulse is present. Figure 6.16 shows the two-dimensional gaussian probability density functions on the x, y plane. Both two-dimensional probability density functions are presented on the same graph, one with the signal absent and one with the signal present.

It is apparent from inspection of Figure 6.16 that, if we knew the phase angle θ, we should rotate the coordinate axes to place the displacement along the new x axis.

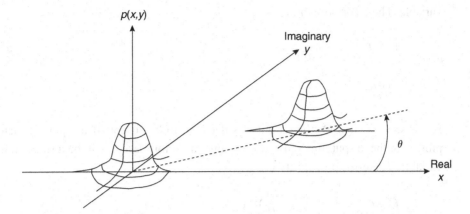

Figure 6.16. Probability density functions for noncoherent OOK.

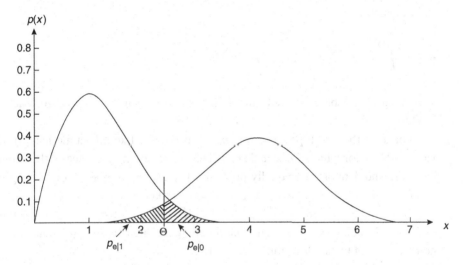

Figure 6.17. Error probabilities for noncoherent OOK.

Then the y component of the data would be superfluous and could be discarded, thereby reducing the problem to coherent OOK.

However, θ is not known. The decision is based not on the displacement along a fixed axis but on the radial distance of the received point (x, y) from the origin. The radial distance $r = \sqrt{x^2 + y^2}$ is a random variable described by a ricean probability density function when the signal is present and a rayleigh probability density function when the signal is absent. The radial coordinate r is compared to the threshold Θ, as shown in Figure 6.17.

If r is larger than Θ, then a pulse is declared to be present. This will be a false detection if the output is actually noise only. In this case, r will be a rayleigh random variable. The error probability is

$$
p_{e|0} = \int_{\Theta}^{\infty} \frac{r}{\sigma^2} e^{-r^2/2\sigma^2} dr
$$
$$
= e^{-\Theta^2/2\sigma^2}.
$$

If r is smaller than Θ, then a pulse is declared to be absent. If the pulse is actually present in the output, this pulse will be missed. In this case, r will be a ricean random variable. The error probability is

$$
p_{e|1} = \int_0^{\Theta} \frac{r}{\sigma^2} e^{-(r^2+A^2)/2\sigma^2} I_0\left(\frac{Ar}{\sigma^2}\right) dr.
$$

To simplify this expression, make the changes in variables $\lambda = A/\sigma$ and $z = r/\sigma$. The integral becomes

$$
p_{e|1} = \int_0^{\Theta/\sigma} z e^{-(z^2+\lambda^2)/2} I_0(\lambda z) dz.
$$

This integral can be evaluated numerically for each value of λ and for each value of Θ/σ.

To balance the two types of errors $p_{e|0}$ and $p_{e|1}$ for fixed λ, numerically find that value of Θ/σ for which $p_{e|0}$ equals $p_{e|1}$. This is the optimum choice of threshold Θ if data zero and data one are equally probable. Then set $p_e = p_{e|0} = p_{e|1}$. This gives p_e as a function of $\lambda = A/\sigma$. Because for a matched filter, $(A/\sigma)^2 = 2E_p/N_0$, $E_b = E_p/2$ for OOK, and $\lambda = A/\sigma = \sqrt{2E_p/N_0} = \sqrt{4E_b/N_0}$, this also gives p_e as a function of E_b/N_0. In this way, using numerical integration, one can plot p_e versus E_b/N_0 for noncoherent OOK, as shown in Figure 6.18.

The second case of noncoherent binary signaling that we treat is noncoherent binary FSK. This calculation requires a different setup. Now there are two matched-filter outputs, denoted x and y. There are four probability density functions corresponding to each of two filter outputs under each of two possible transmitted pulses. First, let $p_{0|0}(x)$ and $p_{1|0}(y)$ denote the two density functions given that pulse $s_0(t)$ was transmitted; the first is ricean and the second is rayleigh. Then let $p_{0|1}(x)$ and $p_{1|1}(y)$ denote the two density functions, given that pulse $s_1(t)$ was transmitted; the first is rayleigh and the second is ricean. When $s_0(t)$ is transmitted, and the filter $s_0(-t)$ has output x, an error occurs if the magnitude of the signal y from the filter $s_1(-t)$ is larger than x. That is, conditional on x, when the data bit is a zero, the error probability is

$$
p_{e|0,x} = \int_x^{\infty} p_{1|0} dy.
$$

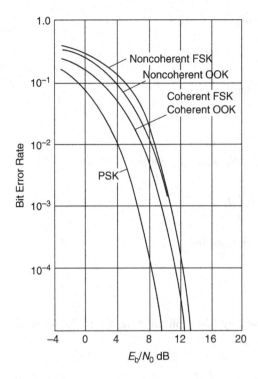

Figure 6.18. Performance of passband binary modulation methods.

The error probability $p_{e|0}$ is then the expectation of this over x

$$p_{e|0} = \int_0^\infty p_{0|0}(x) \left[\int_x^\infty p_{1|0}(y)dy \right] dx.$$

Likewise, when the data bit is a one, the error probability is

$$p_{e|1} = \int_0^\infty p_{1|1}(x) \left[\int_x^\infty p_{0|1}(y)dy \right] dx.$$

Surprisingly, these two probabilities can be expressed in a simple form even though the ricean density in the integrand cannot be expressed in a simple form.

Theorem 6.5.1 *The probability of error p_e of a noncoherent demodulator for a binary, orthogonal, equal-energy waveform alphabet used on an additive white gaussian-noise channel is*

$$p_e = \tfrac{1}{2} e^{-\frac{1}{2} E_b/N_0}.$$

Proof Because $p_e = p_{e|0} = p_{e|1}$ for binary orthogonal, equal-energy pulses in gaussian noise, it is enough to calculate $p_{e|0}$. We begin with the variables x and y, normalized so that $p_{1|0}(y)$ is the rayleigh density function

$$p_{1|0}(y) = ye^{-y^2/2}$$

and $p_{0|0}(y)$ is the ricean density function

$$p_{0|0}(x) = xe^{-(x^2+\lambda^2)/2}I_0(\lambda x)$$

where $\lambda = A/\sigma$ and $\lambda^2 = 2E_b/N_0$. We want to evaluate

$$P_{e|0} = \int_0^\infty p_{0|0}(x)\left[\int_x^\infty p_{1|0}(y)dy\right]dx.$$

The inner integral can be evaluated, resulting in

$$P_{e|0} = \int_0^\infty p_{0|0}(x)e^{-x^2/2}dx.$$

This final integral can be evaluated by first manipulating it into the form of a ricean density in the variable $\sqrt{2}x$. This is

$$P_{e|0} = \frac{1}{2}e^{-\lambda^2/4}\int_0^\infty \sqrt{2}xe^{-(2x^2+\lambda^2/4)/2}I_0\left(\frac{\lambda}{\sqrt{2}}\sqrt{2}x\right)d(\sqrt{2}x).$$

The integral is now a probability density function, so the integral has value one. Therefore,

$$P_{e|0} = \frac{1}{2}e^{-\lambda^2/4}.$$

Because $\lambda^2 = 2E_b/N_0$, the proof is complete.   ∎

6.6   Differential phase-shift keying

Often the phase shift in a channel changes very slowly compared to a bit time. Over a few bit times, the relative phase is constant, although the absolute phase is unknown. *Differential phase-shift keying* (denoted DPSK) is a modulation technique often used with such a channel. (We have already discussed, in Section 2.6, a similar technique at baseband called NRZI.) The reason for using DPSK is to allow the use of a simple demodulator without the need for phase synchronization. In DPSK, as in NRZI, the input binary datastream is differentially encoded into a new binary datastream. The new datastream represents a one by a change in polarity and a zero by the absence of a change in polarity. This operation is made clear by the example of Figure 6.19. In

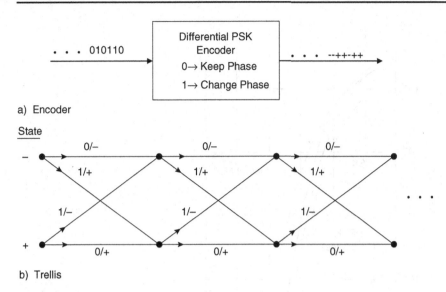

a) Encoder

b) Trellis

Figure 6.19. Differential PSK.

order to establish an initial phase reference, an extra pulse is required at the start. This is an energy overhead inherent in DPSK, but because large message lengths are usually used, this overhead in a single start-up pulse is negligible.

We shall discuss three methods for demodulating DPSK; their performance is summarized in Figure 6.20. The first two methods, labeled "single-bit demodulation" in Figure 6.20, are mathematically equivalent. Either is only optimal for demodulating a single bit received in gaussian noise. Neither of these two methods is an optimal method for demodulating the entire bitstream if the phase is slowly varying, because then the phase can be estimated and the datastream can be coherently demodulated. If the phase is rapidly varying, then DPSK is not suitable.

The third method, labeled "sequence demodulation" in Figure 6.20, is the optimal method. It computes the phase angle of a long sequence of bits, then demodulates that string of bits coherently. If the string of bits is long enough, the residual phase error will be negligible. Even as few as ten bits may be quite enough because then the phase measurement is made on a signal with 10 dB more energy than a single bit. After the phase has been estimated, the waveform can be demodulated as in BPSK. Then a postcoder must be used to invert the differential encoding. Every demodulated bit error becomes two bit errors after the postcoder. Thus the bit error rate of optimally demodulated DPSK is double that of BPSK as is shown in Figure 6.20. The optimal method, however, runs counter to the purpose of transmitting DPSK, which is to eliminate the need for phase synchronization.

Of the "bit-by-bit" methods of demodulating DPSK, the easier to understand is the one depicted in Figure 6.21. In this figure, the implementation is at baseband, but an

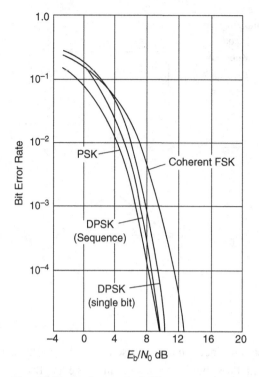

Figure 6.20. Performance of differential PSK.

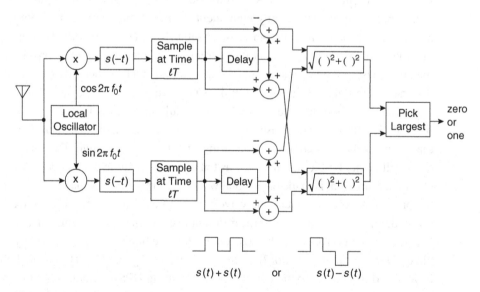

Figure 6.21. Baseband matched-filter demodulator for DPSK.

equivalent passband implementation could be used instead. This demodulator views a pair of bits of the DPSK waveform as a noncoherent FSK waveform with a zero denoted by

$$s_0(t) = s(t) + s(t - T),$$

and a one denoted by

$$s_1(t) = s(t) - s(t - T).$$

The matched filters for the pulses $s_0(t)$ and $s_1(t)$ can be implemented concisely by using a matched filter for the pulselet $s(t)$, followed by a sampler, and an addition and subtraction of delayed samples, as can be seen in Figure 6.21. Each bit is now demodulated as noncoherent FSK by computing the square root of the sum of the squares of the outputs of the matched filters. However, because each individual pulse $s(t)$ contributes to two successive bits, E_b is half of what it would be for noncoherent FSK. Thus we can conclude that the performance of the noncoherent matched-filter demodulator for DPSK has the same performance as noncoherent FSK with twice the E_b/N_0

$$p_e - \tfrac{1}{2}e^{-E_b/N_0}.$$

Simply stated, DPSK is 3 dB better than noncoherent FSK.

An alternative DPSK demodulator, shown in Figure 6.22. looks quite different, and it is somewhat surprising that the performance is the same. In fact, it is mathematically equivalent, which means that the performance must be the same. Let the output of the filter $s(-t)$ at the ℓth sampling instant be denoted $u_R(\ell T) + j u_I(\ell T)$. Then the outputs

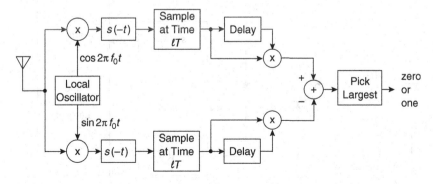

Figure 6.22. Correlation demodulator for DPSK.

of the filters $s_0(-t)$ and $s_1(-t)$ are

$$u_0(\ell T) = [u_R(\ell T) + u_R((\ell - 1)T)] + j[u_I(\ell T) + u_I((\ell - 1)T)]$$
$$u_1(\ell T) = [u_R(\ell T) - u_R((\ell - 1)T)] + j[u_I(\ell T) - u_I((\ell - 1)T)],$$

and the demodulation is based on the sign of $|u_0(\ell T)|^2 - |u_1(\ell T)|^2$. Consequently, a trivial computation gives

$$|u_0(\ell T)|^2 - |u_1(\ell T)|^2 = 4[u_R(\ell T)u_R((\ell - 1)T) + u_I(\ell T)u_I((\ell - 1)T)].$$

Implementation of the right side of this expression gives the alternative demodulator for DPSK, shown in Figure 6.22. We can also understand this demodulator on a more intuitive level. To determine whether the signal component of

$$u(\ell T) = u_R(\ell T) + ju_I(\ell T)$$

is in phase or 180° out of phase with the signal component of

$$u((\ell - 1)T) = u_R((\ell - 1)T) + ju_I((\ell - 1)T),$$

compute $u(\ell T)u^*((\ell - 1)T)$ and detect the sign of the real part.

6.7   Demodulators for M-ary orthogonal signaling

Figure 6.23 shows a coherent demodulator implemented at complex baseband for M-ary orthogonal signaling for a passband pulse alphabet $\{s_m(t)\}$ whose pulses have

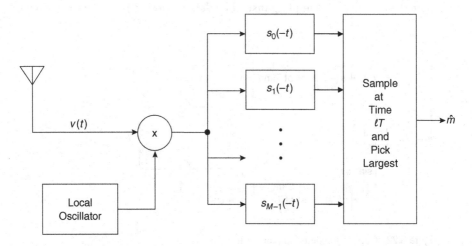

Figure 6.23. A coherent demodulator at baseband for M-ary orthogonal signaling.

only an in-phase component. The quadrature component of each pulse is zero. This means that at complex baseband, the pulse is real. The imaginary component of the complex baseband output contains no signal, and so is not processed.

If the pulses were complex, then the matched filter would be

$$s_m^*(-t) = s_{Rm}(-t) - js_{Im}(-t).$$

For each m, the outputs of the in-phase filter and the quadrature filter would be added to form the output of the matched filter for the full complex pulse. Because the decision of a coherent demodulator is only based on the real part of the matched-filter output, there is no need to develop the imaginary component of the complex matched-filter output.

When the phase of the received signal is not known, a noncoherent demodulator must be used. The complex output samples of the mth matched filter at sampling instant ℓT are

$$u_m(\ell T) = Ae^{j\theta}\delta_{m_\ell m} + n'_{m\ell,R} + jn'_{m\ell,I}$$

and the decision for the ℓth symbol is made based on which of the M magnitudes $|u_m(\ell T)|$ is largest. Figure 6.24 shows a baseband implementation of a noncoherent demodulator for M-ary orthogonal signaling for which the transmitted pulses $s_m(t)$ have only an in-phase component; the quadrature components of the pulses are zero. The received passband waveform is mixed down into a complex baseband waveform which contains the signal $s_m(t)e^{j\theta}$. The complex baseband waveform enters a bank of M matched filters, one filter for each $s_m(t)$, and, for each m, the outputs of the in-phase and quadrature filters are root-sum-squared. If the passband pulses also had a quadrature component, then the matched filters would have both real and imaginary components corresponding to the passband in-phase and quadrature components of each pulse.

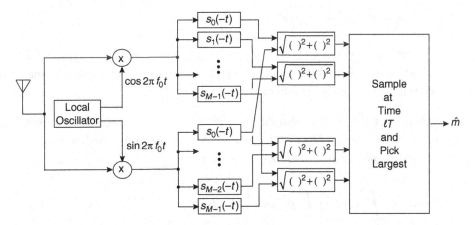

Figure 6.24. A noncoherent demodulator at baseband for M-ary orthogonal signaling.

6.8   Error rates for *M*-ary orthogonal signaling

A demodulator for a passband or complex baseband M-ary orthogonal signaling wave-form passes the received signal through a bank of M matched filters. A coherent demodulator then samples and compares the real parts of the M matched-filter outputs at time ℓT. The performance of this demodulator for real baseband M-ary orthogonal signaling was discussed in detail in Section 3.7. Nothing in that analysis changes if the waveforms are complex. The performance for complex baseband M-ary orthogonal signaling, coherently demodulated, is the same as for real waveforms.

The set of complex M-ary orthogonal signaling waveforms also can be noncoherently demodulated. In fact, if M is moderately large, the degradation in the performance of M-ary orthogonal signaling because of noncoherent demodulation is fairly small. Accordingly, one might prefer a noncoherent demodulator to a coherent demodulator on the grounds of expediency or economics.

We saw in Section 6.7 that, when the phase is unknown, the bank of matched filters has complex output samples given by

$$u_m(\ell T) = Ae^{j\theta}\delta_{m_\ell m} + n'_{m\ell,R} + jn'_{m\ell,I}$$

where θ is an unknown phase angle distributed uniformly on $[0, 2\pi]$, and $n'_{m\ell,R}$ and $n'_{m\ell,I}$ are independent, gaussian random variables of equal variance. The demodula-tion decision is made based on the magnitude $|u_m(\ell T)|$. Because the m_ℓth pulse is transmitted at time ℓT, for $m = m_\ell$, the magnitude

$$|u_m| = \sqrt{(A\cos\theta + n'_{m\ell,R})^2 + (A\sin\theta + n'_{m\ell,I})^2}$$

is a ricean random variable. For $m \neq m_\ell$, the magnitude

$$|u_m| \neq \sqrt{(n'_{m\ell,R})^2 + (n'_{m\ell,I})^2}$$

is a rayleigh random variable. The integer m for which $|u_m|$ is largest is then selected as the estimate $\widehat{m}$ of the modulator input m. An error occurs if the correct matched filter does not have the largest output.

Theorem 6.8.1   *The probability of symbol error p_e of a noncoherent matched-filter demodulator for an M-ary orthogonal waveform alphabet used on an additive white gaussian noise channel at $E_b/N_0 = \gamma$ is given by*

$$p_e = 1 - \int_0^\infty p_1(x)\left[\int_0^x p_0(y)dy\right]^{M-1}dx$$

where

$$p_0(y) = ye^{-y^2/2}$$

and

$$p_1(x) = xe^{-x^2/2}e^{-\gamma \log_2 M}I_0(x\sqrt{2\gamma \log_2 M}).$$

Proof The proof is the same as the proof of Theorem 3.7.1 except that the gaussian probability density functions in that proof must be replaced by rayleigh and ricean probability density functions. If $m \neq m_\ell$, the probability density function of r_m is a rayleigh density function:

$$p_0(r) = \frac{r}{\sigma^2}e^{-r^2/\sigma^2}.$$

If $m = m_\ell$, the probability density function is a ricean density function:

$$p_1(r) = \frac{r}{\sigma^2}e^{-(r^2+A^2)/2\sigma^2}I_0\left(\frac{Ar}{\sigma^2}\right).$$

Change the variables, as in the proof of Theorem 3.7.1, with

$$\left(\frac{A}{\sigma}\right)^2 = \frac{2E_p}{N_0} = 2\gamma \log_2 M$$

to complete the proof of the theorem.   ∎

In contrast to the double integral in Theorem 3.7.1, the double integral in Theorem 6.8.1 can be integrated analytically. First expand the inner integral to write

$$p_e = 1 - \int_0^\infty p_1(x)[1 - e^{-x^2/2}]^{M-1}dx$$

$$= 1 - \int_0^\infty p_1(x)\sum_{\ell=0}^{M-1}(-1)^\ell\binom{M-1}{\ell}e^{-\ell x^2/2}dx$$

$$= 1 - \sum_{\ell=0}^{M-1}(-1)^\ell\binom{M-1}{\ell}\int_0^\infty p_1(x)e^{-\ell x^2/2}dx.$$

The integral at the right, with x now replaced by r, is written out as follows:

$$\int_0^\infty p_1(r)e^{-\ell r^2/2}dx = \int_0^\infty e^{-\ell r^2/2}\int_0^{2\pi}\frac{r}{2\pi}e^{-(r^2+\gamma^2)/2}e^{-\gamma r\cos(\phi-\theta)}d\phi dr$$

$$= \int_0^\infty \int_0^{2\pi}\frac{1}{2\pi}e^{-((\ell+1)r^2+\gamma^2)/2}e^{-\gamma r\cos(\phi-\theta)}rdrd\phi.$$

To recognize a two-dimensional gaussian distribution in the integrand, let $\sigma^2 = 1/(\ell+1)$, then multiply through by $1/(\ell+1)\sigma^2$ and set $a = \gamma\sigma^2$ to write

$$\int_0^\infty p_1(r)e^{-\ell r^2/2}dr = \frac{1}{\ell+1}e^{-\gamma^2/2}e^{a^2/2\sigma^2}\int_0^\infty\int_0^{2\pi}\frac{1}{2\pi\sigma^2}e^{-(r^2+a^2)/2\sigma^2}e^{-ar\cos(\phi-\theta)/\sigma^2}r\,dr\,d\phi.$$

Because the integrand is now a gaussian density function in polar coordinates, the double integral is equal to one. Then

$$p_e = 1 - \sum_{\ell=0}^{M-1}(-1)^\ell\binom{M-1}{\ell}\left[\frac{1}{\ell+1}e^{-\gamma^2/2}e^{a^2/2\sigma^2}\right].$$

This is summarized in the following corollary.

Corollary 6.8.2   *The probability of symbol error for noncoherent demodulation of an M-ary orthogonal signaling waveform is*

$$p_e = \frac{1}{M}\sum_{m=2}^M\binom{M}{m}(-1)^m e^{-[(m-1)/m](E_b/N_0)\log_2 M}.$$

Proof   Because

$$\binom{M-1}{\ell}\frac{1}{\ell+1} = \frac{1}{M}\binom{M}{\ell+1}$$

the expression given prior to the corollary with $\ell+1 = m$ reduces to the statement of the corollary.   ∎

To obtain p_e as a function of E_b/N_0, one can either evaluate the sum in Corollary 6.8.2, or evaluate the integrals of Theorem 6.8.1 numerically. In either case, the result is as shown in Figure 6.25. As M becomes larger, the performance of noncoherent demodulation of M-ary signaling becomes virtually the same as the performance of coherent demodulation of M-ary signaling. In particular, Theorem 3.7.3 also holds for the case of noncoherent demodulation of M-ary orthogonal signaling on an additive gaussian-noise passband channel.

The performance of the three most important M-ary signaling techniques can be compared by a comparison of Figures 3.13, 3.14, and 6.25. When M equals 2, this becomes a comparison of the three most important binary signaling techniques: coherent binary FSK, noncoherent binary FSK, and BPSK, all of which have been studied already in Sections 6.2 and 6.3.

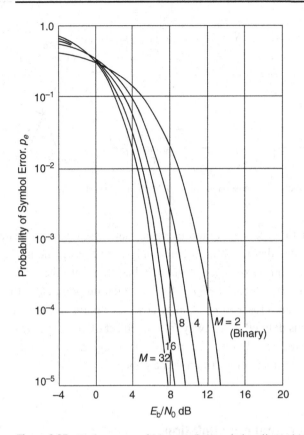

Figure 6.25. Performance of *M*-ary orthogonal signaling with noncoherent demodulation.

6.9   Demodulators for *M*-ary signal constellations

A multilevel signaling waveform using an M-ary signal constellation uses only a single pulse $s(t)$, which is normally real-valued. The received waveform at complex baseband is passed through a filter matched to $s(t)$ and the filter output sampled at ℓT. If $s(t)*s(-t)$ is a Nyquist pulse, there is no intersymbol interference. The sample at ℓT is demodulated into the closest point of the signal constellation.

Figure 6.26 shows a demodulator implemented at complex baseband for an eight-ary PSK signal constellation. The outputs of the real and imaginary matched filters are sampled at time ℓT to define a point u_ℓ in the complex plane. One typical sample of the filter output is denoted in the figure by an asterisk, and the eight points of the eight-ary PSK signal constellation are shown as well. Intuitively, it is easy to see that $u(\ell T)$ should be demodulated into that value of a_ℓ from the signal constellation that is closest to it in euclidean distance. That point then is mapped back into the three-bit data symbol associated with that point of the signal constellation.

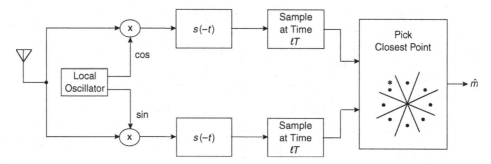

Figure 6.26. A demodulator at complex baseband for a QAM signal constellation.

For any signal constellation used on the additive gaussian-noise channel, the demodulator is similar. Indeed, this demodulator is commonly used even if the noise is not gaussian. Simply map the complex output of the matched filter into the closest point of the signal constellation. The *decision region* for constellation point c_m is the set of all complex numbers that are closer to c_m than to any other point of the signal constellation. Decision regions defined in this way for a fixed set of points in the complex plane are called *Voronoi regions*. Every side of a Voronoi region is a segment of the perpendicular bisector of the line connecting two neighbors.

6.10   Error rates for *M*-ary signal constellations

The calculation of the probability of error for an M-ary signal constellation is quite different from the calculation of the probability of error for an M-ary orthogonal signaling alphabet. The demodulation decision is now based on the output of a single complex matched filter, as discussed in Section 6.9. The output of the matched filter is

$$u(t) = \sum_{\ell=-\infty}^{\infty} a_\ell r(t - \ell T) + n'(t).$$

We continue to suppose that $r(t)$ is a Nyquist pulse, therefore the time samples $u_\ell = u(\ell T)$ of the matched-filter output are

$$u_\ell = a_\ell + n'_\ell$$

where a_ℓ is the point of the complex signal constellation that was transmitted at time ℓT and $n'_\ell = n'(\ell T)$ is a sequence of uncorrelated complex noise samples of variance σ^2 in each component; the samples are independent complex gaussian random variables if the channel noise is gaussian.

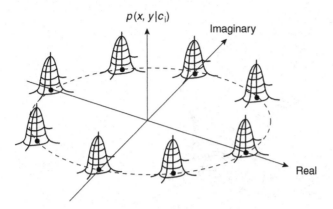

$p(x, y|c_i)$

Figure 6.27. Eight probability density functions for eight-ary PSK.

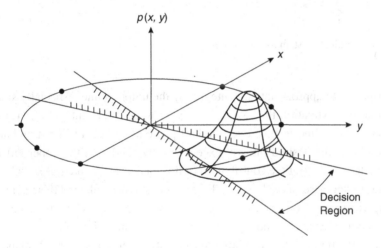

$p(x, y)$

Figure 6.28. A decision region for eight-ary PSK.

For each ℓ, u_ℓ is a complex circular gaussian random variable with a mean equal to the complex number a_ℓ and with variance σ^2 on the real and imaginary axes. Figure 6.27 shows the eight points of the eight-ary PSK signal constellation together with their eight gaussian probability density functions, each one centered on one of the signal points. To calculate the probability of error, on the condition that the transmitted symbol is a_ℓ, we must integrate $p(x, y|a_\ell)$ over all x, y not in the decision region corresponding to a_ℓ.

Figure 6.28 shows the decision region corresponding to one of the points of the signal constellation, and the gaussian probability density function for that point. Figure 6.29 shows the same situation in plan view, now with the gaussian density depicted by contour lines. The probability of error is the integral of the two-dimensional probability density function over that portion of the complex plane not including the decision region. This two-dimensional integral can be evaluated numerically. However, it is

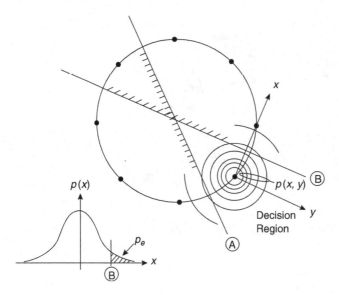

Figure 6.29. A coordinate system for computation.

more informative to approximate the integral by the union bound, especially since the union bound approximation that we give is quite tight. If we integrate $p(x, y)$ over the half plane above line B, then over the half plane to the left of line A and add the results, we obtain the desired integral except that the wedge in the upper left of the plane is counted twice. But the contribution of this wedge to the integral must be small because $p(x, y)$ decreases as $e^{-r^2/2\sigma^2}$ at distance r from the peak, and this wedge is far away compared to some other parts of the region of integration. Counting a negligible contribution twice results in a negligible error in the computation of p_e.

In this way, the union bound can be used to combine integrals over half planes. The integration of the probability density function $p(x, y)$ over a half plane can be expressed concisely with the aid of the following simple theorem.

Theorem 6.10.1   *The integral of the gaussian probability density function*

$$p(x, y) = \frac{1}{2\pi\sigma^2} e^{-[(x-\bar{x})^2+(y-\bar{y})^2]/2\sigma^2}$$

over any half plane not containing $(\bar{x}, \bar{y})$ is

$$p_e = Q\left(\frac{d}{2\sigma}\right)$$

where $d/2$ is the perpendicular distance from $(\bar{x}, \bar{y})$ to the half plane.

Proof   Choose a new coordinate system centered at $(\bar{x}, \bar{y})$, with one coordinate line parallel to the line defining the half plane. Then

$$p_e = \int_{d/2}^{\infty} \int_{-\infty}^{\infty} \frac{1}{2\pi\sigma^2} e^{-(x^2+y^2)/2\sigma^2} \, dx dy,$$

which now has the form of a sequence of two integrals. Evaluating the y integral yields

$$p_e = \int_{d/2}^{\infty} \frac{1}{\sqrt{2\pi\sigma^2}} e^{-x^2/2\sigma^2} \, dx$$

$$= Q\left(\frac{d}{2\sigma}\right),$$

which completes the proof of the theorem.   ■

We can now find the probability of symbol error for eight-ary PSK signaling as follows. All signal points have an identical pattern of neighbors, so the average probability of error is equal to the conditional probability of error given any signal point. If the radius of the constellation is A, then a simple calculation gives

$$\frac{d}{2} = A \sin \frac{\pi}{8} = .3827A.$$

Finally, because $(A/\sigma)^2 = 2E_p/N_0$, and with the aid of Theorem 6.10.1, we can write the bound as

$$p_e \leq 2Q\left(\sqrt{(.38^2)2E_p/N_0}\right).$$

The factor of two multiplying the Q function comes about because a signal point has two nearest neighbors, which leads to an integration over two (overlapping) half planes. The overlap is incorrectly included twice, but because the overlap is small, this bound on probability of symbol error will be tight for large values of E_b/N_0. If a Gray-coded signal constellation is used, then a single symbol error will normally lead to one bit error in three bits. Hence the bit error rate is approximated by

$$p_{eb} \approx \frac{2}{3}Q\left(\sqrt{\frac{0.88E_b}{N_0}}\right)$$

where we have also made the substitution $E_p = 3E_b$. This should be compared to the expression for BPSK

$$p_{eb} = Q\left(\sqrt{\frac{2E_b}{N_0}}\right).$$

For a first approximation, we look only at the argument of Q and see that for eight-ary PSK, E_b must be larger by 2 divided by 0.88 in order to have the same performance as BPSK. Thus, we say that eight-ary PSK uses about 3.6 dB more energy per bit than does binary PSK.

A similar analysis, using Theorem 6.10.1, can be used to approximate the probability of symbol error for any signal constellation by an expression of the form

$$p_e \approx N_{d_{\min}} Q \left(\frac{d_{\min}}{2\sigma} \right)$$

where $d_{\min}$ is the minimum of the euclidean distances between all pairs of points of the signal constellation, and $N_{d_{\min}}$ is the average number of nearest neighbors, defined as neighbors at minimum distance. This can be rewritten by using the following substitutions

$$\left(\frac{d_{\min}}{2\sigma} \right)^2 = \frac{d_{\min}^2 \, A^2}{4A^2 \, \sigma^2} = \frac{d_{\min}^2 \, 2E_p}{4E_p \, N_0} = \frac{d_{\min}^2 \, 2E_b}{4E_b \, N_0}.$$

Consequently,

$$p_e \approx N_{d_{\min}} Q \left(\frac{\rho_{\min}}{2} \sqrt{\frac{2E_b}{N_0}} \right)$$

where $\rho_{\min} = d_{\min}/\sqrt{E_b}$ is the normalized minimum distance of the signal constellation. Within the accuracy of this approximation, the signal constellation is adequately described by the two parameters, $N_{d_{\min}}$ and $\rho_{\min}$.

In particular, the probability of symbol error of 2^k-ary PSK is approximated by

$$p_e \approx 2Q \left(\sin \frac{\pi}{M} \sqrt{\frac{2E_b}{N_0}} \right)$$

because $d_{\min} = A \sin(\pi/M)$ if there are $M = 2^k$ uniformly spaced points on the circle of radius A. With Gray coding, there is usually only one bit error per symbol error, so the bit error rate is approximated by

$$p_e \approx \frac{2}{k} Q \left(\sin(\pi 2^{-k}) \sqrt{\frac{2E_b}{N_0}} \right).$$

Comparing the argument of the Q function with the comparable terms for BPSK, we say that 2^k-ary PSK requires approximately $\sin^2(\pi 2^{-k})$ more transmitted energy per bit to achieve a specified bit error rate. The reason for expending this extra energy per bit is to increase the data rate from one bit per sampling instant to k bits per sampling instant.

Problems for Chapter 6

6.1. a. Explain the underlying intuitive reason why, for fixed E_b/N_0, M-ary orthogonal signaling has a smaller bit error rate than binary signaling.

   b. Explain the underlying intuitive reason why, for large M, noncoherent M-ary orthogonal signaling has nearly the same bit error rate as coherent M-ary orthogonal signaling.

   c. What are the disadvantages of M-ary orthogonal signaling?

6.2. a. An MSK waveform is used for coherent binary signaling in white gaussian noise. Give an expression for the probability of bit error as a function of E_b and N_0.

   b. If QPSK is used instead of MSK, how will the probability of bit error change? How will the spectrum change?

6.3. The QPSK signal constellation is assigned pairs of data bits in counterclockwise sequence as 00, 01, 11, 10. (This sequence is known as a two-bit Gray code.)

   a. Prove that the (Gray-coded) QPSK waveform received in white gaussian noise can be demodulated with no loss in performance by demodulating the in-phase and quadrature channels as independent BPSK waveforms to obtain the two bits.

   b. Prove that if instead the QPSK signal constellation is assigned pairs of data bits in the counterclockwise sequence 00, 01, 10, 11, then each data bit cannot be obtained independently from either the in-phase or the quadrature bit. How does the bit error rate degrade at fixed E_b/N_0?

6.4. Suppose that a BPSK waveform is received with a phase error of θ, but nevertheless is demodulated by using a coherent demodulator. By how much is the signal-to-noise ratio reduced because of the phase error? If $E_b/N_0 = 10$ dB, by how much will the bit error rate degrade due to a $10°$ phase error? How much phase error can be accepted in a coherently demodulated binary FSK before it is better to use a noncoherent demodulator?

6.5. The noncoherent demodulator for binary FSK uses as the decision statistic the square root of the sum of the squares of the noisy outputs of the in-phase and quadrature matched filters.

   a. A co-worker proposes that instead of using a noncoherent demodulator, one could instead estimate phase from the matched-filter outputs and correct for it. Specifically, if $u_R(t)$ and $u_I(t)$ are the matched-filter outputs, let

$$\widehat{\theta} = \tan^{-1}\left(\frac{u_I(0)}{u_R(0)}\right)$$

and let

$$\widehat{u}(t) = u_R(t)\cos\widehat{\theta} + u_I(t)\sin\widehat{\theta}.$$

Prove that this gives a demodulator that is mathematically equivalent to the noncoherent demodulator.

b. Now the co-worker proposes that θ be estimated on two successive bit times, as above, and averaged. Will this give a better or worse demodulator if phase error is independent from bit to bit? Will this give a better or worse demodulator on a partially-coherent channel? A *partially-coherent channel* is one on which phase error changes very slowly compared to one bit duration.

6.6. a. Construct a four-ary orthogonal signaling waveform alphabet of duration T by using a half-cosine pulse of pulsewidth $T/4$ as a "chip".

b. Given a matched filter for a half-cosine pulse as an existing component, sketch the design of a coherent demodulator.

c. Under the same condition, sketch the design of a noncoherent demodulator.

6.7. a. Sketch a noncoherent binary FSK demodulator at complex baseband for signaling pulses that have both a real part and an imaginary part.

b. Sketch an equivalent demodulator at passband.

6.8. A passband PSK waveform is passed through its passband matched filter. Derive an expression for the signal at the output of the filter.

6.9. Show that the expression for the probability of symbol error of a coherent demodulator for M-ary orthogonal signaling reduces to the expression for the probability of bit error of coherent binary FSK when M equals 2.

6.10. Prove that the probability of symbol error of a QPSK waveform on an additive white gaussian-noise channel is given exactly by

$$p_e = 2Q\left(\sqrt{\frac{2E_b}{N_0}}\right) - Q^2\left(\sqrt{\frac{2E_b}{N_0}}\right).$$

How does this compare with the approximate expression based on the union bound?

6.11. a. Prove that the probability of error of a coherently demodulated 32-ary biorthogonal alphabet of waveforms with pulse energy E_p is strictly smaller than the probability of error of a coherently demodulated 32-ary orthogonal family with the same E_p, and strictly larger than a coherently demodulated 31-ary orthogonal family with the same E_p.

b. Set up an integral expression for the probability of symbol error of an M-ary biorthogonal family of waveforms coherently demodulated.

6.12. a. Prove that on an additive gaussian-noise channel, as M goes to infinity, the probability of error of a biorthogonal family of waveforms, coherently demodulated, behaves as one of the two cases

(i) $p_e \to 0$ if $E_b/N_0 > \log_e 2$

(ii) $p_e \to 1$ if $E_b/N_0 < \log_e 2$.

(**Hint:** Show that when M is large, the biorthogonal family is negligibly different from an orthogonal family.)

b. Prove that as M goes to infinity, an M-ary orthogonal family of waveforms noncoherently demodulated behaves as one of the two cases

   (i) $p_e \to 0$ if $E_b/N_0 > \log_e 2$
   (ii) $p_e \to 1$ if $E_b/N_0 < \log_e 2$.

6.13. A QPSK demodulator is designed to put out a null signal, called an erasure, when the decision is ambivalent. Specifically, the decision regions are modified as shown in the accompanying diagram.

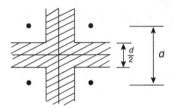

When the received sample lies in the hatched region, the demodulator output is a special symbol denoting an erasure. When the received sample lies in one of the other regions, the appropriate two bits of demodulated data are the output of the demodulator. Use the union bound to set up approximate expressions for the probability of erasure and the probability of symbol error.

6.14. Orthogonal matrices whose elements are all ± 1 are called *Hadamard matrices*. When n is a power of two, an n by n Hadamard matrix can be defined by the recursion

$$H_2 = \begin{bmatrix} 1 & 1 \\ 1 & -1 \end{bmatrix}, \quad H_{2n} = \begin{bmatrix} H_n & H_n \\ H_n & -H_n \end{bmatrix}.$$

(The $2n \times 2n$ matrix H_{2n} is called the *Kronecker product* of H_2 and H_n.) Hadamard sequences of blocklength n are the rows of H_n. A communication system uses eight-ary Hadamard sequences of blocklength 8 and half-cosine shaped chips as orthogonal waveforms.

a. Sketch a functional block diagram for a noncoherent receiver for this system.

b. Repeat the design, this time decomposing the matched filter into two filters: the "chip filter" matched to the chip, and the "sequence filter" matched to the Hadamard sequences. The use of two local oscillators is suggested with the chip filter at an intermediate frequency (IF) and the sequence filter at baseband.

c. Repeat the design, this time decomposing the n sequence filters simultaneously based on the structure of the Kronecker product. (This structure is called a fast-Hadamard-transform receiver.)

6.15. Derive an approximate expression for the probability of symbol error as a function of E_b/N_0 of a sixteen-ary QAM (square pattern) signal constellation, shown in Problem 5.15. The energy E_b is the average energy per bit over the sixteen points of the constellation.

6.16. Derive an approximate expression for the probability of symbol error as a function of E_b/N_0 of a sixteen-ary PSK signal constellation, shown in Problem 5.16.

6.17. a. Use a series expansion on a term of the form $\sqrt{(A+x)^2 + y^2}$ to explain why the ricean probability density function looks like a gaussian probability density function when A is large.

   b. The modified Bessel function of the first kind, $I_0(x)$, can be approximated when x is large as

$$I_0(x) \approx \frac{e^x}{\sqrt{2\pi x}}$$

   Using this approximation as a starting point, show again that the ricean probability density function looks like a gaussian probability density function when A is large.

6.18. Two eight-ary signal constellations are shown below.

   a. If the points in the QAM constellation are spaced as shown (nearest neighbors separated by two units), what should the radius of the PSK constellation be so that E_b is the same in both constellations?

   b. By studying the distance structure of the two constellations, decide which constellation will have a lower probability of error in white gaussian noise, and explain why.

   c. Set up approximate expressions for the error probabilities. Describe where approximations are made.

6.19. Let $s(t)$ be a "half-cosine chip". That is,

$$s(t) = \begin{cases} \cos \pi t/T & |t| \leq \frac{T}{2} \\ 0 & \text{otherwise.} \end{cases}$$

   Two "MSK pulses", each with four chips, are given by

$$s_0(t) = s(t) + js\left(t - \tfrac{1}{2}T\right) + s(t - T) + js\left(t - \tfrac{3}{2}T\right)$$

$$s_1(t) = s(t) - js\left(t - \tfrac{1}{2}T\right) - s(t - T) + js\left(t - \tfrac{3}{2}T\right).$$

a. Show that $s_0(t)$ and $s_1(t)$ are orthogonal where the definition of orthogonal is

$$\int_{-\infty}^{\infty} s_0(t)s_1^*(t)dt = 0.$$

b. Using $s_0(t)$ and $s_1(t)$ as a binary orthogonal signaling alphabet, sketch a coherent demodulator. Sketch a noncoherent demodulator for the same waveform.

c. In place of $s(t)$, now use the alternative pulse

$$s_1'(t) = s(t) + js\left(t - \tfrac{1}{2}T\right) - s(t - T) - js\left(t - \tfrac{3}{2}T\right).$$

Show that $s_0(t)$ and $s_1(t)$ are not orthogonal. Will the use of $s_1'(t)$ instead of $s_1(t)$ degrade the performance if a coherent demodulator is used? What if a noncoherent demodulator is used?

6.20. Prove that the correlation demodulator for DPSK is equivalent to the noncoherent matched-filter demodulator.

6.21. Keeping E_b constant, draw an eight-ary PSK signal constellation and a sixteen-ary PSK signal constellation to a common scale. By comparing distances in the two signal constellations, determine which signal constellation will have a smaller bit error rate in the same background of white gaussian noise, and explain how this conclusion is reached. By approximately how much must E_b be changed in the second constellation to make the bit error rate the same?

6.22. A noncoherent demodulator for an M-ary orthogonal signaling alphabet flags the demodulated output symbol as "unreliable" whenever the largest matched-filter output magnitude is not at least twice as large as the second largest matched-filter output magnitude. Give an expression for the probability that the output symbol is wrong but is not flagged as unreliable.

6.23. An eight-ary signal constellation in the complex plane places four points equi-spaced symmetrically in angle on each of two circles of radii r_1 and r_2 as illustrated:

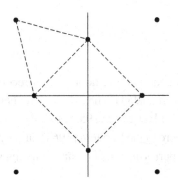

The radii are chosen so that each interior point has four nearest neighbors.

a. Using the union bound (and mentioning all major details), give an expression for the probability of symbol error versus d/σ when the modem is used on an additive white gaussian-noise channel. Explain how to convert d/σ to E_b/N_0.

b. If the modem is now used instead on a channel that makes only phase errors $e^{j\theta_e}$, describe the graph of probability of symbol error versus θ_e. If θ_e is now a gaussian random variable with variance σ_θ^2, what is p_e as a function of σ_θ^2?

6.24. Let $s(t)$ be a "half-cosine chip". That is,

$$s(t) = \begin{cases} \cos \pi t/T & |t| \leq \frac{T}{2} \\ 0 & \text{otherwise.} \end{cases}$$

Two "MSK pulses", each with four chips, are given by

$$s_0(t) = s(t) + js\left(t - \tfrac{1}{2}T\right) + s(t - T) + js\left(t - \tfrac{3}{2}T\right)$$

$$s_1(t) = s(t) + js\left(t - \tfrac{1}{2}T\right) - s(t - T) - js\left(t - \tfrac{3}{2}T\right).$$

a. Using $s_0(t)$ and $s_1(t)$ as a binary orthogonal signaling alphabet, sketch a functional diagram of a noncoherent demodulator for a communication system that sends one bit in time $4T$. Include down conversion and clearly indicate the complex structure of the matched filters.

b. Now describe how to decompose the matched filters into chip filters and sequence filters.

c. Describe how to implement the chip filter at an intermediate frequency. Is a single-chip matched filter adequate at intermediate frequency for both the in-phase and quadrature modulation components?

d. Describe the chip filter output. How can the sampler be implemented at intermediate frequency?

Notes for Chapter 6

The earliest derivation of the optimum coherent and noncoherent receivers for binary signaling appears in the work of Woodward (1954). The calculation of error probability for noncoherent binary signaling is due to Helstrom (1955), based on the noise studies of Rice (1945). Further contributions were made by Stein (1964), and by Doelz, Heald, and Martin (1957). Early surveys of the performance of binary signaling were published by Lawton (1958) and Cahn (1959).

The performance of M-ary coherent demodulation was studied by Kotel'nikov (1959) and by Viterbi (1961). The earliest study of the performance of M-ary noncoherent demodulation was by Reiger (1958), with later work by Nuttall (1962). Turin (1959) first derived the asymptotic behavior for noncoherent demodulation as M goes to infinity.

7 Principles of Optimal Demodulation

This chapter approaches the theory of modem design starting from well-accepted basic principles of inference. In particular, we will study the maximum-likelihood principle and the maximum-posterior principle. By studying the optimum demodulation of passband sequences, we shall develop an understanding of the maximum-likelihood principle and its application. In Chapter 8, we will also treat the topic of synchronization as applied to both carrier recovery and time recovery by using the maximum-likelihood principle.

It is appropriate at this point to develop demodulation methods based on the likelihood function. The maximum-likelihood principle will enable us to derive optimal methods of demodulating in the presence of intersymbol interdependence and, as a side benefit, to establish the optimality of the demodulators already discussed in Chapter 3.

The *maximum-likelihood principle* is a general method of inference applying to many problems of decision and estimation besides those of digital communications. The development will proceed as follows. First we will introduce the maximum-likelihood principle as a general method to form a decision under the criterion of minimum probability of decision error when given a finite set of measurements. Then we will approximate the continuous-time waveform $v(t)$ by a finite set of discrete-time samples to which we apply the maximum-likelihood principle. Finally, we will take the limit as the number of samples of $v(t)$ goes to infinity to obtain the maximum-likelihood principle for the waveform measurement $v(t)$.

7.1   The likelihood function

We begin with the general decision problem, not necessarily the problem of demodulation, of deciding between M hypotheses when given a measurement. Suppose that we are given a finite set of data $(x_1, \ldots, x_n)$ and M probability density functions $p_m(x_1, \ldots, x_n)$ on the data space and indexed $m = 0, \ldots, M - 1$. We are to decide which probability density function was used to generate the observed data. A decision rule is a function of the data, denoted $\widehat{m}(x_1, \ldots, x_n)$, where $\widehat{m}$ is the estimated value of m when given data vector $(x_1, \ldots, x_n)$. A decision error occurs whenever the mth probability density function was used to generate the observed data, but the decision $\widehat{m}(x_1, \ldots, x_n)$ is not equal to m. The conditional probability of error, given that m is true, is denoted $p_{e|m}$. The optimality criterion we shall use for forming the decision rule is that the average probability of decision error is minimized, given that each m is equally likely to be true. At the end of this chapter, we will see that this criterion is not always as compelling as it seems to be.

The following theorem gives optimum decision rules for the case where each hypothesis occurs with probability $1/M$.

Theorem 7.1.1 *For the M-ary decision problem with equiprobable hypotheses, the optimum decision rule when given the vector measurement $(x_1, \ldots, x_n)$ is to choose that m for which $p_m(x_1, \ldots, x_n)$ is largest.*

Proof The probability of decision error, given that m is true, is

$$p_{e|m} = \int_{\mathcal{U}_m^c} p_m(x_1, \ldots, x_n) dx_1 \ldots dx_n$$

where $\mathcal{U}_m$ is the set of data vectors $(x_1, \ldots, x_n)$ for which the mth hypothesis is the output of the decision rule. The average probability of decision error is

$$p_e = \sum_{m=0}^{M-1} \frac{1}{M} \int_{\mathcal{U}_m^c} p_m(x_1, \ldots, x_n) dx_1 \ldots dx_n$$

$$= \sum_{m=0}^{M-1} \frac{1}{M} \left[1 - \int_{\mathcal{U}_m} p_m(x_1, \ldots, x_n) dx_1 \ldots dx_n \right].$$

Consequently, to minimize p_e, the decision set $\mathcal{U}_m$ should contain the measurement $(x_1, \ldots, x_n)$ whenever $p_m(x_1, \ldots, x_n)$ is larger than $p_{m'}(x_1, \ldots, x_n)$ for all $m' \neq m$. (A tie for the largest probability can be broken by any arbitrary rule, say, break a tie in favor of the smallest index.) This concludes the proof. ∎

Corollary 7.1.2 *If, for each m, the random data vector $(X_{m1}, \ldots, X_{mn})$ is a gaussian vector random variable, corresponding to the mth hypothesis, and with independent components of means $c_{m1}, \ldots, c_{mn}$ and identical variances σ^2, then the optimum decision rule given observed data $(x_1, \ldots, x_n)$ is to choose the hypothesis indexed by m for which the euclidean distance $\sum_{\ell=1}^{n} (x_\ell - c_{m\ell})^2$ is smallest.*

Proof   Suppose that, for each m, $(X_{m1}, \ldots, X_{mn})$ is a gaussian vector random variable with independent components of means $c_{m1}, \ldots, c_{mn}$ and identical variances σ^2. Then

$$p_m(x_1, \ldots, x_n) = \prod_{\ell=1}^{n} \frac{1}{\sqrt{2\pi}\,\sigma} e^{-(x_\ell - c_{m\ell})^2/2\sigma^2}$$

$$= \frac{1}{(\sqrt{2\pi}\,\sigma)^n} e^{-\Sigma_\ell (x_\ell - c_{m\ell})^2/2\sigma^2}.$$

An optimum decision rule decides on that m for which $p_m(x_1, \ldots, x_n)$ is maximum. To maximize $p_m(x_1, \ldots, x_n)$ over m, one should minimize $\Sigma_\ell (x_\ell - c_{m\ell})^2$ over m. This completes the proof of the corollary.   ∎

Corollary 7.1.3   *If $\sum_{\ell=1}^{n} c_{m\ell}^2$ is independent of m, then the optimum decision rule is to choose that m for which the correlation coefficient $\sum_{\ell=1}^{n} x_\ell c_{m\ell}$ is largest.*

Proof   Expand the square and note that the term $\Sigma_\ell x_\ell^2$ and, by the condition of the corollary, the terms $\Sigma_\ell c_{m\ell}^2$ are independent of m. The proof of the corollary follows.   ∎

The function $p_m(x_1, \ldots, x_m)$ appearing in Theorem 7.1.1 arises as a probability density function, one such probability density function for each m. However, the theorem regards m as the variable and the data $(x_1, \ldots, x_m)$ as given. This is the view that one adopts after the data is observed. Then we think of $p_m(x_1, \ldots, x_m)$ as a function of m. In this role $p_m(x_1, \ldots, x_m)$ is called a *likelihood function*, and its logarithm, denoted

$$\Lambda(m) = \log p_m(x_1, \ldots, x_n)$$

is called the *log-likelihood function*.

In general, the unknown index m can be replaced by a vector of parameters γ, and the probability density function on the vector of measurements $p(x_1, \ldots, x_n \mid \gamma)$ is conditional on γ. Possibly, some components of the vector γ are discrete parameters such as data components and some are continuous parameters, such as carrier phase. As a function of the vector γ the *log-likelihood function* is defined as

$$\Lambda(\gamma) = \log p(x_1, \ldots, x_n \mid \gamma).$$

The logarithmic form of the likelihood function is preferred because it replaces products of likelihoods by sums of log-likelihoods, and sums are more convenient to deal with.

7.2   The maximum-likelihood principle

We are usually interested in a log-likelihood function only for the purpose of finding where it achieves its maximum, but not in the value of the maximum. Constants added to or multiplying the log-likelihood function do not affect the location of the maximum, so it is common practice to suppress such constants when they occur by redefining $\Lambda(\gamma)$.

For example, the gaussian distribution with unknown mean $\bar{x}$ has a log-likelihood function

$$\Lambda(\bar{x}) = -\log\sqrt{2\pi\sigma^2} - (x - \bar{x})^2/2\sigma^2.$$

For economy, we will commonly discard the constants and write

$$\Lambda(\bar{x}) = -(x - \bar{x})^2.$$

We will also call this function a likelihood "statistic" to recall that is not precisely the log-likelihood function. A likelihood statistic is an example of a sufficient statistic. In general, a *statistic* is any function of the received data, and a *sufficient statistic* is a statistic from which the likelihood function can be recovered by reintroducing known constants; no essential information contained in the data is lost.

We shall be interested in problems in which the number of data measurements n goes to infinity. In such a case, the limit as n goes to infinity of $\Lambda(\gamma)$ may be infinite for all (or many) values of γ. It then would be meaningless to deal with the maximum over γ of the limit of $\Lambda(\gamma)$. But we are not interested in the value of the maximum. We are interested only in the value of γ where the maximum occurs, or, more precisely, in the limit as n goes to infinity of the sequence of values of γ that achieve the maximum for each n. Therefore, to avoid divergence to infinity, we may discard terms from the log-likelihood function that do not affect the location of the maximum.

A satisfying way to do this is to use a form called the *log-likelihood ratio*, also denoted by the symbol Λ, and given by the general form

$$\Lambda(\gamma, \gamma') = \log\frac{p(x_1, \ldots, x_n \mid \gamma)}{p(x_1 \ldots, x_n \mid \gamma')},$$

or perhaps

$$\Lambda(\gamma) = \log\frac{p(x_1, \ldots, x_n \mid \gamma)}{p(x_1, \ldots, x_n)}$$

where $p(x_1 \ldots, x_n)$ is a convenient reference probability distribution, sometimes corresponding to noise only. The purpose in introducing either form of the log-likelihood ratio is to have a form that remains finite as n goes to infinity. The maximum-likelihood estimate is then that value of γ maximizing $\Lambda(\gamma)$, or equivalently, that value for which $\Lambda(\gamma, \gamma')$ is nonnegative for all γ'.

We are now ready to consider the important instance of a log-likelihood function of a received noisy waveform $v(t)$, given by

$$v(t) = c(t, \gamma) + n(t),$$

where γ is an unknown vector parameter and $n(t)$ is additive gaussian noise. The log-likelihood function that we will derive for this waveform often will be written formally as

$$\Lambda(\gamma) = -\frac{1}{N_0} \int_{-\infty}^{\infty} [v(t) - c(t, \gamma)]^2 dt.$$

However, as it is written, the integral is infinite because the noise has infinite energy, so the formula with infinite limits can only be understood symbolically. We shall derive the formula only for finite observation time and only for white noise. Nonwhite noise can be converted to white noise by the use of a whitening filter, but not without violating the assumption of a finite observation interval. The use of a whitening filter can be justified under the assumption that the observation interval is very long in comparison with the response time of the whitening filter, so the transient effects at the edges of the interval have a negligible effect. A more elegant, and more advanced treatment of the case of nonwhite noise would use the methods of functional analysis to reach the same conclusion rigorously.

We shall want to replace the waveform $v(t)$ with a finite-dimensional vector of samples having independent noise so that Theorem 7.1.1 can be used. In creating this vector, we suppose that $v(t)$ has been passed through a whitening filter to whiten the noise.

Proposition 7.2.1 *Given the reference signal $c(t, \gamma)$ depending on the parameter γ and the observed signal $v(t)$ supported on $[-T/2, T/2]$ received in white gaussian noise $n(t)$, the log-likelihood function for γ is the function of γ given by*

$$\Lambda(\gamma) = -\frac{1}{N_0} \int_{-T_0/2}^{T_0/2} [v(t) - c(t, \gamma)]^2 dt.$$

Proof   The received signal is given by

$$v(t) = c(t, \gamma) + n(t)$$

for some actual value of γ. Consider the Fourier series expansion of $v(t)$ on the interval $[-T/2, T/2]$. The complex Fourier expansion coefficients $\ldots, V_{-1}, V_0, V_1, \ldots$ are an infinite number of gaussian random variables with means $C_k, \ldots, C_{-1}, C_0, C_1, \ldots$ given by the Fourier coefficients of $c(t, \gamma)$ on the interval $[-T/2, T/2]$. Because the noise is white, the random variables $V_{-K}, \ldots, V_K$ are independent and identically distributed. The probability density function of these $2K + 1$ random variables is

$$p(V_{-K}, \ldots, V_K \mid c(t, \gamma)) = [2\pi\sigma^2]^{-(2K+1)} \prod_{k=-K}^{K} e^{-[V_k - C_k]^2/2\sigma^2}$$

and $\sigma^2 = N_0/2$. For a set of samples, we use as the likelihood statistic

$$\Lambda(\boldsymbol{\gamma}) = \log B p(V_{-K}, \ldots, V_K \mid c(t, \boldsymbol{\gamma}))$$

conditional on the expected complex value $c(t, \boldsymbol{\gamma})$ of the received signal, which may depend on the vector parameter γ, and where we use the constant

$$B = [2\pi\sigma^2]^{2K+1}$$

to normalize the log-likelihood function.

With this choice of normalizing constant,

$$\Lambda(C_{-K}, \ldots, C_K) = -\sum_{k=-K}^{K} [V_k - C_k]^2 / N_0.$$

Now let K go to infinity and use the energy theorem for Fourier series[1] to write

$$\lim_{K \to \infty} \Lambda(C_{-K}, \ldots, C_K) = -\frac{1}{N_0} \int_{-T_0/2}^{T_0/2} |v(t) - c(t, \gamma)|^2 dt$$

which completes the proof of the theorem.   ■

The log-likelihood function given in the theorem goes to infinity as T_0 goes to infinity. From a practical point of view, it is always possible to find a way around this mathematical difficulty by regarding the expression symbolically and its consequences as a guide.

7.3   Maximum-likelihood demodulation

In this section, we shall derive the maximum-likelihood demodulator for various modulation waveforms in the absence of intersymbol interference. The maximum-likelihood demodulator minimizes the probability of demodulated block error. It is

---

[1] The energy theorem for Fourier series

$$\int_{-T/2}^{T/2} |v(t)|^2 dt = \sum_{k=-\infty}^{\infty} |V_k|^2$$

is a special case of Parseval's formula for Fourier series

$$\int_{-T/2}^{T/2} v(t) u^*(t) dt = \sum_{k=-\infty}^{\infty} V_k U_k^*.$$

not the demodulator that minimizes the bit error rate. This is because the maximum-likelihood demodulator minimizes the probability of block (or message) error, not the probability of bit (or symbol) error. It may be that one could increase the message error rate, but reduce the number of bit errors in each incorrect message in such a way that the bit error rate decreases.

We shall defer the detailed study of maximum-likelihood demodulation when there is intersymbol interference to Section 7.4 where we shall see that the maximum-likelihood demodulator for sequences need not be the same as many of the demodulators studied in earlier chapters. In this section, we shall specialize the maximum-likelihood demodulator to the simple case in which the channel symbols do not interfere and the noise is additive white gaussian noise. We shall see that the maximum-likelihood demodulator then reduces to the matched-filter demodulator. Hence one result of this section is to establish that some demodulators based on the matched filter that were studied in earlier chapters are optimal in gaussian noise.

The maximum-likelihood principle will now be applied to the task of demodulation. As usual, the model of a digital communication waveform that we shall study next, in Theorem 7.3.1, is

$$v(t) = c(t) + n(t)$$

$$= \sum_{\ell=0}^{n-1} a_\ell s(t - \ell T) + n(t)$$

where $s(t)$ is a general pulse and a_ℓ takes values in some real or complex signal constellation. For the moment, the pulse is not necessarily one that satisfies a Nyquist condition. If there is dispersion in the channel, it is included in the definition of the pulse $s(t)$. The received pulse may exhibit intersymbol interference, perhaps because it was generated that way in the transmitter, or perhaps because of dispersion in the channel.

Theorem 7.3.1 *The sequence of (complex) samples at times ℓT at the output of the matched filter $s^*(-t)$ is a sufficient statistic for demodulation of any waveform of the form*

$$c(t) = \sum_{\ell=0}^{n-1} a_\ell s(t - \ell T)$$

in additive white gaussian noise.

Proof   It suffices to prove that the likelihood statistic can be reconstructed from the stated sequence of matched-filter outputs. As implied by Corollary 7.1.2, maximizing the likelihood function $\Lambda(a_0, \ldots, a_{n-1})$ in gaussian noise is equivalent to minimizing the euclidean distance $d(v(t), c(t))$. Thus,

$$d(v(t), c(t)) = \int_{-\infty}^{\infty} |v(t) - c(t)|^2 dt$$

$$= \int_{-\infty}^{\infty} |v(t)|^2 dt - 2\text{Re}\left[\int_{-\infty}^{\infty} v(t)c^*(t)dt\right] + \int_{-\infty}^{\infty} |c(t)|^2 dt$$

$$= \int_{-\infty}^{\infty} |v(t)|^2 dt - 2\text{Re}\left[\sum_{\ell=0}^{n-1} a_\ell^* \int_{-\infty}^{\infty} v(t)s^*(t - \ell T)dt\right]$$

$$+ \sum_{\ell=0}^{n-1}\sum_{\ell'=0}^{n-1} a_\ell a_{\ell'}^* \int_{-\infty}^{\infty} s(t - \ell T)s^*(t - \ell' T)dt$$

$$= \int_{-\infty}^{\infty} |v(t)|^2 dt - 2\text{Re}\left[\sum_{\ell=0}^{n-1} a_\ell^* u_\ell\right] + \sum_{\ell=0}^{n-1}\sum_{\ell'=0}^{n-1} a_\ell a_{\ell'}^* r_{\ell'-\ell}$$

where

$$u_\ell = u(\ell T) = \int_{-\infty}^{\infty} v(t)s^*(t - \ell T)dt$$

and

$$r_\ell = r(\ell T) = \int_{-\infty}^{\infty} s(t)s^*(t - \ell T)dt$$

$$= r^*(-\ell T) = r_{-\ell}^*.$$

Given a received $v(t)$, the first term of the last line on the right is a constant that is independent of the hypothesized data sequence so it need not be included. A suitable redefinition of the likelihood statistic to be maximized is

$$\Lambda(a_0, \ldots, a_{n-1}) = 2\text{Re}\left[\sum_{\ell=0}^{n-1} a_\ell^* u_\ell\right] - \sum_{\ell=0}^{n-1}\sum_{\ell'=0}^{n-1} a_\ell a_{\ell'}^* r_{\ell'-\ell},$$

which depends on the received signal $v(t)$ only through the samples u_ℓ, so these samples form a sufficient statistic. This completes the proof of the theorem. ∎

The theorem only provides a sufficient method of forming a sufficient statistic, not a necessary method. The theorem does not preclude the possibility of other statistics that are just as good, but none is better. Proposition 7.2.1 indicates that, to achieve an optimal demodulator, the matched filter should be designed for the received pulse (which may include the effects of dispersion in the channel), rather than for the transmitted pulse.

We shall now apply the maximum-likelihood principle to the case of a BPSK waveform in white noise in the absence of intersymbol interference. In the next section we shall repeat the discussion in the presence of intersymbol interference. For the demodulation of BPSK, there is no purpose in dealing with the quadrature component, so we may treat the signal as a real baseband signal. The a_ℓ are unknown parameters equal to $\pm A$. The log-likelihood function is maximized by the sequence of values of a_ℓ that maximize the likelihood statistic that was given at the end of the preceding proof.

Theorem 7.3.2 *In additive white gaussian noise, the maximum-likelihood demodulator for the* BPSK *waveform with no intersymbol interference is to pass the received signal through the matched filter and detect the sign of the output at each sampling instant ℓT.*

Proof For this special case, the double summation in the proof of Theorem 7.3.1 reduces to a single summation because of the assumption of the theorem. That is,

$$d(v(t), c(t)) = \int_{-\infty}^{\infty} v^2(t)dt - 2\sum_{\ell=0}^{n-1} a_\ell u_\ell + \sum_{\ell=0}^{n-1} a_\ell^2 \int_{-\infty}^{\infty} s^2(t)dt.$$

The first term on the right does not depend on a_ℓ. Because $a_\ell = \pm A$, the last term does not depend on a_ℓ either. To maximize $\Sigma_\ell a_\ell u_\ell$, the estimate of a_ℓ should have the same sign as u_ℓ for each ℓ. Because any sequence of data bit values is allowed, this is the maximum-likelihood demodulator. ∎

Notice that Theorem 7.3.2 derives, as an optimal demodulator, the same coherent matched-filter demodulator that was studied in Chapter 3, but now starting from a stronger definition of optimality. The matched-filter demodulator was derived in Chapter 3 under a simpler setting for the problem. There the demodulator was specified to have the structure of a linear filter followed by a threshold, and the matched filter was shown to maximize signal-to-noise ratio over the class of demodulators with this special structure. In Theorem 7.3.2, the structure itself is developed during the proof of optimality. On the other hand, Theorem 7.3.2 applies only to white gaussian noise, while Theorem 3.1.2 applies to any stationary noise.

Theorem 7.3.3 *In additive white gaussian noise, the maximum-likelihood demodulator for QAM signaling with complex signal constellation $\{c_m : m = 0, \ldots, M-1\}$, and with no intersymbol interference, is to pass the signal through a matched filter and to choose that m for which c_m is closest in euclidean distance to the output of the filter at the sampling instant.*

Proof As in the proof of Corollary 7.1.2, maximizing the likelihood function $\Lambda(a_0, \ldots, a_{n-1})$ is equivalent to minimizing the distance

$$d(v(t), c(t)) = \int_{-\infty}^{\infty} |v(t)|^2 dt - 2\mathrm{Re}\left[\sum_{\ell=0}^{n-1} a_\ell^* u_\ell\right] + \sum_{\ell=0}^{n-1} |a_\ell|^2 \int_{-\infty}^{\infty} s^2(t)dt$$

by choice of the sequence of a_ℓ where, for each ℓ, $a_\ell \in \{c_m : m = 0, \ldots, M - 1\}$. But $\int_{-\infty}^{\infty} |v(t)|^2 dt$ is a constant throughout the calculation and can be replaced by the constant $\sum_{\ell=0}^{n-1} |u_\ell|^2$ without affecting which m achieves the minimum. Thereafter, with the implicit understanding that the pulse $s(t)$ is normalized so that $\int_{-\infty}^{\infty} s^2(t) dt = 1$, this is the same as maximizing

$$d(\boldsymbol{u}, \boldsymbol{a}) = \sum_{\ell=0}^{n-1} |u_\ell|^2 - 2\mathrm{Re}\left[\sum_{\ell=0}^{n-1} a_\ell^* u_\ell\right] + \sum_{\ell=0}^{n-1} |a_\ell|^2$$

$$= \sum_{\ell=0}^{n-1} |u_\ell - a_\ell|^2.$$

This is maximized by choosing, for each ℓ, the estimate $\widehat{a}_\ell$ equal to that element of the signal constellation that minimizes $|u_\ell - a_\ell|^2$.   ∎

The maximum-likelihood principle also can be used to provide an optimality condition for a noncoherent demodulator of an M-ary orthogonal waveform. The assumption is that the phase error is independent from symbol to symbol and there is no intersymbol interference, and therefore nothing is to be gained by demodulating more than one symbol at a time. The following is the optimality theorem for noncoherent demodulation of a single symbol of an M-ary orthogonal signaling waveform.

Theorem 7.3.4 *In additive white gaussian noise, when the carrier phase is unknown, the maximum-likelihood demodulator for a single symbol of an M-ary orthogonal signaling waveform with no intersymbol interference is a bank of matched filters, followed by a threshold on the magnitude of the output at the sampling instant.*

Proof Because, by assumption, there is no intersymbol interference, and orthogonal signaling pulses are defined to be orthogonal to all of their translates by ℓT, it suffices to consider only a single transmitted pulse. Let $s_m(t)$ for $m = 0, \ldots, M - 1$ denote the M orthogonal pulses. For a single transmitted pulse, the received signal at complex baseband is

$$v(t) = s_m(t)e^{j\theta} + n(t).$$

The proof consists of writing down the log-likelihood function as a function of the unknown index m and the unknown phase θ, and then maximizing over both m and θ. But this reduces to choosing m and θ so as to minimize the squared euclidean distance:

$$d^2(v(t), s_m(t)e^{j\theta}) = \int_{-\infty}^{\infty} |v(t) - s_m(t)e^{j\theta}|^2 dt$$

$$= \int_{-\infty}^{\infty} |v(t)|^2 dt - 2\mathrm{Re}\left[e^{-j\theta}\int_{-\infty}^{\infty} v(t)s_m^*(t) dt\right] + \int_{-\infty}^{\infty} |s_m(t)|^2 dt.$$

The first term does not depend on m or θ. Because we have defined an M-ary orthogonal alphabet to have the same energy in each of its pulses, the last term does not depend on m, nor on θ. The middle term has the form of the number $\text{Re}[ze^{-j\theta}]$, which is maximized over θ by choosing $e^{-j\theta} = z^*/|z|$. Consequently,

$$\min_{\theta} d^2(v(t), s_m(t)e^{j\theta}) = \int_{-\infty}^{\infty} |v(t)|^2 dt - 2\left|\int_{-\infty}^{\infty} v(t)s_m^*(t)dt\right| + E_p,$$

and the maximum-likelihood demodulator chooses m to minimize the right side. That is, choose that m for which the magnitude of the mth matched-filter output is largest. ∎

7.4   Maximum-likelihood sequence demodulation

Maximum-likelihood demodulation of a BPSK waveform in the absence of intersymbol interference was studied in the previous section. Now we turn to the demodulation of BPSK in the presence of intersymbol interference. Now, the log-likelihood function for a block of length n cannot itself be separated in a simple way into a sum of noninteracting log-likelihood functions, as was done in the proof of Theorem 7.3.2. In its place, we will formulate an alternative likelihood statistic that can be so separated in a simple way into an appropriate sum of terms.

We begin with the likelihood statistic for the general case of a QAM waveform

$$c(t) = \sum_{\ell=0}^{n-1} a_\ell s(t - \ell T)$$

received in additive white gaussian noise, which was developed within the proof of Theorem 7.3.1 as

$$\Lambda(a_0, \ldots, a_{n-1}) = 2\text{Re}\left[\sum_{\ell=0}^{n-1} a_\ell^* u_\ell\right] - \sum_{\ell=0}^{n-1}\sum_{\ell'=0}^{n-1} a_\ell a_{\ell'}^* r_{\ell'-\ell}$$

where the first term involves the matched-filter output u_ℓ, given by

$$u_\ell = u(\ell T) = \int_{-\infty}^{\infty} v(t)s^*(t - \ell T)dt$$

and the second term involves r_ℓ, given by

$$r_\ell = r(\ell T)$$

$$= \int_{-\infty}^{\infty} s(t)s^*(t - \ell T)dt$$

$$= r^*(-\ell T) = r_{-\ell}^*$$

which does not involve the received signal $v(t)$. Because $\Lambda(a_0, \ldots, a_{n-1})$ can be recovered from the sequence of the matched-filter output samples, the sequence of the matched-filter output samples is a sufficient statistic for the demodulation of a QAM waveform, whether or not there is intersymbol interference.

The maximum-likelihood demodulator finds the vector $\widehat{a} = (a_0, \ldots, a_{n-1})$, of block-length n, each of whose components is in the QAM signal constellation, that maximizes $\Lambda(a_0, \ldots, a_{n-1})$. For the special case of BPSK, all of the components of the n-vector $(a_0, \ldots, a_{n-1})$ must be $\pm A$. One way of finding this maximum-likelihood solution is to substitute, in turn, each of the 2^n such binary vectors (or, in the general case, each of the M^n such M-ary vectors) into $\Lambda(a_0, \ldots, a_{n-1})$ and then search for the largest value of $\Lambda(a_0, \ldots, a_{n-1})$. However, this process is rather tedious, even intractable, for large n. Further, in most applications, n is not fixed but is ever-increasing. In such applications, one does not want a block-organized computational structure for the maximum-likelihood demodulator. One wants a recursive computational structure that will demodulate each bit (or symbol) in turn after a fixed computational delay. The Viterbi algorithm has this kind of structure, sliding along the incoming senseword and producing output bits in sequence after some fixed latency. We shall show that, in the presence of intersymbol interference and white gaussian noise, the maximum-likelihood principle can be formulated in such a way that the Viterbi algorithm provides a fast computational algorithm with the desired properties. The only condition we impose is that the intersymbol interference has finite duration. That is, for some integer v, called the *constraint length*, $r_\ell = 0$ for $\ell > v$ and for $\ell < 0$. This means that a symbol will have nonnegligible intersymbol interference only from the v previous data symbols. Consequently, we shall refer to the value of the previous v symbols as the *state* of the channel. Specifically, define the state as the v-tuple consisting of the previous v data symbols

$$\sigma_\ell = (\sigma_{\ell-v}, \sigma_{\ell-v+1}, \ldots, \sigma_{\ell-1}).$$

For QAM signaling, each symbol takes values in a finite signal constellation with $M = 2^k$ points so there are a finite number, 2^{kv}, of states. As ℓ increases by one, the state σ_ℓ changes to state $\sigma_{\ell+1}$. The state transitions define a trellis on the 2^{kv} states. The cumulative likelihood statistic for the subblock $(a_0, \ldots, a_\ell)$ is nondecreasing as ℓ increases.

The next theorem provides a formulation of the likelihood statistic that has a regular additive structure and can be fitted to the structure of the Viterbi algorithm. Let $\lambda_{\ell+1}(\sigma_\ell, \sigma_{\ell+1})$ denote the increase in the likelihood statistic in going from state σ_ℓ to state $\sigma_{\ell+1}$. That is, define $\lambda_{\ell+1}(\sigma_\ell, \sigma_{\ell+1})$ by

$$\Lambda(a_0, \ldots, a_{n-1}) = \Lambda(a_0, \ldots, a_{n-2}) + \lambda_{\ell+1}(\sigma_\ell, \sigma_{\ell+1}).$$

Theorem 7.4.1 *The likelihood statistic for a QAM signaling waveform in additive white gaussian noise with modulation levels denoted a_ℓ, can be written as the sum*

$$\Lambda(a_0, \ldots, a_{n-1}) = \sum_{\ell=0}^{n-1} \lambda_{\ell+1}(\sigma_\ell, \sigma_{\ell+1})$$

where

$$\lambda_{\ell+1}(\sigma_\ell, \sigma_{\ell+1}) = 2\text{Re}[a_\ell^* u_\ell] - 2\text{Re}\left[a_\ell^* \sum_{k=\ell-v}^{\ell-1} a_k r_{\ell-k}\right] - |a_\ell|^2 r_0,$$

with the understanding that $a_\ell = 0$ for ℓ less than zero and where u_ℓ and r_ℓ are the actual and expected values of the matched-filter output, respectively.

Proof   The equation for $\Lambda(a_0, \ldots, a_{n-1})$ was given at the start of the section. We will expand the equation for $\Lambda(a_0, \ldots, a_\ell)$ by writing the sum from 0 to ℓ as a sum from 0 to $\ell - 1$, then adding the terms at ℓ explicitly. This is

$$\Lambda(a_0, \ldots, a_\ell) = 2\text{Re}\left[\sum_{k=0}^{\ell} a_k^* u_k\right] - \sum_{k=0}^{\ell} \sum_{k'=0}^{\ell} a_k a_{k'}^* r_{k'-k}$$

$$= 2\text{Re}\left[\sum_{k=0}^{\ell-1} a_k^* u_k\right] - \sum_{k=0}^{\ell-1} \sum_{k'=0}^{\ell-1} a_k a_{k'}^* r_{k'-k}$$

$$+ 2\text{Re}[a_\ell^* u_\ell] - a_\ell \sum_{k'=0}^{\ell-1} a_{k'}^* r_{k'-\ell} - a_\ell^* \sum_{k=0}^{\ell-1} a_k r_{\ell-k} - |a_\ell|^2 r_0$$

$$= \Lambda(a_0, \ldots, a_{\ell-1}) + 2\text{Re}[a_\ell^* u_\ell] - 2\text{Re}\left[a_\ell^* \sum_{k=\ell-v}^{\ell-1} a_k r_{\ell-k}\right] - |a_\ell|^2 r_0.$$

Now use the facts that $r_k = r_{-k}^*$ and $r_k = 0$ for $k > v$ to rewrite this as

$$\Lambda(a_0, \ldots, a_\ell) = \Lambda(a_0, \ldots, a_{\ell-1}) + 2\text{Re}[a_\ell^* u_\ell] - 2\text{Re}\left[a_\ell^* \sum_{k=\ell-v}^{\ell-1} a_k r_{\ell-k}\right] - |a_\ell|^2 r_0$$

$$= \Lambda(a_0, \ldots, a_{\ell-1}) + \lambda_\ell(\sigma_\ell, \sigma_{\ell+1}).$$

Hence the same reduction applies to $\Lambda(a_0, \ldots, a_{\ell+1})$ and so forth. Because $\lambda_{\ell+1}(\sigma_\ell, \sigma_{\ell+1}) = 0$ for ℓ less than zero, the theorem is proved.   ∎

The likelihood statistic of Theorem 7.4.1 takes the form of a sum of n terms, the ℓth of which depends only on the states at times ℓ and $\ell + 1$ and the ℓth sample u_ℓ

of the matched-filter output. This is the structure that is needed to apply the Viterbi algorithm. To this end, we use the terminology of "distance". With a change in sign, we refer to the negative of the likelihood statistic as a "distance" between an observed sequence and a code sequence. On the trellis with states σ_ℓ, define the *branch distance* as $d_{\ell+1} = -\lambda_{\ell+1}(\sigma_\ell, \sigma_{\ell+1})$ on the branch between states $\sigma_\ell = (a_{\ell-\nu}, \ldots, a_{\ell-1})$ and $\sigma_{\ell+1} = (a_{\ell+1-\nu}, \ldots, a_\ell)$. The *path distance*, then, between any path through the trellis and the received sequence of matched-filter samples is the sum of branch distances on the traversed path. To maximize the likelihood statistic, we must minimize the path distance by the choice of a path through the trellis.

The likelihood statistic is expressed as a sum of branch distances in Theorem 7.4.1. Then the term $|a_\ell^2|r_0$ in Theorem 7.4.1 is a constant and need not be included in the branch distance. We shall also drop a factor of two to obtain the modified branch distance

$$d_{\ell+1}(\sigma_\ell, \sigma_{\ell+1}) = \mathrm{Re}\left[a_\ell \left(-u_\ell + \sum_{k=\ell-\nu}^{\ell-1} a_k r_{\ell-k} \right) \right]$$

which conveys all relevant information, and is more compact.

We shall illustrate the use of Theorem 7.3.1 by developing the trellis shown in Figure 7.1 for the case of BPSK using amplitudes $a_\ell = \pm A$ with intersymbol interference of constraint length one. This is similar to the example studied in Section 4.3. In that section, we used the Viterbi algorithm to demodulate a BPSK waveform in the presence of intersymbol interference caused by channel dispersion, but we made no claims about the optimality of this procedure. Now, we shall describe an alternative demodulator for BPSK, also using the Viterbi algorithm, that is optimal in the sense of the maximum-likelihood principle. Because $\nu = 1$ and the data is real, the branch distance becomes

$$d_{\ell+1} = a_\ell(a_{\ell-1}r_1 - u_\ell).$$

The sum of the branch distances is to be minimized by the choice of path through the trellis, as defined by the a_ℓ.

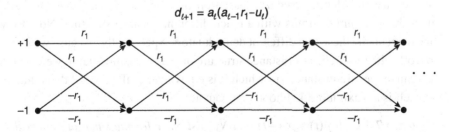

Figure 7.1. A trellis for a maximum-likelihood demodulator.

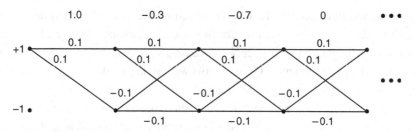

Figure 7.2. A trellis for a specific example.

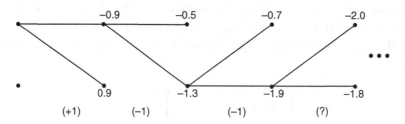

Figure 7.3. Example of a Viterbi demodulator.

The trellis shown in Figure 7.1 is for a maximum-likelihood demodulator for a BPSK waveform with constraint length ν equal to one. The term $a_{\ell-1}r_\ell$ labels the branches, and the branch distance is computed from u_ℓ by the indicated equation. Notice in this example that the branch distances are to be combined by direct addition. This is in marked contrast to the example of Figure 4.5 in which squares of distances were added.

Figure 7.2 shows a specific numerical example using $r_1 = 0.1$. Figure 7.3 shows the surviving paths for that example at each recursion of the Viterbi algorithm.

Let us pause here to summarize our present position before we move on to conclude the chapter. We saw earlier that if $p(t) * p^*(-t)$ is a Nyquist pulse where $p(t)$ is the transmitted pulse, then the sampled output of the matched filter $p^*(-t)$ will have uncorrelated noise components, but if there is dispersion in the channel, there will be intersymbol interference. This intersymbol interference can be accommodated by using the Viterbi demodulator as described in Section 4.3 to search for the minimum euclidean distance path. However, in Theorem 7.4.1, we saw that the optimum demodulator actually has a different structure, based on the matched filter $s^*(-t)$ and a Viterbi algorithm searching a trellis with a much different distance structure. Now we would like to reconcile these two different demodulators. Specifically, because we may prefer to work with the euclidean-distance structure of the first demodulator, we would like to determine the circumstances in which it is equivalent to the second demodulator, and so itself is a maximum-likelihood demodulator.

Theorem 7.4.2 *If $p(t) * p^*(-t)$ is a Nyquist pulse for the sampling interval T, and $s(t) = \sum_{i=0}^{\nu} g_i p(t - iT)$, then the Nyquist samples of the output of the filter $p^*(-t)$ are* *a sufficient statistic for estimating the data symbols a_ℓ of the QAM waveform*

$$c(t) = \sum_{\ell=-\infty}^{\infty} a_\ell s(t - \ell T)$$

in additive white gaussian noise.

Proof   Because we know that the samples $u(\ell T)$ of the output of the filter $s^*(-t)$ form a sufficient statistic, it suffices to show that these samples can be computed from the sampled outputs of the filter $p^*(-t)$. Thus,

$$
\begin{aligned}
u(\ell T) &= \int_{-\infty}^{\infty} v(t)s^*(t - \ell T)dt \\
&= \int_{-\infty}^{\infty} v(t) \sum_{\ell=0}^{\nu} g_i p(t - \ell T - iT)dt \\
&= \sum_{i=0}^{\nu} g_i \int_{-\infty}^{\infty} v(t)p(t - \ell T - iT)dt \\
&= \sum_{i=0}^{\nu} g_i u'(\ell T + iT)
\end{aligned}
$$

where $u'(\ell T)$ is the ℓth output sample of the filter $p^*(-t)$. Thus $u(\ell T)$ can be computed from the sequence $u'(\ell T)$.   ∎

Because of this theorem, for such a received pulse $s(t)$, we are free to use the outputs of the matched filter $p^*(-t)$ with no loss of information. Then instead of following Theorem 7.3.1, we can formulate the maximum-likelihood demodulator by using these alternative statistics. This maximum-likelihood demodulator will be the Viterbi demodulator. Even though the structure may look different, Theorem 7.4.2 allows us to conclude that this demodulator is equivalent to the demodulator of Theorem 7.3.1, and usually it is to be preferred because of its more convenient structure.

The essential thought behind Theorem 7.4.2 is also important when there is a precoder between the datastream and the channel. For such an application, there may be a modulo-q operation in the precoder. The following theorem confirms that this modulo-q operation does not invalidate the conclusion of Theorem 7.4.2.

Theorem 7.4.3   *Let*

$$c_\ell = \sum_{i=0}^{\nu} g_i a_{\ell-i} \qquad (\bmod\ q)$$

where g_i and a_i are complex integers modulo q, and

$$c(t) = \sum_{\ell=-\infty}^{\infty} c_\ell p(t - \ell T)$$

where $p(t) * p^*(-t)$ is a Nyquist pulse for the sampling interval T. Then in additive white noise, the output samples of the matched filter $p^*(-t)$ are a sufficient statistic for demodulating the data sequence a_ℓ.

Proof   This is essentially the same as the proof of Theorem 7.4.2.   ∎

7.5   Noncoherent combining of diversity signals

A *diversity communication system* sends several copies of the same message to a user through several different channels so that even if all channels but one are broken, the message will still arrive at the receiver. Diversity may be at the level of an entire message with L diversity channels. The same full message is sent independently through the L diversity channels (or through one channel at L different times). Figure 7.4 shows a system with two diversity channels. If neither channel is broken or impaired, then the user receives the same message twice. The duplicate copy of the message gives no additional value.

Diversity transmission can also be at the symbol level; then it is called *symbol splitting*. Each symbol is transmitted through each of L diversity channels. The demodulator sees the noisy signal coming out of all L channels and, by appropriately combining the multiple copies, must decide which symbol was transmitted at time ℓT. There are many ways to design this demodulator, and the choice will depend on the kinds of things that might go wrong on the channel. One may individually demodulate all L channel outputs and vote on the correct symbol, but this can be a poor approach if all channels are equally noisy. Alternatively, one may integrate the set of all L channel output signals into a single statistic and demodulate based on that statistic. For example, the L

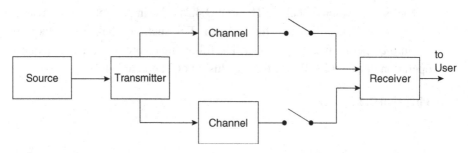

Figure 7.4. Diversity transmission.

matched-filter output samples could be added and the sum applied to a threshold, but this can be a poor approach if most channels are relatively noise-free and a few are very noisy. Between these two approaches are alternative methods. One prefers to use an approach that is robust. This is an approach whose performance depends only weakly on the disparity in the noise strengths of the channels.

The energy per symbol is E_s/L, and the output of each subchannel matched filter has a signal-to-noise ratio E_s/LN_0. If the probability of symbol error of the selected modulation waveform is written $p_{es}(E_b/N_0)$ as a function of E_b/N_0, then the probability of demodulation error in a single subchannel is $p_{es}(E_b/LN_0)$. When one expects that more than half of the subchannels are relatively noise free, one should demodulate every channel output and vote on the correct channel symbol. The vote is correct whenever more than half of the L subchannels are demodulated correctly.

If every channel has additive gaussian noise of the same power density spectrum $N_0/2$, the use of voting would have a severe energy penalty. In this case, the probability of demodulation failure after a majority vote is given by

$$
p_e = \sum_{\ell=\lfloor L/2 \rfloor}^{L} \binom{L}{\ell} p_{es}\left(\frac{E_b}{LN_0}\right)^{\ell} \left(1 - p_{es}\left(\frac{E_b}{LN_0}\right)\right)^{L-\ell}.
$$

If the demodulator accepts a plurality decision, then the probability of demodulation error will not be smaller than this expression. To get a general sense of the consequence of this equation, let L be even and approximate the sum by its dominant term. It then becomes

$$
p_e \approx \binom{L}{L/2} p_{es}\left(\frac{E_b}{LN_0}\right)^{(L-1)/2}.
$$

Now recall that for many of the kinds of modulation that were studied in Chapter 3, the probability of symbol error for large E_b/N_0 can be written as

$$
p_{es}\left(\frac{E_b}{N_0}\right) = Q\left(\sqrt{\beta\frac{E_b}{N_0}}\right)
$$

for some constant β that depends on the kind of modulation. Consequently, using the inequality $Q(x) < \frac{1}{2}e^{-x^2/2}$ as a coarse approximation, we can manipulate the approximate terms as follows:

$$
\left[p_{es}\left(\frac{E_b}{LN_0}\right)\right]^{(L-1)/2} \approx \left(e^{-\beta E_b/2LN_0}\right)^{(L-1)/2} \approx e^{-\beta E_b/4N_0}
$$

$$
\approx p_{es}\left(\frac{E_b}{2N_0}\right).
$$

Consequently,

$$
p_e \sim \binom{L}{L/2} p_{es}\left(\frac{E_b}{2N_0}\right).
$$

The appearance of a factor of two in the argument of p_{es} suggests that making a hard decision in each individual subchannel results in an energy loss of at least 3 dB. The first term indicates even more degradation in the probability of error. Indeed, an exact analysis would show an even larger degradation.

For binary modulation, because the total energy expenditure is E_b joules per bit, each of the L subchannels expends E_b/L joules per bit. If every subchannel were a white gaussian-noise channel with noise power density spectrum $N_0/2$ watts per hertz, one would hope to communicate with the same E_b/N_0 as would be needed by a single channel. Indeed, this can be done simply by adding the L outputs of the L matched filters to form a single statistic. This is referred to as *coherent integration*. Because the L noise samples are independent, their variances add when the samples are added, so the noise power in the composite statistic will increase by L while the signal amplitude increases by L. The composite signal amplitude is $L\sqrt{E_b/L}$, so the energy of the composite signal is LE_b. Because the signal-to-noise ratio of the sum is just what it would be if a single channel were used, there is no energy loss in using coherent integration whenever all subchannel outputs are available and equally noisy. However, in such cases, there is no point in using a diversity system. It would be enough to use all of the available energy on a single channel.

When the diversity channels are passband channels with unknown and independent phase angles, the decision statistic is formed by the method called *noncoherent integration*. The absolute value, or a monotonic function of the absolute value, of the output of each matched filter is computed, and these values are added to form the decision statistic.

The optimum monotonic function for this decision statistic for L-ary diversity, M-ary orthogonal, noncoherent signaling will be derived from the maximum-likelihood principle. Let $s_{m\ell}(t)$ for $m = 0, \ldots, M - 1$ be a set of M orthogonal pulses of equal energy and unknown phase. At each sample time, a particular pulse $s_m(t)$, representing k data bits, is transmitted on all L channels. The waveform on the ℓth diversity channel is $c_\ell(t) = \sum_{i=-\infty}^{\infty} s_{m_i}(t - iT)$, which does not depend on ℓ. Let $x_{m\ell}$ be the magnitude of the output of the matched filter for the pulse $s_{m\ell}(t)$. Let $p_N(x_{m\ell})$ be the probability density function on $x_{m\ell}$, given that the input to the matched filter is noise only, and let $p_S(x_{m\ell})$ be the probability density function, given that the input to the matched filter is signal plus noise. Because the pulses have equal energy, these probability density functions do not depend on m or ℓ. The log-likelihood function is

$$
\Lambda(m) = \log\left[\prod_{\ell=0}^{L-1} p_S(x_{m\ell}) \prod_{m'\neq m}\prod_{\ell=0}^{L-1} p_N(x_{m'\ell})\right]
$$

$$
= \log\prod_{\ell=0}^{L-1}\frac{p_S(x_{m\ell})}{p_N(x_{m\ell})}\left[\prod_{m'=0}^{M-1}\prod_{\ell=0}^{L-1} p_N(x_{m'\ell})\right].
$$

The bracketed term in the second line is independent of m, so it is enough to maximize the redefined likelihood statistic

$$
\Lambda(m) = \log\prod_{\ell=0}^{L-1}\frac{p_S(x_{m\ell})}{p_N(x_{m\ell})}
$$

by choice of m.

Theorem 7.5.1 *Given an M-ary orthogonal signaling scheme with L-ary nonco-herent diversity, all pulses having the same energy $E_p = A^2$, and all channels having additive white gaussian noise of two-sided power density spectrum $N_0/2$, the maximum-likelihood demodulator chooses that m for which*

$$
\Lambda(m) = \sum_{\ell=0}^{L-1}\log I_0\left(\frac{2Ax_{m\ell}}{N_0}\right)
$$

is largest, where $x_{m\ell}$ is the absolute value of the output sample of the m_ℓth matched filter.

Proof To evaluate the maximum-likelihood demodulator, we should write the likeli-hood function with the set of phase angles included as unknown parameters. However, because each phase term occurs in only one pulse and the pulses are orthogonal, the maximization with respect to phases will separate into individual maximizations lead-ing to the prescription that only the absolute value of each matched filter is used. Consequently, we begin the mathematical treatment at this point.

Let $p_N(x_{m\ell})$ for $m = 0, \ldots, M-1$ and $\ell = 0, \ldots, L-1$ be the rayleigh density function describing the magnitude of the matched-filter output when m is not the trans-mitted symbol; let $p_S(x_{m\ell})$ for $m = 0, \ldots, M-1$ and $\ell = 0, \ldots, L-1$ be the ricean density function describing this output when m is the transmitted symbol. Then given $x_{m\ell}$ for $m = 0, \ldots, M-1$, and $\ell = 0, \ldots, L-1$, we choose $\widehat{m}$ as the value of m for which

$$\Lambda(m) = \log \prod_{\ell=0}^{L-1} \frac{p_S(x_{m\ell})}{p_N(x_{m\ell})}$$

$$= \log \prod_{\ell=0}^{L-1} \left[\frac{x_{m\ell} e^{-(x_{m\ell}^2+A^2)/2\sigma^2}}{\sigma^2} I_0\left(\frac{Ax_{m\ell}}{\sigma^2}\right) \right] \Bigg/ \frac{x_{m\ell} e^{-x_{m\ell}^2/2\sigma^2}}{\sigma^2}$$

$$= \log \left[\prod_{\ell=0}^{L-1} e^{-A^2/2\sigma^2} I_0\left(\frac{Ax_{m\ell}}{\sigma^2}\right) \right]$$

is maximized. Replacing the logarithm of a product by the sum of logarithms and σ^2 by $N_0/2$ completes the proof of the theorem.   ■

An optimal receiver for a four-ary orthogonal alphabet with noncoherent two-way diversity is shown in Figure 7.5, in which $I_0(x)$ is the modified Bessel function. The structure of this receiver, using log-Bessel integration, is an immediate consequence of Theorem 7.5.1.

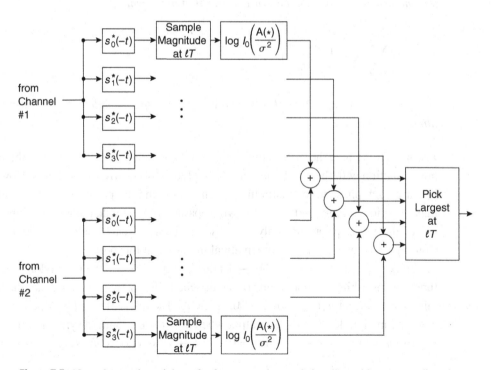

Figure 7.5. Noncoherent demodulator for four-ary orthogonal signaling with two-way diversity.

For small values of x, we may consider the series expansion

$$\log I_0(x) = \frac{1}{4}x^2 - \frac{1}{64}x^4 + \cdots .$$

Therefore the function $\log I_0(x)$ can be well approximated by a quadratic function of x when x is small. The approximation

$$\log I_0(x) \equiv \frac{1}{4}x^2$$

is sometimes used for all values of x because it does not much matter that the demodulator is suboptimal when the signal strength is large.

The performance of a noncoherent L-ary diversity system can be evaluated numerically using the diagram of Figure 7.6. Figure 7.7 shows the performance of an eight-ary orthogonal modulation system with L-ary symbol splitting and square-law combining in additive white gaussian noise. If the square-law combining nonlinearity were replaced by the optimal $\log I_0(x)$ combining nonlinearity, then the curves for $L = 2, 3$, and 4 would move slightly to the left. An inspection of Figure 7.7 shows that, to achieve a given probability of error when all channels have the same noise power, symbol splitting requires about 1 dB more energy than would simple orthogonal signaling. This loss of 1 dB against white gaussian noise is the cost for using a diversity system with noncoherent combining to protect against channel loss or disparate noise distributions.

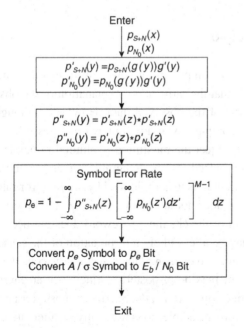

Figure 7.6. Computing the performance of a diversity system.

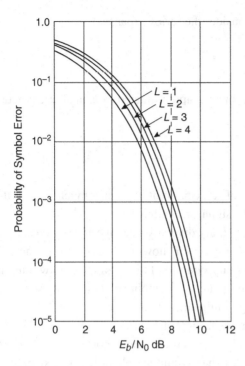

Figure 7.7. Performance of noncoherent eight-ary modulation with L-ary symbol splitting and square-law combining.

7.6   The maximum-posterior principle

The maximum-likelihood principle is a compelling principle of optimality. The fundamental premise of this principle is that the most likely explanation of the observed data is the most desirable explanation of the observed data. However, although this premise may be compelling, it is not compulsory. It does ignore other possible considerations of merit. One may choose to reject the maximum-likelihood principle, or to use it only when it fits one's purpose. Indeed, although the message error rate may be more meaningful, a communication system is usually judged by the bit error rate, and the maximum-likelihood demodulator does not attempt to minimize the bit error rate because the most likely interpretation of the channel senseword does not necessarily have the fewest bit errors. There may be a less likely interpretation of the senseword that has a lower bit error rate or possibly favors some other, secondary, consideration.

In this section, we shall introduce the maximum-posterior principle as an alternative method of inference. Often, for routine applications, the maximum-posterior principle is equivalent to the maximum-likelihood principle, so is not really an alternative. This is so, for example, in the case of decisions at the block level when all blocks are equally likely. However, when applied at the bit level, the maximum posterior principle can be quite different.

Discussion of the maximum-posterior principle makes repeated reference to the notions of marginal probability distributions and conditional probability distributions, which are defined as follows. Let $p(x, y)$ be any bivariate probability distribution[2] on two random variables X and Y. Then the *marginal distributions* on X and Y are given by

$$p(x) = \sum_y p(x, y)$$

$$p(y) = \sum_x p(x, y).$$

The *conditional distributions* on X and Y are given by

$$p(x|y) = p(x, y)/p(y)$$
$$p(y|x) = p(x, y)/p(x).$$

It follows from these definitions that $p(x, y) = p(x|y)p(y) = p(y|x)p(x)$ from which the Bayes formula

$$p(y|x) = \frac{p(x|y)p(y)}{\sum_y p(x|y)p(y)}$$

follows immediately.

Similar definitions apply to the multivariate probability distribution $p(x_1, \ldots, x_n)$ on the block of random variables $(X_1, \ldots, X_n)$. The block has the marginal distribution

$$p(x_\ell) = \sum_{x_1} \cdots \sum_{x_{\ell-1}} \sum_{x_{\ell+1}} \cdots \sum_{x_n} p(x_1, \ldots, x_n)$$

on random variable X_ℓ. Marginalization to several random variables, such as $p(x_\ell, x_{\ell'})$, is defined in a similar way. A variety of conditionals can then be defined, such as $p(x_1, \ldots, x_{\ell-1}, x_{\ell+1}, \ldots, x_n | x_\ell)$ or $p(x_1, x_2 | x_\ell)$.

The conditional probability of observation x when given the parameter γ is $p(x|\gamma)$. The task of inference of a parameter γ from an observation x when given the conditional probability $p(x|\gamma)$ led us to introduce the maximum-likelihood principle. However, if a probability distribution $p(\gamma)$ on γ is known, then one can write the joint distribution $p(x, \gamma) = p(\gamma)p(x|\gamma)$. By the Bayes formula, the posterior probability of parameter γ when given x is

$$p(\gamma|x) = \frac{p(\gamma)p(x|\gamma)}{\sum_\gamma p(\gamma)p(x|\gamma)}.$$

---

[2] As is customary, we carelessly use p to denote all of these probability distributions even though they are actually many different functions.

The probability $p(\gamma|\boldsymbol{x})$ is called the *posterior probability* of γ, and $p(\gamma)$ is called the *prior probability* of γ.

The maximum-posterior principle is a principle of inference that states that one should estimate γ so as to maximize $p(\gamma|\boldsymbol{x})$. Thus $\widehat{\gamma} = \max_\gamma p(\gamma|\boldsymbol{x})$. The maximum-posterior principle can be used only if $p(\gamma)$ is known and this is often not known. It is easy to see that the maximum-posterior principle is the same as the maximum-likelihood principle if $p(\gamma)$ is the uniform distribution. However, if $\boldsymbol{y}$ is a block of multiple parameters, then marginalization to one parameter may produce a fundamental change in the structure of the maximum-posterior estimate.

For example, if there are two parameters γ_1 and γ_2, then

$$p(\gamma_1, \gamma_2|\boldsymbol{x}) = \frac{p(\gamma_1, \gamma_2)p(\boldsymbol{x}|\gamma_1, \gamma_2)}{\sum_{\gamma_1\gamma_2} p(\gamma_1, \gamma_2)p(\boldsymbol{x}|\gamma_1, \gamma_2)}$$

is the posterior probability of the pair (γ_1, γ_2). The marginalization of $p(\gamma_1, \gamma_2|\boldsymbol{x})$ to γ_1 is the marginal posterior probability

$$p(\gamma_1|\boldsymbol{x}) = \sum_{\gamma_2} p(\gamma_1, \gamma_2|\boldsymbol{x}) = \frac{\sum_{\gamma_2} p(\gamma_1, \gamma_2)p(\boldsymbol{x}|\gamma_1, \gamma_2)}{\sum_{\gamma_1\gamma_2} p(\gamma_1, \gamma_2)p(\boldsymbol{x}|\gamma_1, \gamma_2)}$$

$$p(\gamma_2|\boldsymbol{x}) = \sum_{\gamma_1} p(\gamma_1, \gamma_2|\boldsymbol{x}) = \frac{\sum_{\gamma_1} p(\gamma_1, \gamma_2)p(\boldsymbol{x}|\gamma_1, \gamma_2)}{\sum_{\gamma_1\gamma_2} p(\gamma_1, \gamma_2)p(\boldsymbol{x}|\gamma_1, \gamma_2)}.$$

The componentwise maximum-posterior estimate then is

$$(\widehat{\gamma}_1, \widehat{\gamma}_2) = (\mathrm{argmax}_{\gamma_1} p(\gamma_1|\boldsymbol{x}), \mathrm{argmax}_{\gamma_2} p(\gamma_2|\boldsymbol{x}))$$

which, in general, is not equal to the blockwise maximum-posterior estimate

$$(\widehat{\gamma_1, \gamma_2}) = \left(\mathrm{argmax}_{\gamma_1\gamma_2} p(\gamma_1, \gamma_2|\boldsymbol{x})\right).$$

The comparison between blockwise and componentwise maximum-posterior estimation is at the heart of many applications of this principle. Indeed, even if γ_1 and γ_2 are independent so that $p(\gamma_1, \gamma_2) = p(\gamma_1)p(\gamma_2)$, a change in $p(\gamma_2)$, the prior on γ_2, does change the marginal $p(\gamma_1|\boldsymbol{x})$, and so the estimate $\widehat{\gamma}_1$ may change.

7.7   Maximum-posterior sequence demodulation

The *maximum-posterior demodulator* involves a form of demodulation that is straightforward to explain at the conceptual level, although it can be burdensome to implement. To recover the transmitted dataword, it demodulates each bit or symbol of the received senseword individually, but within the context of the entire senseword. It does this by marginalizing the blockwise posterior probability distribution to the individual bitwise (or symbolwise) posterior probability distribution. It computes this marginalization for each bit (or symbol) of the block, and then forms a componentwise maximum-posterior estimate of the transmitted dataword.

A maximum-posterior demodulator at the block level is not popular for the following reason. The maximum-posterior block estimate is the maximum of the posterior $p(a|v)$ over all data sequences a. However, if all data sequences are equally probable, this is the same as the maximum-likelihood demodulator because then $p(a|v)$ is a constant multiple of $p(v|a)$. Only when the block probabilities are unequal, and known, is the maximum-posterior demodulator at the block level useful.

The maximum-posterior demodulator for an intersymbol interference channel forms the componentwise posteriors on the n data symbols when given the channel output matched-filter samples, which have the form

$$v_i = \sum_\ell g_\ell a_{i-\ell} + n_i$$

$$= c_i + n_i$$

where c is the noise-free channel output and v is the noisy channel output. If the additive noise vector n is memoryless gaussian noise, then the conditional probability on v is given by

$$p(v|a) = \prod_{\ell=1}^n \frac{1}{\sqrt{2\pi}\sigma} e^{-(v_\ell - c_\ell)^2 / 2\sigma^2}$$

where $c = g * a$. To compute the marginal $p(a_\ell|v)$, we must compute the marginal component $p(a_\ell = s_m|v)$ for $m = 0, 1, \ldots, M-1$, where s_m is the mth point of the signal constellation S. It will prove convenient to also express the marginal imprecisely as $p(a_\ell = m|v)$ regarding a_ℓ as taking on the data values m rather than the symbols s_m representing the data values.

When the block posterior probability, given by

$$p(a|v) = \frac{p(v|a)p(a)}{\sum_a p(v|a)p(a)}$$

is marginalized to the single letter a_ℓ of the message sequence a, it becomes the conditional

$$p(a_\ell|v) = \sum_{a_0}\sum_{a_1}\cdots\sum_{a_{\ell-1}}\sum_{a_{\ell+1}}\cdots\sum_{a_{n-1}} p(a|v)$$

in which all components except a_ℓ are summed out, and a_ℓ is equal to a fixed value, say $a_\ell = m$. Each sum, say the sum on a_j, designates a sum over all possible values of component a_j. We will abbreviate this formula for marginalization as

$$p(a_\ell|v) = \sum_{a/a_\ell} p(a|v)$$

or more specifically, as

$$p(a_\ell = m|v) = \sum_{a/a_\ell} p(a|v)$$

where the notation a/a_ℓ denotes the vector a punctured by a_ℓ, so that

$$a/a_\ell = (a_0, \ldots, a_{\ell-1}, a_{\ell+1}, \ldots, a_{n-1}).$$

Then we may write

$$a = (a/a_\ell, a_\ell)$$

to reinsert a_ℓ into a/a_ℓ, with the understanding that a_ℓ is to be inserted in its proper place.

For the simplest case of BPSK, each a_j ranges over zero and one. Thus, the sum on only the jth component is

$$p(a/a_j|v) = \sum_{a_j} p(a|v)$$

$$= p(a|a_j = 0, v) + p(a|a_j = 1, v).$$

This sum on only one component provides the marginalization to the punctured block a/a_ℓ, which has $n-1$ components. In contrast, to marginalize to a single component is to compute the marginal $p(a_\ell|v)$ by summing over $n-1$ components. To compute this marginal, one must sum out over all values of a/a_ℓ. For BPSK, this requires a sum over 2^{n-1} such terms, one summand for each value of the sequence a/a_ℓ. For an M-ary signal constellation, the sum over each a_j ranges over the M elements of the signal constellation S. The sum over the entire punctured data vector a/a_ℓ then requires a sum over M^{n-1} terms. This componentwise posterior $p(a_\ell|v)$ must be computed for each of n values of ℓ.

The maximum-posterior estimate of the ℓth symbol is now given by

$$\widehat{a}_\ell = \text{argmax}_{a_\ell \in S} p(a_\ell|v)$$

for $\ell = 0, \ldots, n-1$. If S contains only two values, as for a BPSK waveform, then the ℓth binary data bit is recovered as a one or a zero according to whether the posterior component $p(a_\ell = 1|v)$ is greater than or less than $p(a_\ell = 0 \mid v)$. This can be decided by noting whether $p(a_\ell = 1 \mid v)$ is greater than or less than $1/2$.

If there is no intersymbol interference and the a_ℓ are independent, then $p(a|v) = \prod_\ell p(a_\ell|v_\ell)$. In this case, the marginals satisfy

$$p(a_\ell|v) = p(a_\ell|v_\ell) \qquad \ell = 0, \ldots, n-1.$$

Accordingly, there is no need then to explicitly marginalize $p(a|v)$ to find $p(a_\ell|v)$ because, in this situation, $p(a_\ell|v)$ is already given as $p(a_\ell|v_\ell)$. This means that only the ℓth component of v gives information about a_ℓ. Maximum-posterior demodulation at the bit (or symbol) level then differs from maximum-likelihood demodulation at the bit level only if the zeros and ones are not equiprobable.

To compute the marginals of $p(a|v)$ for a vector a of blocklength n in the general case can be a formidable task because $n-1$ summations are required to compute each marginal, and a marginal is needed for each of the M values of each of the n symbols. Each summation, itself, is a sum of M terms. This means that there are $M(M-1)(n-1)n$ additions, where M is the size of the signal constellation. In addition, to first compute $p(a|v)$ for an intersymbol interference channel can be a substantial task. Because, however, the symbol interdependence has the form of a trellis, fast algorithms to compute the collection of all branch probabilities are available, and will be explained next. This trellis description is always available for intersymbol interference. Indeed, intersymbol interference is described by a trellis with each branch leaving the same node labeled by a unique data symbol and each symbol is repeated in that frame on branches leaving other nodes.

The *two-way* (or *frontward–backward*) *algorithm* is a fast algorithm for calculating all of the posterior probabilities on the nodes of a trellis of finite length. The algorithm starts partial computations at each end of the trellis. It works backwards along the trellis from one end, and frontwards along the trellis from the other. The task is to compute, starting with the sequence of matched-filter output samples, denoted v, the posterior probability for each node of the trellis that was visited by the encoder. The demodulator can then determine, for each data symbol, the received componentwise maximum-posterior estimate of that symbol.

The two-way algorithm and the Viterbi algorithm both move through the frames of a trellis by a sequential process, so it is not surprising that both are similar at the logical flow level. Both have a similar pattern of walking the trellis. However, they are very different at the computational level. Whereas the fundamental computational unit of the Viterbi algorithm is an "add–compare–select" step, the fundamental computational unit of the two-way algorithm is a "multiply–sum" (or sum of products) step. Moreover,

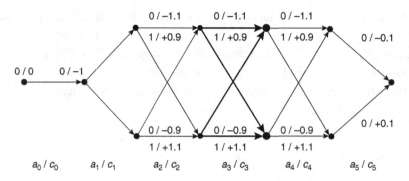

Figure 7.8. Trellis for explaining the two-way algorithm.

the two-way algorithm makes two passes through the trellis, one forward pass and one backward pass.

In contrast to the Viterbi algorithm, the two-way algorithm does not make a hard decision for each symbol. Instead, as its output the algorithm provides soft information. Accordingly, the demodulation decision is not regarded as part of the two-way algorithm; it comes later. Upon receiving the output of the algorithm, the demodulator may choose to use other information that was not given to the two-way algorithm. In Section 10.9, for example, the availability of soft information at the output of the algorithm leads to the possibility of iterative decoding, which makes the performance of turbo codes so attractive. For the present section, it is enough to posit that the symbols are demodulated by simple threshold methods.

The motivation for the two-way algorithm is presented in the simple example in Figure 7.8. This is a trellis for a BPSK waveform of blocklength four preceded and followed by zeros, with intersymbol interference described by $g_0 = 1$, $g_1 = 0.1$, and $g_\ell = 0$ otherwise. The branches in the third frame are highlighted in order to discuss the computation of the posterior of the third bit. To marginalize to a_3, one must sum the probabilities of all paths of the trellis that have $a_3 = 0$; these are the paths that go through the top highlighted node. Then one must sum the probabilities of all paths that have $a_3 = 1$; these are the paths that go through the bottom highlighted node. Because these two probabilities sum to one, only one of them actually needs to be computed directly. However, it may be more convenient to ignore common constants during the computation, and to rescale the result after the computation so that the two probabilities sum to one.

To compute $p(a_3 = 0|v)$, referring to Figure 7.8, first sum the path probabilities over all paths that start at the beginning of the trellis and reach the top node at the end of the frame of a_3, then sum over all paths from that same top node of the trellis that reach the end of the trellis. Multiply these together. Do the same for the paths that go through the bottom highlighted node at the end of the frame of a_3. More generally, for a constraint length of v, there are 2^v nodes at the end of a frame.

Now the recursive computation is obvious. Let $\beta_{i\ell}$ for $i = 0, 1$ be the sum of the probabilities of all paths from the beginning of the trellis to the ith node at the end of frame ℓ. Let $\gamma_{i\ell}$ for $i = 0, 1$, be the sum of the probabilities of all paths from the ith node of the ℓth frame to the end of the trellis. To establish the recursion, write $\beta_{0\ell}$ and $\beta_{1\ell}$ in terms of $\beta_{0,\ell-1}$ and $\beta_{1,\ell-1}$, which have already been computed in the previous iteration, then

$$\beta_{0\ell} = \beta_{0,\ell-1}p(v_\ell|a_{\ell-1} = 0, a_\ell = 0)p(a_\ell = 0) + \beta_{1,\ell-1}p(v_\ell|a_{\ell-1} = 1, a_\ell = 0)p(a_\ell = 0)$$

$$\beta_{1\ell} = \beta_{0,\ell-1}p(v_\ell|a_{\ell-1} = 0, a_\ell = 1)p(a_\ell = 1) + \beta_{1,\ell-1}p(v_\ell|a_{\ell-1} = 1, a_\ell = 1)p(a_\ell = 1).$$

The probability of the set of all paths leaving the node with $a_3 = 0$ or with $a_3 = 1$ and going to the end of the trellis can be computed in the same way simply by viewing the trellis backwards. The backward recursion is

$$\gamma_{0\ell} = \gamma_{0,\ell+1}p(v_{\ell+1}|a_{\ell+1} = 0, a_\ell = 0)p(a_{\ell+1} = 0) + \gamma_{1,\ell+1}p(v_{\ell+1}|a_{\ell+1} = 1, a_\ell = 0)p(a_{\ell+1} = 0)$$

$$\gamma_{1\ell} = \gamma_{0,\ell+1}p(v_{\ell+1}|a_{\ell+1} = 0, a_\ell = 1)p(a_{\ell+1} = 1) + \gamma_{1,\ell+1}p(v_{\ell+1}|a_{\ell+1} = 1, a_\ell = 1)p(a_{\ell+1} = 1).$$

More generally, if the constraint length of the intersymbol interference is v, there will be 2^v terms in each of these sums. The posterior $p(a_3 = 0|v)$ is the product of two terms. Then

$$\beta_{i\ell} = \sum_{j=1}^{2^v} \beta_{j,\ell-1}p(v_\ell|a_{\ell-1} = j, a_\ell = i)$$

$$\gamma_{i\ell} = \sum_{j=1}^{2^v} \gamma_{j,\ell+1}p(v_{\ell+1}|a_{\ell+1} = j, a_\ell = 1).$$

The two-way algorithm performs two sets of partial computations: one partial computation starts at the beginning of the trellis and works forward until it reaches the end of the trellis, storing as it goes the result of all partial computations in an array as shown in Figure 7.9. The other partial computation starts at the end of the trellis and works backward until it reaches the beginning of the trellis, again storing partial results as it goes. After the two arrays of partial computations are complete, they are combined to form the vector of branch posteriors.

After the symbol posterior is computed by the two-way algorithm, it can be converted to a hard decision or used in some other manner. The symbol value with the largest posterior probability is the appropriate hard decision.

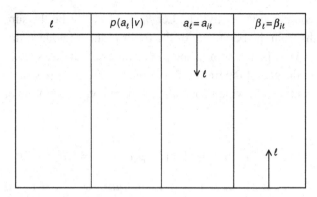

Figure 7.9. Partial computations for the two-way algorithm.

Problems for Chapter 7

7.1. A received baseband waveform for digital communication in white noise is

$$v(t) = \left[\sum_{\ell=0}^{n-1} a_\ell s(t - \ell T) \right] * g(t) + n(t)$$

where $s(t)$ is the transmitted pulse such that $s(t) * s(-t)$ is a Nyquist pulse, and $g(t)$ is the channel impulse response. The received signal is passed through a filter matched to $s(t)$ and sampled at ℓT. We are given the samples as our received data but we cannot observe the received signal at any point prior to the output of the sampler.

a. Derive the maximum-likelihood demodulator based on the observed samples, and describe an efficient implementation.

b. Would it be better to observe the received signal at the input to the matched filter?

7.2. A baseband BPSK waveform for digital communication is dispersed by a linear channel with an impulse response $g(t)$ and is received in white noise as

$$v(t) = \sum_{\ell=0}^{n-1} a_\ell s(t - \ell T) * g(t) + n(t).$$

a. Prove that a sufficient statistic for demodulation is given by the set of r_ℓ given by

$$r_\ell = \int_{-\infty}^{\infty} V(f) S^*(f) G^*(f) e^{j2\pi f \ell T} \, df$$

for all ℓ.

b. Use the necessary and sufficient conditions of a Fourier series to show that the sequence of samples

$$r'_\ell = \int_{-\infty}^{\infty} V(f)S^*(f)e^{j2\pi f \ell T}\,df$$

for all ℓ is not a sufficient statistic for demodulation, that is, show that the r_ℓ need not be recoverable from the r'_ℓ.

7.3. a. Derive the maximum-likelihood demodulator for M-ary orthogonal signaling with no intersymbol interference.

b. By averaging out phase as a random nuisance parameter, derive the maximum-likelihood demodulator for noncoherent M-ary orthogonal signaling with no intersymbol interference.

7.4. Given a channel response

$$v(t) = \sum_{\ell=-\infty}^{\infty} a_\ell s(t - \ell T) + n(t)$$

where

$$s(t) = \frac{T \sin 2\pi Wt}{\pi t(T - t)}$$

and $n(t)$ is white gaussian noise.

a. Is $s(t)$ a Nyquist pulse for this symbol rate?

b. Let $u(t) = v(t) * g(t)$ where

$$g(t) = 2W \operatorname{sinc} 2Wt.$$

Does the sequence $u(\ell T)$ form a sufficient statistic for demodulating a_ℓ?

7.5. Given the sequence of received samples

$$v_\ell = c_\ell + n_\ell$$

where the sequence n_ℓ is a sequence of white (independent) gaussian-noise samples, and the sequence c_ℓ is one of a given set of fixed code sequences, prove that the maximum-likelihood estimate of the sequence c_ℓ is that sequence for which the euclidean distance $d(v, c)$ between sequences v and c is minimized.

Notes for Chapter 7

The maximum-likelihood principle is a central principle of statistics and of classical estimation theory. The development of its properties and appeal as a principle of inference are largely due to Fisher (1922). This principle is basic to much of modern communication theory. Forney (1972) developed the maximum-likelihood demodulator in the presence of noise and intersymbol interference as a matched filter, followed by a whitening transversal filter to make the noise samples independent, and finally followed by the Viterbi algorithm to find the maximum-likelihood explanation for the data that has thus been reduced to discrete-time data in white gaussian noise. Implicit in this structure is the fact that the whitening filter can be in discrete time. Ungerboeck (1974) showed how to eliminate the transversal filter by setting up the trellis so that the Viterbi algorithm could be used as a maximum-likelihood demodulator even in the presence of any nonwhite noise. In his paper, Forney also derived bounds on the probability of error of a maximum-likelihood sequence demodulator.

The notion of diversity communication is very natural, and was used early on by many. A systematic survey of combining diversity signals was published by Brennen (1955, 1959).

Maximum-posterior algorithms have long received little attention in comparison to maximum-likelihood algorithms, but this is no longer the case. The two-way forward–backward, or BCJR algorithm, developed from the earlier Baum–Welch algorithm (1966), was published by Bahl, Cocke, Jelinek, and Raviv (1974) and was largely ignored for several decades. Essentially the same algorithm has also appeared independently in other fields. The introduction of turbo codes created a need for the BCJR algorithm, and it was rediscovered for that purpose.

8 Synchronization

A channel may introduce unpredictable changes into the waveform passing through it. In a passband channel, such as a radio frequency channel, unpredictable phase shifts of the carrier may occur in the atmosphere, in antennas, and in other system elements or because of uncertainty in the time of propagation. In order to demodulate a digital waveform coherently, a coherent replica of the carrier is needed in the receiver. Because the receiver does not know the carrier phase independently of the received signal, the receiver must locally regenerate a coherent replica of the carrier. Uncertainty in the phase of the received waveform introduces the task of phase synchronization in the receiver.

Uncertainty in the time of propagation also introduces problems of time synchronization. The local clock must be synchronized with the incoming datastream so that incoming symbols and words can be correctly framed and assigned their proper indices. Time synchronization may be subdivided into two tasks: symbol synchronization, and block or frame synchronization. These two kinds of time synchronization are quite different. Symbol synchronization is a fine time adjustment that adjusts the sampling instants to their correct value. It exploits the shape of the individual pulses making up the waveform to adjust the time reference. The content of the datastream itself plays no role in symbol synchronization. Block synchronization takes place on a much longer time scale. It looks for special patterns embedded in the datastream so that it can find the start of a message or break the message into constituent parts.

8.1 Estimation of waveform parameters

A received passband waveform may have an unknown phase and an unknown time origin. The complex representation of a typical received signal with these unknowns is

$$v(t) = \sum_{\ell=-\infty}^{\infty} a_\ell s(t - \ell T - \alpha(t))e^{j\theta(t)} + n(t)$$

where $\theta(t)$ is a random process called the *phase noise*, $\alpha(t)$ is a random process called the *timing jitter*, and $n(t)$ is additive stationary noise, which is usually gaussian. Whenever $\theta(t)$ and $\alpha(t)$ are varying very slowly compared to the duration of a block of the message of blocklength n, we can approximate them as constants and write the approximation

$$v(t) = \sum_{\ell=0}^{n-1} a_\ell s(t - \ell T - \alpha) e^{j\theta} + n(t)$$

where θ is an unknown parameter in the interval $[0, 2\pi]$, representing the phase offset; α is an unknown parameter in the interval $[0, T]$, representing the symbol timing offset; and $n(t)$ is additive noise, which is usually gaussian. Often there is no need to model θ and α as random variables, but when we want to do so, we may model them as uniform random variables in their respective intervals. Otherwise, we model them simply as unknown parameters.

Estimation of the parameter θ is called *carrier synchronization*. Estimation of the parameter α is called *symbol synchronization*. There may also be an uncertainty in the symbol index ℓ due to uncertainty in the message start time. This is dealt with under the heading *block synchronization*, or *message synchronization*, using techniques that may be quite different from those used for carrier or symbol synchronization.

Given the received signal $v(t)$, the task is to estimate θ and α. Let us consider what this involves. A common transmitted signal is a quadrature-amplitude-modulated signal, given by

$$c(t) = \sum_{\ell=-\infty}^{\infty} a_\ell s(t - \ell T),$$

where each a_ℓ is a complex number taken from the QAM signal constellation. If the modulation system is to achieve a data rate that is very nearly equal to the channel capacity, then we may expect that the signal constellation must appropriately discretize the complex plane around the origin, and the pulse $s(t)$ must have a spectrum $S(f)$ closely suitable for the channel bandwidth. This leads to a waveform with a lot of irregular fluctuations. Indeed, the subject of information theory tells us that a good waveform $c(t)$ will appear to be very similar to complex gaussian noise; the more nearly the modulation achieves the capacity of an additive gaussian-noise channel, the more nearly $c(t)$ will mimic gaussian noise. Consequently, there will be little structure in $v(t)$ from which to form an estimate of θ or α. To extract its needed information, an estimator of θ or of α requires some residual inefficiency in the communication waveform. If the communication waveform is very inefficient, the estimation of θ and α can be rather trivial. However, because communication systems must now transmit at data rates ever closer to the channel capacity, the development of synchronization techniques has been hard pressed to keep up. Sophisticated techniques are being used, and continue to evolve.

Simple estimators can be designed in an ad hoc way by using heuristic arguments. More advanced estimators can be derived by using the maximum-likelihood principle. The exact maximum-likelihood estimator may be too complicated to use, so one finds approximations to it.

Consider a received waveform corresponding to a block of symbols of length n,

$$v(t) = c(t - \alpha)e^{j\theta} + n(t)$$

$$= \sum_{\ell=0}^{n-1} a_\ell s(t - \ell T - \alpha)e^{j\theta} + n(t).$$

We want to estimate a_ℓ for $\ell = 0, \ldots, n - 1$. The secondary parameters θ and α are nuisance parameters that affect the structure of the estimator but themselves are not of independent interest. In principle, the best way to minimize the probability of demodulation block error is to simultaneously estimate all the unknowns θ, α, and a_ℓ for $\ell = 0, \ldots, n-1$ by means of a single maximum-likelihood estimator. The estimates of a_ℓ provide the demodulator output; the estimates of θ and α are not of lasting interest and can be discarded. The log-likelihood function is of the form

$$\Lambda(\theta, \alpha, a_0, \ldots, a_{n-1}) = \int_{-\infty}^{\infty} \log p(v(t)|\theta, \alpha, a_0, \ldots, a_{n-1}) dt$$

where, for each t, $p(v(t)|\theta, \alpha, a_0, \ldots, a_{n-1})$ is the probability density function of $v(t)$ conditional on θ, $\alpha, a_0, \ldots, a_{n-1}$. The maximum-likelihood estimator computes the argument over $\theta, \alpha, a_0, \ldots, a_{n-1}$ of the maximum of the log-likelihood function. By Proposition 7.2.1, when the noise is white and gaussian, this is equivalent to minimizing the squared distance

$$d^2(v(t), c(t - \alpha)e^{j\theta}) = \int_{-\infty}^{\infty} |v(t) - c(t - \alpha)e^{j\theta}|^2 dt$$

between the actual received signal $v(t)$ and each possible noise-free received signal $c(t - \alpha)e^{j\theta}$. The minimum of this distance occurs at $(\widehat{\theta}, \widehat{\alpha}, \widehat{a}_0, \ldots, \widehat{a}_{n-1})$, of which the n-tuple $(\widehat{a}_0, \ldots, \widehat{a}_{n-1})$ provides the desired estimate of the data.

We shall see that when the blocklength n is equal to one and α is known, the maximum-likelihood demodulator reduces to a noncoherent demodulator. This is a meaningful demodulator for a binary waveform, such as frequency-shift keying or on–off keying, but is meaningless for binary phase-shift keying when n is equal to one. For n larger than one, the maximum-likelihood demodulator for BPSK will have a sign ambiguity; it will not be able to distinguish between the true message and the complement of that message.

The maximum-likelihood demodulator for large blocklengths will usually be pro-hibitively expensive to implement directly. One usually uses a suboptimal procedure whose performance is a satisfactory approximation of the optimal performance. A common technique is to use a modular demodulator structure, possibly with feed-back between the modules. For example, first estimate θ, then estimate α, and then demodulate the data in the manner discussed in Chapter 6. The demodulated data may be fed back to aid in refining the estimate of θ or α. We will study phase and time synchronization for a succession of simple problems evaluating and discussing the maximum-likelihood estimator for each of them.

8.2   Estimation of phase

We begin our study of phase estimation with the study of estimating the phase angle of a pure sinusoid of known frequency and unknown phase received in additive white gaussian noise. The received signal is given by

$$\tilde{v}(t) = \cos(2\pi f_0 t + \theta) + n(t),$$

and observed on the interval from $-T_0/2$ to $T_0/2$. The complex baseband representation of this signal is

$$v(t) = e^{j\theta} + n_R(t) + jn_I(t) \qquad |t| \le T_0/2$$

where $n_R(t)$ and $n_I(t)$ are independent, white, gaussian noise processes. The log-likelihood decision statistic, as stated in Proposition 7.2.1, is

$$\Lambda(\theta) = \int_{-T_0/2}^{T_0/2} |v(t) - e^{j\theta}|^2 dt$$

$$= \int_{-T_0/2}^{T_0/2} [(v_R(t) - \cos\theta)^2 + (v_I(t) - \sin\theta)^2] dt.$$

Expanding the squares and discarding irrelevant constants gives

$$\Lambda(\theta) = -\cos\theta \int_{-T_0/2}^{T_0/2} v_R(t) dt - \sin\theta \int_{-T_0/2}^{T_0/2} v_I(t) dt$$

as an expression to be minimized by the choice of θ. The derivative of $\Lambda(\theta)$ with respect to θ, when set equal to zero, gives

$$\sin\theta \int_{-T_0/2}^{T_0/2} v_R(t) dt - \cos\theta \int_{-T_0/2}^{T_0/2} v_I(t) dt = 0,$$

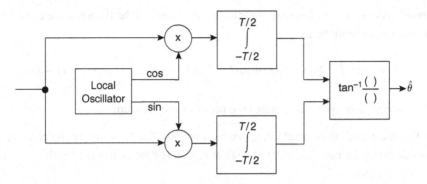

Figure 8.1. Estimating the phase of a sinusoid.

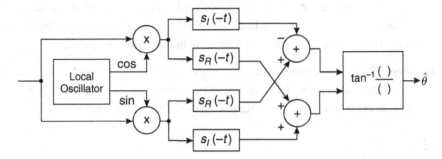

Figure 8.2. Estimating the phase of a known pulse.

and the estimate of θ is the solution of this equation:

$$\widehat{\theta} = \tan^{-1}\left[\frac{\int_{-T_0/2}^{T_0/2} v_I(t)dt}{\int_{-T_0/2}^{T_0/2} v_R(t)dt}\right].$$

Figure 8.1 shows a block diagram of this maximum-likelihood estimator of phase.

There is nothing in this analysis that requires the waveform $c(t)$ to have a constant amplitude, or even a constant phase. The more general case is described by the following theorem, and illustrated in Figure 8.2.

Theorem 8.2.1 *Suppose that a received signal at complex baseband is*

$$v(t) = s(t)e^{j\theta} + n_R(t) + jn_I(t)$$

where $n_R(t)$ and $n_I(t)$ are independent, gaussian, white noise processes, and $s(t)$ is a finite energy pulse, possibly complex. The maximum-likelihood estimator of θ is

$$\widehat{\theta} = \tan^{-1}\left[\frac{\int_{-\infty}^{\infty}[s_R(t)v_I(t) - s_I(t)v_R(t)]dt}{\int_{-\infty}^{\infty}[s_R(t)v_R(t) + s_I(t)v_I(t)]dt}\right].$$

Proof As in the development prior to the statement of the theorem, we can formulate the likelihood statistic as

$$\Lambda(\theta) = -\cos\theta \int_{-\infty}^{\infty} [s_R(t)v_R(t) + s_I(t)v_I(t)] dt - \sin\theta \int_{-\infty}^{\infty} [s_R(t)v_I(t) - s_I(t)v_R(t)] dt.$$

Maximizing this over θ completes the proof of the theorem. ■

The estimated phase angle $\widehat{\theta}$, given by Theorem 8.2.1, is simply the phase angle of the output of the matched filter $s^*(-t)$ at $t = 0$. A block diagram for this estimator is shown in Figure 8.2.

Now that we know the maximum-likelihood estimator of the phase of an otherwise known pulse, we need to determine the accuracy of the phase estimate by finding the variance of the phase error. This can be difficult to analyze for low values of E_p/N_0. When the signal-to-noise ratio is high, however, the error can be found by a linearized analysis of the output of the matched filter, as in the following theorem.

Theorem 8.2.2 *Asymptotically, for high signal-to-noise ratio, the maximum-likelihood estimator of the received phase of a known pulse $s(t)$ of energy E_p in white gaussian noise has a phase error variance (in radians) satisfying*

$$\sigma_\theta^2 = \frac{1}{2E_p/N_0}.$$

Proof The condition for the estimate $\widehat{\theta}$ as given in Theorem 8.2.1 can be written as

$$\sin\widehat{\theta} u_R - \cos\widehat{\theta} u_I = 0$$

where

$$u_R + ju_I = \int_{-\infty}^{\infty} [v_R(t) + jv_I(t)][s_R(t) + js_I(t)]^* dt.$$

Then, because

$$v(t) = s(t)e^{j\theta} + n(t),$$

this reduces to

$$u_R + ju_I = e^{j\theta} \int_{-\infty}^{\infty} |s(t)|^2 dt + \int_{-\infty}^{\infty} n(t)s^*(t) dt.$$

For high signal-to-noise ratio, the error in the estimate $\widehat{\theta}$ may be expressed by considering the first-order infinitesimals satisfying

$$\delta\widehat{\theta}(\cos\widehat{\theta} u_R + \sin\widehat{\theta} u_I) = -\sin\widehat{\theta} du_R + \cos\widehat{\theta} du_I$$

where all terms but the infinitesimals are to be evaluated at their undisturbed values. Therefore

$$\delta\widehat{\theta}E_P = -\sin\theta \int_{-\infty}^{\infty} [s_R(t)n_R(t) + s_I(t)n_I(t)]dt - \cos\theta \int_{-\infty}^{\infty} [s_I(t)n_R(t) - s_R(t)n_I(t)]dt$$

$$= -\int_{-\infty}^{\infty} [s_R(t)n_R'(t) + s_I(t)n_I'(t)]dt$$

where

$$n_R'(t) = n_R(t)\sin\theta - n_I(t)\cos\theta$$
$$n_I'(t) = n_R(t)\cos\theta + n_I(t)\sin\theta.$$

By the properties of circular complex gaussian random processes, these are also independent, gaussian, random processes, each with power density spectrum $N_0/2$. Take the expected value of the square of $\delta\widetilde{\theta}E_p$ to obtain

$$E_p^2\sigma_\theta^2 = \int_{-\infty}^{\infty}\int_{-\infty}^{\infty}\left[s_R(t)s_R(t')\frac{N_0}{2}\delta(t-t') + s_I(t)s_I(t')\frac{N_0}{2}\delta(t-t')\right]dtdt'$$

$$= \frac{N_0}{2}\int_{-\infty}^{\infty}[s_R^2(t) + s_I^2(t)]dt$$

from which the theorem follows. ∎

A similar analysis can be used if the gaussian noise is not white. In that case the asymptotic expression for the phase variance becomes

$$\sigma_\theta^2 = \left[\int_{-\infty}^{\infty}\frac{|S(f)|^2}{N(f)}df\right]^{-1}.$$

The denominator inside the integral arises due to a whitening filter $1/N(f)$ that becomes part of the matched filter.

8.3   Recursive phase estimation

As a practical matter when the communication waveform has a very long duration, the carrier phase angle θ is actually a slowly-varying function of time, $\theta(t)$. One then wants to replace the block structure for estimating a constant phase θ with a recursive structure for estimating a time-varying phase $\theta(t)$. A way to do this for a received sinusoid in noise, given by

$$v(t) = \cos(2\pi f_0 t + \theta(t)) + n(t),$$

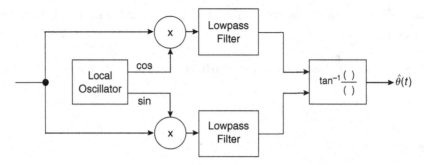

Figure 8.3. Estimating a varying phase of a sinusoid.

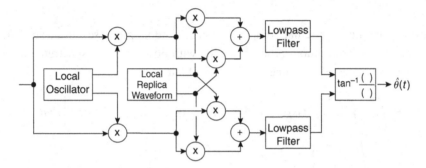

Figure 8.4. Estimating a varying phase of a known waveform.

is shown in Figure 8.3. Now, the integration of Figure 8.1 is replaced with a lowpass filter, and the output of the estimator is a time-varying function $\widehat{\theta}(t)$ that is an estimate of the time-varying phase angle $\theta(t)$. Figure 8.4 shows an estimator of a varying phase on a more general complex waveform $v(t)$, for which presumably the individual data symbols have a short duration in comparison to the fluctuations of $\theta(t)$.

There is an alternative to the use of the lowpass filters shown in Figures 8.3 and 8.4. This is the *phase-locked loop*, which is often used in applications because it forms $\cos\widehat{\theta}$ and $\sin\widehat{\theta}$ rather than $\widehat{\theta}$. We shall study this loop initially for a pure sinusoidal carrier received in noise. Later, in Sections 8.5 and 8.6, we shall extend the discussion to phase locking of the carrier in the presence of modulation.

Suppose that a sinusoid of an unknown slowly-varying phase angle $\theta(t)$ is received in the presence of additive passband noise

$$v(t) = \cos(2\pi f_0 t + \theta(t)) + n_R(t)\cos 2\pi f_0 t - n_I(t)\sin 2\pi f_0 t$$

where $n_R(t)$ and $n_I(t)$ are independent, identically distributed noise processes. The amplitude of the signal has been set to one rather than A. Accordingly, we regard the received signal to be normalized in which case, the noise signal is divided by A and the noise variance is $N_0/2A^2$. The carrier phase will be recovered from $v(t)$ by generating a noise-free local carrier

$$q(t) = \sin(2\pi f_0 t + \widehat{\theta}(t))$$

where $\widehat{\theta}(t)$ is nearly equal to $\theta(t)$. This local carrier is generated by a device called a *controlled local oscillator*, abbreviated CLO,[1] which is a device with input $e(t)$, with at least one output given by

$$q(t) = \sin\left(2\pi f_0 t + \int_0^t e(t)dt\right)$$

and possibly a second output in phase quadrature given by

$$q'(t) = \cos\left(2\pi f_0 t + \int_0^t e(t)dt\right).$$

In many applications, in which the second of the two outputs is not needed, it is suppressed.

A *phase-locked loop* is a feedback loop containing a controlled local oscillator that is used to estimate phase angle. The phase-locked loop controls the controlled oscillator so that the oscillator's output provides the desired sinusoid for the local replica of the carrier. The simplest form of the phase-locked loop is shown in Figure 8.5. Many variations of this basic loop are in use.

When the gain parameter K shown in Figure 8.5 is a constant, the loop is called a *first-order phase-locked loop*. In practical applications, it is common for the constant K to be replaced by a filter to create a more desirable loop transient response. We shall study only the elementary first-order phase-locked loop.

A phase-locked loop is a nonlinear feedback loop driven by noise and described by a nonlinear differential equation. The simple case of a first-order loop can be solved in

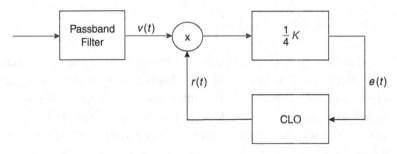

Figure 8.5. A basic phase-locked loop.

---

[1] Often a controlled oscillator is called a *voltage-controlled oscillator*, abbreviated VCO.

closed form. The estimate is

$$\widehat{\theta}(t) = \int_0^t e(t)dt.$$

The feedback signal is given by

$$e(t) = -\frac{1}{2}Kq(t)v(t)$$

$$= -\frac{1}{2}K\sin(2\pi f_0 t + \widehat{\theta}(t))\cos(2\pi f_0 t + \theta(t))$$

$$\qquad - \frac{1}{2}Kq(t)[n_R(t)\cos 2\pi f_0 t - n_I(t)\sin 2\pi f_0 t]$$

and $\dot{\widehat{\theta}}(t) = e(t)$. The trigonometric product in the first term is expanded into a sum and difference term, and the composite noise term is denoted $n'(t)$ to give

$$\dot{\widehat{\theta}}(t) = -K\sin(\widehat{\theta}(t) - \theta(t)) - K\sin(4\pi f_0 t + \widehat{\theta}(t) + \theta(t)) + n'(t).$$

The second term of this expression is at frequency $2f_0$ and is easily rejected by a simple lowpass filter within the loop, so that term is ignored in our analysis. The bandwidth of that lowpass filter can be so large that it does not otherwise affect the loop behavior; in fact, the phase-locked loop itself acts as a lowpass filter rejecting the unwanted term at double frequency. Accordingly, this term will be ignored.

We now have the differential equation

$$\dot{\widehat{\theta}}(t) + K\sin(\widehat{\theta}(t) - \theta(t)) = n'(t)$$

where the noise $n'(t)$ here is related to the incoming channel noise as follows:

$$n'(t) = -\frac{1}{2}K[n_R(t)\cos 2\pi f_0 t - n_I(t)\sin 2\pi f_0 t]\sin(2\pi f_0 t + \widehat{\theta}(t))$$

$$= -K[n_R(t)\sin\widehat{\theta}(t) - n_I(t)\cos\widehat{\theta}(t)].$$

If the channel noise is gaussian, then so is $n'(t)$ and the noise variances are related by the factor K^2. Because $\widehat{\theta}(t)$ varies in time very slowly in comparison to $n_R(t)$ and $n_I(t)$, $n'(t)$ can be considered to be white noise with power density spectrum related to that of the channel noise $n(t)$ by $N'(f) = K^2 N_0/2$ if $A = 1$, or $K^2 N_0/2A^2$ in general.

In the absence of noise, the loop behavior is described by the solution of the first-order differential equation

$$\dot{\psi} + K\sin\psi = 0$$

where $\theta_e = \widehat{\theta} - \theta$ is the phase error. It is now easy to find that

$$\tan\frac{\theta_e(t)}{2} = \tan\frac{\theta_{e0}}{2}e^{-Kt}$$

is the transient solution to the differential equation.

In the steady state, $\theta_e(t)$ will consist of a fluctuating error signal called phase noise whose variance depends on the noise $n'(t)$. We will calculate this error variance under the assumption that $\sin(\widehat{\theta} - \theta)$ can be approximated by $\widehat{\theta} - \theta$ in the steady state. In this approximation, the differential equation is

$$\dot{\theta}_e + K\theta_e = n'(t),$$

with the inhomogeneous solution

$$\theta_e(t) = \int_0^\infty h(\xi)n'(t-\xi)d\xi$$

where the impulse-response function is $h(\xi) = e^{-K\xi}$ for $\xi \geq 0$.

The correlation function of the phase noise is

$$\begin{aligned}
E[\theta_e(t)\theta_e(t+\tau)] &= E\int_0^\infty h(\xi)n'(t-\xi)d\xi\int_0^\infty h(\xi')n'(t+\tau-\xi')d\xi' \\
&= \int_0^\infty\int_0^\infty e^{-K\xi}e^{-K\xi'}E[n'(t-\xi)n'(t+\tau-\xi')]d\xi d\xi' \\
&= \frac{K^2N_0}{2}\int_0^\infty\int_0^\infty e^{-K\xi}e^{-K\xi'}\delta(\tau-(\xi'-\xi))d\xi d\xi'
\end{aligned}$$

if $A = 1$. The range of integration requires that both ξ and ξ' are nonnegative, while τ can be negative or positive. If τ is positive, then the impulse occurs at $\xi' = \xi + |\tau|$. In this case,

$$\begin{aligned}
E[\theta_e(t)\theta_e(t+\tau)] &= \frac{K^2N_0}{2}\int_0^\infty e^{-K\xi}e^{-K(\xi+|\tau|)}d\xi \\
&= \frac{KN_0}{4}e^{-K|\tau|}.
\end{aligned}$$

If τ is negative, then the impulse occurs at $\xi = \xi' + |\tau|$. In this case

$$\begin{aligned}
E[\theta_e(t)\theta_e(t+\tau)] &= \frac{K^2N_0}{2}\int_0^\infty e^{-K(\xi'+|\tau|)}e^{-K\xi'}d\xi' \\
&= \frac{KN_0}{4}e^{-K|\tau|}
\end{aligned}$$

which is the same integral as before. In general, N_0 should be replaced by N_0/A^2. Thus, the correlation function of the phase error is

$$\phi_{\theta_e\theta_e}(\tau) = \mathrm{E}[\theta_e(t)\theta_e(t+\tau)]$$
$$= \frac{KN_0}{4A^2}e^{-K|\tau|}.$$

The variance in the phase error is found by setting $\tau = 0$. Then

$$\sigma_\theta^2 = \frac{KN_0}{4A^2}$$
$$= \frac{1}{2E_{\mathrm{eff}}/N_0}$$

where the effective energy E_{eff} is set to $2A^2/K$ so that this equation can be compared to the variance of the optimal estimator of phase of a finite-energy pulse, which was given as

$$\sigma_\theta^2 = \frac{1}{2E_p/N_0}$$

in Theorem 8.2.2. To decrease the phase-noise variance, the constant K should be made smaller, but this means that time variations in $\theta(t)$ are not tracked as closely.

8.4   Estimation of delay

Next, we turn to synchronization of the sampling interval. The estimation of a time delay α in a received waveform is very similar to the estimation of phase offset θ. We will structure the development to underscore this parallel: first by finding an estimator for a known pulse, then giving a recursive structure for tracking a time-varying time delay $\alpha(t)$.

The estimation of the pulse arrival time is based on a received signal of the form

$$v(t) = s(t-\alpha) + n(t)$$

where $s(t)$ is a known pulse, possibly complex, with finite energy E_p, and $n(t)$ is stationary gaussian noise, possibly complex, whose correlation function $\phi(\tau)$ and power density spectrum $N(f)$ (per component) are known. The estimation problem is to determine the unknown delay α from $v(t)$. We will develop the maximum-likelihood estimator of α.

Theorem 8.4.1 *Suppose that a received signal at complex baseband is*

$$v(t) = s(t - \alpha) + n_R(t) + jn_I(t)$$

where $n_R(t)$ and $n_I(t)$ are independent, identically distributed, white gaussian-noise processes, and $s(t)$ is a known differentiable finite energy pulse, possibly complex. The maximum-likelihood estimator of the unknown α is a value of α satisfying

$$\text{Re} \left[\int_{-\infty}^{\infty} \frac{ds(t - \alpha)}{d\alpha} v^*(t) dt \right] = 0.$$

Proof The log-likelihood function in additive gaussian noise can be written as the distance

$$\Lambda(\alpha) = - \int_{-\infty}^{\infty} |v(t) - s(t - \alpha)|^2 dt$$

$$= -2 \int_{-\infty}^{\infty} [v_R(t)s_R(t - \alpha) + v_I(t)s_I(t - \alpha)] dt + \text{constant}$$

where the constant does not depend on α. The maximum-likelihood estimate occurs where the derivative with respect to α equals zero, which is equivalent to the statement of the theorem. ■

To calculate the variance of the estimate, which is given next, we will assume that the signal-to-noise ratio is large enough so that the noise causes only a small perturbation in $\Lambda(\alpha)$, which can be analyzed by treating only the dominant term in a Taylor series expansion. Because $\widehat{\alpha}$ is the point at which $\Lambda'(\alpha) = 0$, we will write up to first-order terms

$$(\delta\alpha)\Lambda''(\alpha) = \delta\Lambda'(\alpha)$$

and

$$\sigma_\alpha^2 = \frac{E[\delta\Lambda'(\alpha)]^2}{[\Lambda''(\alpha)]^2}.$$

Proposition 8.4.2 *Asymptotically, for high signal-to-noise ratio, the variance of the maximum-likelihood estimator of the arrival time of a known pulse $s(t)$ of energy E_p in white gaussian noise is*

$$\sigma_\alpha^2 = \frac{1}{(2\pi)^2 \overline{f^2}(2E_p/N_0)}$$

where

$$\overline{f^2} = \frac{\int_{-\infty}^{\infty} f^2 |S(f)|^2 df}{\int_{-\infty}^{\infty} |S(f)|^2 df}.$$

Proof For a high signal-to-noise ratio, the condition of Theorem 8.4.1 may be used to form an equation involving first-order infinitesimals

$$\text{Re}\left[\int_{-\infty}^{\infty}\frac{d^2s(t-\alpha)}{dt^2}(\delta\alpha)v^*(t)dt - \int_{-\infty}^{\infty}\frac{ds(t-\alpha)}{dt}\delta v^*(t)dt\right] = 0.$$

Replace the infinitesimal $\delta v^*(t)$ by $n(t)$ and $v^*(t)$ by its expectation $s^*(t)$, leaving

$$\text{Re}\left[\delta\alpha\int_{-\infty}^{\infty}\frac{d^2s(t-\alpha)}{dt^2}s^*(t)dt - \int_{-\infty}^{\infty}\frac{ds(t-\alpha)}{dt}n^*(t)dt\right] = 0.$$

Using Parseval's inequality on the first term and squaring leads to

$$(\delta\alpha)^2\left[\int_{-\infty}^{\infty}|2\pi fS(f)|^2df\right]^2 = \left[\int_{-\infty}^{\infty}\left[\frac{ds_R(t-\alpha)}{dt}n_R(t) + \frac{ds_I(t-\alpha)}{dt}n_I(t)\right]dt\right]^2.$$

The expectation of this equation gives

$$\sigma_\alpha^2\left[\int_{-\infty}^{\infty}|2\pi fS(f)|^2df\right]^2 = \frac{N_0}{2}\int_{-\infty}^{\infty}\left|\frac{ds(t)}{dt}\right|^2dt.$$

Using Parseval's formula on the right side gives

$$\sigma_\alpha^2\left[\int_{-\infty}^{\infty}|2\pi fS(f)|^2df\right]^2 = \frac{N_0}{2}\int_{-\infty}^{\infty}|2\pi fS(f)|^2df,$$

and, because $E_p = \int_{-\infty}^{\infty}|S(f)|^2df$, the theorem follows. ∎

More generally, when the gaussian noise is not white, the expression for the asymptotic error variance is replaced by

$$\sigma_\alpha^2 = \left[(2\pi)^2\int_{-\infty}^{\infty}f^2\frac{|S(f)|^2}{N(f)}df\right]^{-1}.$$

The appearance of $N(f)$ in the denominator can be thought of as a whitening filter $1/N(f)$ applied to $s(t)$.

As a practical matter, the time delay will be a slowly varying function of time, $\alpha(t)$, throughout the duration of a long waveform $c(t)$. One wishes to replace the block structure for estimating constant delay with an iterative structure for estimating the time-varying delay $\alpha(t)$. This can be accomplished for a sequence of pulses by using the *delay-locked loop*, shown in Figure 8.6, in which the error signal is used to drive the controlled clock.

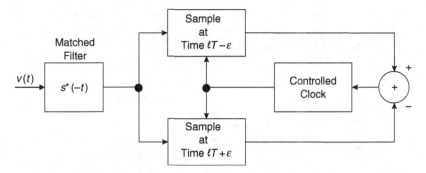

Figure 8.6. A basic delay-locked loop.

A delay-locked loop is a discrete-time version of the phase-locked loop. The purpose of a delay-locked loop is to adjust the time reference of a controlled clock. This is a version of a controlled oscillator whose output is a sequence of timing impulses rather than a sinusoidal signal. Whereas a phase-locked loop slews the phase of a controlled oscillator continuously based on the phase error signal, a delay-locked loop increments the reference time of a controlled clock based on the timing error signal. The delay-locked loop of Figure 8.6 obtains its error signal by sampling the output of the matched filter at each side of the presumed peak, called the *early-gate sample* and the *late-gate sample*, and computing the difference. The difference gives an approximation to the derivative, and provides the error signal. The error signal is multiplied by a loop gain constant (not shown) and adjusts the controlled clock. The trade between response time and noise sensitivity is determined by choice of loop gain constant.

8.5   Carrier synchronization

The phase-locked loop studied in Section 8.3 is used to acquire the phase of a pure sinusoid. However, a pure sinusoid conveys no information. It is usually necessary to apply the phase-locked loop in more complicated situations in which the received signal contains modulation. We shall study the phase-locked loop embedded into larger structures containing more of the receiver functions.

A large variety of techniques can be employed to recover the carrier phase of a data-modulated waveform. We shall first discuss methods that devote a portion of the transmitted energy to a data-free segment of the waveform. This method imposes an energy overhead on the system and so reduces the overall E_b/N_0 of the system. The overhead may not be acceptable in a high-performance system. We shall also study techniques that recover the carrier phase from a fully modulated waveform. In general, the more sophisticated the waveform modulation, the more difficult it is to extract the carrier phase.

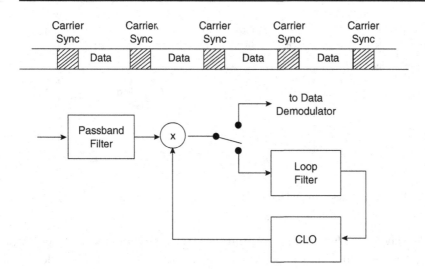

Figure 8.7. A time-sampled phase-locked loop.

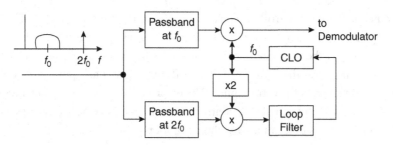

Figure 8.8. Locking to a displaced carrier.

The simplest method of carrier phase recovery is to periodically interrupt the modulation to transmit an unmodulated segment of the sinusoidal carrier. Figure 8.7 shows a periodic segment of the clear sinusoid interspersed with the fully modulated waveform. The clear carrier segment is used to lock a time-sampled phase-locked loop by using the sampled phase error as a feedback to a controlled local oscillator. The output of the loop's local oscillator provides a continuous local replica of the carrier with proper phase and frequency.

Alternatively, the clear carrier might be transmitted as a pilot signal by a frequency side tone, say a sine wave whose frequency is an integer multiple of the carrier frequency. Figure 8.8 shows how a side tone of frequency $2f_0$ can be used to drive a phase-locked loop at frequency f_0.

In either of these methods, the clear carrier contains transmitted power but no message bits. The power in the pilot signal divided by the bit rate is the energy overhead per data bit. The power in the side tone increases the required E_b/N_0 because, in calculating E_b, the fraction of the total power devoted to the side tone must be apportioned among the data bits.

More general methods of carrier phase recovery will recover the phase of a fully modulated waveform directly from the waveform. Then one should use the maximum-likelihood principle to simultaneously estimate both the data and the phase. This process, however, is usually far too complicated to use for a practical application. Somehow one must decouple the task of phase estimation from the task of data demodulation. We shall study a suboptimal technique based on the so-called *generalized likelihood function*, in which the data is regarded as a random *nuisance parameter* in the likelihood function for the phase, and averaged out. This is justified primarily by pragmatism, not by a deeper theory. Thus, the method of averaging out nuisance parameters gives a tainted – but not discredited – form of the maximum-likelihood principle.

The starting point for forming the generalized likelihood function lies in the fact that if a measurement, x, is governed by the probability density function $p_0(x|\theta)$ with probability $\frac{1}{2}$, and is governed by the probability density function $p_1(x|\theta)$ with probability $\frac{1}{2}$, then its unconditional probability density function is

$$p(x) = \tfrac{1}{2}p_0(x|\theta) + \tfrac{1}{2}p_1(x|\theta).$$

This statement says that the likelihood functions (rather than the log-likelihood functions) are to be averaged. Therefore the expected likelihood function e^Λ can be related to the conditional log-likelihood functions Λ_0 and Λ_1 by

$$e^\Lambda = \tfrac{1}{2}e^{\Lambda_0} + \tfrac{1}{2}e^{\Lambda_1}.$$

Thus, rather than maximize the log-likelihood function, we maximize the log of the expected likelihood function.

Let us first see what this principle tells us about acquiring the phase of a single bit of a BPSK waveform. When the single bit of a BPSK waveform is a one, the received signal is

$$v(t) = As(t)e^{j\theta} + n_R(t) + jn_I(t),$$

and the log-likelihood function is

$$\Lambda_0(\theta) = -\frac{1}{N_0}\int_{-\infty}^{\infty} |v(t) - As(t)e^{j\theta}|^2 dt.$$

When the single bit is a zero, then the received signal is

$$v(t) = -As(t)e^{j\theta} + n_R(t) + jn_I(t)$$

and the log-likelihood function is

$$\Lambda_1(\theta) = -\frac{1}{N_0} \int_{-\infty}^{\infty} |v(t) + As(t)e^{j\theta}|^2 dt.$$

The expected likelihood function is

$$e^{\Lambda(\theta)} = \tfrac{1}{2}e^{\Lambda_0(\theta)} + \tfrac{1}{2}e^{\Lambda_1(\theta)}.$$

Expand the squares in the exponents and collect the terms that do not depend on θ into a constant C. This gives

$$e^{\Lambda(\theta)} = C \left[e^{(2A/N_0)\mathrm{Re}[\int_{-\infty}^{\infty} v^*(t)s(t)e^{j\theta} dt]} + e^{-(2A/N_0)\mathrm{Re}[\int_{-\infty}^{\infty} v^*(t)s(t)e^{j\theta} dt]} \right]$$

$$= C \cosh \left[(2A/N_0)\mathrm{Re} \left[\int_{-\infty}^{\infty} v^*(t)s(t)e^{j\theta} dt \right] \right]$$

$$= C \cosh \left[(2A/N_0) \left[\cos\theta \int_{-\infty}^{\infty} v_R(t)s(t)dt + \sin\theta \int_{-\infty}^{\infty} v_I(t)s(t)dt \right] \right]$$

$$= C \cosh \left[\frac{u_R(0)\cos\theta + u_I(0)\sin\theta}{N_0/2A} \right].$$

Therefore, the (generalized) maximum-likelihood estimate is

$$\widehat{\theta} = \mathrm{argmax}_\theta \, \log\cosh \left[\frac{u_R(0)\cos\theta + u_I(0)\sin\theta}{N_0/2A} \right].$$

To maximize the quantity $\cosh[\mathrm{Re}[ze^{j\theta}]]$ where z is any complex number, we can choose the estimate $\widehat{\theta}$ equal to the negative of the angle of z. However, because the hyperbolic cosine is an even function, there will be another maximum at $\widehat{\theta} + 180°$. We say that there is a 180° phase ambiguity in acquiring the phase of BPSK. Because of this ambiguity, a single bit cannot be demodulated if phase is not known.

We shall now derive a more general procedure for estimating the phase from the entire BPSK waveform of finite length. For a randomly modulated BPSK waveform with n data bits, and with no intersymbol interference, the probability of output sequence v is

$$p(v|\theta) = \prod_{\ell=0}^{n-1} \left[\tfrac{1}{2}p_0(v_\ell|\theta) + \tfrac{1}{2}p_1(v_\ell|\theta) \right].$$

Theorem 8.5.1 *The log-likelihood function for an unknown phase θ in a randomly modulated BPSK waveform of blocklength n with no intersymbol interference is*

$$\Lambda(\theta) = \sum_{\ell=1}^{n} \log\cosh \left[\frac{u_R(\ell T)\cos\theta + u_I(\ell T)\sin\theta}{N_0/2A} \right]$$

where

$$u_R(t) = \int_{-\infty}^{\infty} v_R(\xi)s(\xi - t)d\xi$$

and

$$u_I(t) = \int_{-\infty}^{\infty} v_I(\xi)s(\xi - t)d\xi$$

are the components of the matched-filter output.

Proof The BPSK data block of blocklength n is $a = (a_0, \ldots, a_{n-1})$, where $a_\ell = \pm A$. Each block occurs with probability 2^{-n}. The log-likelihood function is

$$\Lambda(\theta, a) = -\frac{1}{N_0} \int_{-\infty}^{\infty} \left| v(t) - e^{j\theta} \sum_{\ell=0}^{n-1} a_\ell s(t - \ell T) \right|^2 dt.$$

When the square inside the integral is opened, there will be a term in $|v(t)|^2$ that does not depend on θ, and a double sum on k that also does not depend on θ. Because there is no intersymbol interference and a_ℓ^2 is independent of the data, the latter term will not depend on the data either. By collecting terms that do not depend on θ or a into a constant, we have the log-likelihood function

$$\Lambda(\theta, a) = -\frac{2}{N_0} \sum_{\ell=0}^{n-1} \text{Re}\left[a_\ell u(\ell T)e^{-j\theta} \right] + \log C$$

where $u(\ell T)$ is the ℓth sample of the matched-filter output. Therefore

$$e^{\Lambda(\theta, a)} = C \prod_{\ell=0}^{n-1} e^{-(2/N_0)\text{Re}[a_\ell u(\ell T)e^{-j\theta}]}.$$

Now average over all data vectors a, each of which occurs with probability 2^{-n}. Thus

$$e^{\Lambda(\theta)} = C \sum_{a} \frac{1}{2^n} \prod_{\ell=0}^{n-1} e^{-(2/N_0)\text{Re}[a_\ell u(\ell T)e^{-j\theta}]}$$

$$= C \prod_{\ell=0}^{n-1} \left[\frac{1}{2} e^{(2A/N_0)\text{Re}[u(\ell T)e^{-j\theta}]} + \frac{1}{2} e^{-(2A/N_0)\text{Re}[u(\ell T)e^{-j\theta}]} \right]$$

$$= C \prod_{\ell=0}^{n-1} \cosh \left[\frac{\text{Re}[u(\ell T)e^{-j\theta}]}{N_0/2A} \right],$$

and the theorem follows. ■

Because the hyperbolic cosine is an even function, it is clear that if $\widehat{\theta}$ maximizes $\Lambda(\theta)$, then $\widehat{\theta} + 180°$ does also. Thus BPSK has a $180°$ phase ambiguity even when the phase is recovered from an entire waveform. This was obvious at the outset because changing phase by $180°$ affects the waveform in the same way as replacing the datastream by its complement.

To maximize the log-likelihood statistic $\Lambda(\theta)$, we differentiate $\Lambda(\theta)$ in Theorem 8.5.1 with respect to θ and drive the derivative to zero. The value of θ satisfying

$$\sum_{\ell=1}^{n} \left[\frac{-u_R(\ell T)\sin\theta + u_I(\ell T)\cos\theta}{N_0/2A} \right] \tanh \left[\frac{u_R(\ell T)\cos\theta + u_I(\ell T)\sin\theta}{N_0/2A} \right] = 0,$$

is the output $\widehat{\theta}$ of the block estimator.

For a single pulse at time ℓT, $n = 1$, and the equation is satisfied when either of the two bracketed terms is equal to zero. Of these two, we choose the solution that gives the maximum. But the maximum cannot occur when the argument of the hyperbolic tangent equals zero because $\cosh x$ has a minimum at $x = 0$. Hence the maximum occurs when

$$u_R(\ell T)\sin\theta - u_1(\ell T)\cos\theta = 0.$$

This leads to the maximum-likelihood estimate

$$\widehat{\theta} = \tan^{-1}\frac{u_I}{u_R},$$

as was also the case for an unmodulated carrier.

8.6   Recursive carrier synchronization

The derivative of the log-likelihood function given in Theorem 8.5.1 tells how the matched-filter samples should be weighted, combined, and set equal to zero to find the maximum. Because this implicit equation cannot be solved explicitly, one can use a recursive procedure on the entire block to find the solution, which we write as

$$\theta_{r+1} = \theta_r + \frac{K}{n}\sum_{\ell=1}^{n} \left[\frac{-u_R(\ell T)\sin\theta_r + u_I(\ell T)\cos\theta_r}{N_0/2A} \right] \tanh \left[\frac{u_R(\ell T)\cos\theta_r + u_I(\ell T)\sin\theta_r}{N_0/2A} \right]$$

where K is a constant that can be chosen to control convergence. This expression, as written, requires all n samples to be collected before the iterations can begin. Moreover, all terms of the equation must be updated every iteration.

To make the estimator into a recursive estimator, the update increment $\Delta\widehat{\theta}$ is made proportional to the summand at each sampling instant. If $\Delta\widehat{\theta}$ is small at each bit time, the incremental phase update can be approximated as the differential equation

$$\frac{d\widehat{\theta}}{dt} = -K[u_R(t)\sin\widehat{\theta} - u_I(t)\cos\widehat{\theta}]\tanh\left[\frac{u_R(t)\cos\widehat{\theta} + u_I(t)\sin\widehat{\theta}}{N_0/2A}\right]$$

$$= Ku_I'(t)\tanh\left(\frac{2A}{N_0}u_R'(t)\right)$$

where K is a constant that controls the rate of the convergence, and

$$u_R'(\ell T) = u_R(\ell T)\cos\widehat{\theta} + u_I(\ell T)\sin\widehat{\theta}$$
$$u_I'(\ell T) = -u_R(\ell T)\sin\widehat{\theta} + u_I(\ell T)\cos\widehat{\theta}.$$

This equation suggests a recursive phase estimator at complex baseband, which can be reformulated at an intermediate frequency to obtain the estimator shown in Figure 8.9. The double-loop structure is composed of two matched filters implemented at passband with inputs that are in phase quadrature. The sampled outputs (of the amplitude modulation) of the two passband matched filters are exactly the two terms u_R' and u_I' that are required. The product of u_I' and the hyperbolic tangent of u_R' provides the feedback signal to the controlled local oscillator to drive the loop. This means that if the input to a controlled oscillator is held constant, the output frequency is changed by an amount proportional to this constant.

The feedback circuit of Figure 8.9 can be simplified by approximating the hyperbolic tangent in either of two ways. Each approximation results in a different popular version

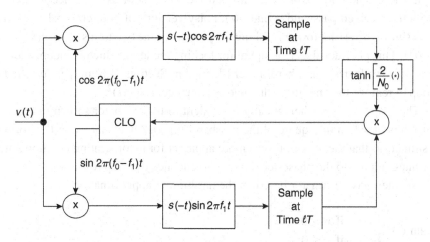

Figure 8.9. A recursive phase estimator for a BPSK waveform.

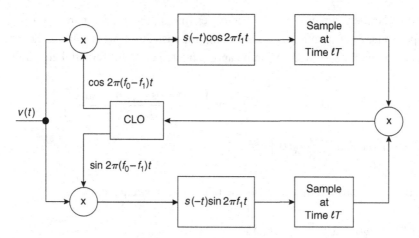

Figure 8.10. The Costas loop.

of the estimator. An estimator based on the approximation,

$$\tanh x \sim x,$$

shown in Figure 8.10, is most appropriate at low signal-to-noise ratios. This estimator, called a *Costas loop*, is used for estimating phase not only of a BPSK waveform but also of a variety of other amplitude-modulated carriers, even in situations where it is not an approximation to the maximum-likelihood phase estimator. In one application of the Costas loop, $c(t)$ is an arbitrary pulse-amplitude-modulated waveform received in noise, and no time reference is available for sampling the matched-filter output. Therefore a weaker form of the Costas loop is used. The loop may be driven by the raw received signal lightly filtered, without full benefit of a matched filter. By using both sine and cosine outputs of the controlled local oscillator as reference signals to mix with the received passband signal $v(t)$, noisy versions of both $c(t)\cos[\theta - \theta'(t)]$ and $c(t)\sin[\theta - \theta'(t)]$ are formed. Taking the product of the two terms gives $c(t)^2 \sin 2(\theta - \theta'(t))$. This provides the error signal for locking the loop, which is independent of the sign of $c(t)$. If desired, a hardlimiter (defined in Section 11.8) may be inserted in the loop to remove the amplitude fluctuations that occur in $c(t)^2$.

The loop now is mathematically equivalent to the estimator shown in Figure 8.11, which is based on the square of the passband signal. An advantage of this form of the estimator is that the opportunity is more apparent for incorporating additional filtering or integration into the phase-locked loop to enhance the loop response.

An alternative estimator based on the hardlimiter approximation

$$\tanh x \approx \begin{cases} 1 & \text{if } x > 0 \\ -1 & \text{if } x \leq 0, \end{cases}$$

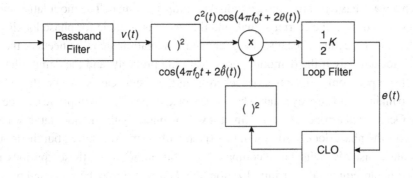

Figure 8.11. Phase-locking to a squared signal.

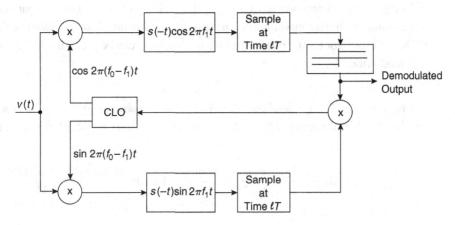

Figure 8.12. A decision-directed phase estimator.

as shown in Figure 8.12, is most appropriate at high signal-to-noise ratios. Notice that the nonlinearity can be viewed as a threshold that is demodulating the BPSK datastream to form ±1. The multiplication by ±1 can be viewed as an inversion of those negative pulses in the waveform corresponding to $-As(t)$. For this reason, this form of the estimator is called a *decision-directed phase estimator*.

In a data-directed phase estimator, the estimated data samples $\hat{a}_\ell$ are used to strip the modulation from the modulated carrier. From the demodulated output, a copy of the estimated data, called a *local replica sequence*, and synchronous with the received signal, is generated in the receiver. This replica multiplies the received signal, as shown in Figure 8.12. After multiplication, the signal is

$$u'(\ell T) = a_\ell u(\ell T) \sin \theta$$
$$= a_\ell^2 \sin \theta + n_\ell,$$

which always has a positive modulation A^2 multiplying $\sin \theta$. The modulation has now been stripped from the carrier signal, and the signal is suitable for phase locking.

With this procedure, because a regenerated modulation signal is needed, the phase can be locked only if the demodulator is operating properly, and the demodulator will operate properly only if the phase is locked. Thus the system can be used with confidence only to maintain an already established phase lock, or possibly to improve a crude phase lock. Occasional erroneous data estimates will be made by the demodulator. When this happens, the phase-locked loop will be driven with a spurious input, but the loop time constant is usually quite large compared to a bit duration, so these spurious inputs will be inconsequential. For initialization, the datastream may be preceded by a fixed preamble, called a *training sequence*, which is known to the receiver and so simplifies the problem of initial phase acquisition.

More generally, a receiver must acquire and lock the phase of a signal that has both in-phase and quadrature modulation. The simplest such example of a QAM waveform is QPSK. An estimator for the phase of a QPSK waveform can be found by the maximum-likelihood principle.

Theorem 8.6.1 *A log-likelihood function for an unknown phase in a randomly modulated* QPSK *waveform of blocklength n, with no intersymbol interference, is*

$$\Lambda(\theta) = \sum_{\ell=1}^{n} \left[\log \cosh \frac{u_R(\ell T) \cos \theta + u_I(\ell T) \sin \theta}{N_0/2A} + \log \cosh \frac{u_R(\ell T) \sin \theta - u_I(\ell T) \cos \theta}{N_0/2A} \right].$$

Proof   We will carry through the steps of the proof for $n = 1$. Because there is no intersymbol interference, it is clear that the solution for general n is a sum of such terms.

There are now four log-likelihood functions corresponding to the four QPSK points. They are

$$\Lambda_{00}(\theta) = -\frac{1}{N_0} \int_{-\infty}^{\infty} |v(t) + (-1 - j)As(t)e^{j\theta}|^2 dt$$

$$\Lambda_{01}(\theta) = -\frac{1}{N_0} \int_{-\infty}^{\infty} |v(t) + (1 - j)As(t)e^{j\theta}|^2 dt$$

$$\Lambda_{10}(\theta) = -\frac{1}{N_0} \int_{-\infty}^{\infty} |v(t) + (1 + j)As(t)e^{j\theta}|^2 dt$$

$$\Lambda_{11}(\theta) = -\frac{1}{N_0} \int_{-\infty}^{\infty} |v(t) + (-1 + j)As(t)e^{j\theta}|^2 dt.$$

Consequently,

$$e^{\Lambda(\theta)} = \tfrac{1}{4}e^{\Lambda_{00}(\theta)} + \tfrac{1}{4}e^{\Lambda_{01}(\theta)} + \tfrac{1}{4}e^{\Lambda_{10}(\theta)} + \tfrac{1}{4}e^{\Lambda_{11}(\theta)},$$

which can be reduced to

$$e^{\Lambda(\theta)} = C \cosh\left[\frac{u_R(0)\cos\theta + u_I(0)\sin\theta}{N_0/2A}\right]\cosh\left[\frac{u_R(0)\sin\theta - u_I(0)\cos\theta}{N_0/2A}\right].$$

For a train of QPSK pulses, indexed by ℓ, this becomes the equation in the theorem, with $\Lambda(\theta)$ redefined to suppress the constant C. ∎

The derivative of $\Lambda(\theta)$ is

$$\frac{d\Lambda}{d\theta} = \sum_{\ell=1}^{n}\left[\frac{-u_R(\ell T)\sin\theta + u_\ell(\ell T)\cos\theta}{N_0/2A}\tanh\frac{u_R(\ell T)\cos\theta + u_I(\ell T)\sin\theta}{N_0/2A}\right.$$
$$\left.+\frac{u_R(\ell T)\cos\theta + u_I(\ell T)\sin\theta}{N_0/2A}\tanh\frac{u_R(\ell T)\sin\theta - u_I(\ell T)\cos\theta}{N_0/2A}\right],$$

which leads to a recursive feedback estimator for θ, as shown in Figure 8.13. As for BPSK, each hyperbolic tangent can be approximated, either by its argument to obtain a circuit known as an extended Costas loop, or by a hardlimiter to obtain a decision-directed phase estimator. The extended Costas loop can be manipulated into the form of a phase-locked loop on the fourth power of the received signal.

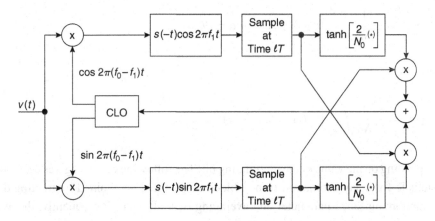

Figure 8.13. A sequential phase estimator for a QPSK waveform.

8.7   Symbol synchronization

The task of time synchronization is divided into *symbol synchronization* (studied in this section) and *block synchronization* (studied in the next section). The task of symbol synchronization is to synchronize the local clock to the incoming datastream so that each incoming symbol can be detected at the proper time. Symbol synchronization will not necessarily provide block synchronization or frame synchronization. This is because there can be ambiguities in a frame synchronization. An analogy can be seen in setting the second hand of a clock. Even though the second hand is set precisely, the minute and hour may be completely unknown. Similarly, a receiver may learn to sample precisely at the center of each data symbol, and yet not know the index of that symbol within a block. Usually, the method of block synchronization will be completely separate from the method of symbol synchronization. Symbol synchronization is more closely related to carrier recovery than it is to block synchronization. Just as carrier recovery techniques are built around the idea of a phase-locked loop, so symbol synchronization is built around the idea of a delay-locked loop.

Symbol synchronization works directly from the modulated datastream corrupted by noise as it is seen by the receiver. One method of synchronization is to extract a harmonic of the symbol frequency from the received signal. Then a local symbol clock can be synchronized by methods that are very similar to the phase-locked loops used to recover the carrier phase. If necessary, a start-up procedure, such as one that uses a slow search, can be used for initialization. Synchronization is then maintained by locking a feedback loop to a clock signal that is extracted from the modulated waveform.

We shall develop the techniques for symbol time synchronization more formally by starting from the maximum-likelihood principle. The log-likelihood function for a single bit of a BPSK waveform with unknown delay α is either

$$\Lambda_0(\alpha) = -\frac{1}{N_0} \int_{-\infty}^{\infty} |v_R(t) + As(t - \alpha)|^2 dt$$

or

$$\Lambda_1(\alpha) = -\frac{1}{N_0} \int_{-\infty}^{\infty} |v_R(t) - As(t - \alpha)|^2 dt,$$

depending on the value of the modulating bit. The following theorem gives a suboptimal statistic for delay estimation from a block of data that is obtained by treating data as a random nuisance parameter and averaging over all data. Consequently, the average log-likelihood statistic is given by

$$e^{\Lambda(\alpha)} = \frac{1}{2}e^{\Lambda_0(\alpha)} + \frac{1}{2}e^{\Lambda_1(\alpha)}$$

$$= C \cosh\left[\frac{2A}{N_0}\int_{-\infty}^{\infty} v_R(\xi)s(\xi - \alpha)d\xi\right].$$

The constant C does not depend on α, and can be suppressed by redefining $\Lambda(\alpha)$. Therefore, the log-likelihood statistic is

$$\Lambda(\alpha) = \log \cosh\left[\frac{2A}{N_0}u_R(\alpha)\right]$$

where $u_R(\alpha)$ is the real component of the matched filter output sampled at time α.

Theorem 8.7.1 *A log-likelihood statistic for an unknown delay in a randomly modulated BPSK waveform of blocklength n is*

$$\Lambda(\alpha) = \sum_{\ell=1}^{n} \log \cosh\left[\frac{2A}{N_0}u_R(\ell T + \alpha)\right]$$

where

$$u_R(t) = \int_{-\infty}^{\infty} v_R(\xi)s(\xi - t)d\xi.$$

Proof For a single pulse, we have seen that

$$\Lambda(\alpha) = \log \cosh\left[\frac{2A}{N_0}u_R(\alpha)\right].$$

For a train of BPSK pulses, the log-likelihood function becomes the sum

$$\Lambda(\alpha) = \sum_{\ell=1}^{n} \log \cosh\left[\frac{2A}{N_0}\int_{-\infty}^{\infty} v_R(\xi + \ell T)s(\xi - \alpha)d\xi\right],$$

and the theorem follows. ∎

The log-likelihood function is to be maximized by the choice of α. The maximum over α occurs where the derivative with respect to α is zero. The derivative of $\Lambda(\alpha)$, set equal to zero, gives

$$\sum_{\ell=1}^{n} \frac{2A}{N_0}\frac{du_R(\ell T + \alpha)}{d\alpha} \tanh\frac{u_R(\ell T + \alpha)}{N_0/2A} = 0.$$

This implicit equation in α must be solved by computing the left side for sufficiently many α to find the value of α that solves the equation.

Alternatively, one can work directly with the derivative to develop a sequential estimator. To solve this implicit equation for α, one can use an iterative procedure. A *delay tracker* is a delay-locked loop consisting of an iterative procedure that drives the term on the right side to zero by processing one term of the sum at a time. To form the iteration, write the feedback equation

$$\Delta\alpha(\ell T) = K \left[\frac{du_R(\ell T + \alpha)}{d\alpha} \tanh \frac{u_R(\ell T + \alpha)}{N_0/2A} \right].$$

If a large value of K is chosen for the feedback signal adjusting α, then the likelihood statistic $\Lambda(\alpha)$ would quickly be driven to its maximum provided there is no noise. However by choosing a large value of K, the loop will be sensitive to noise. By choosing a small value of K, the expected response is slowed, and the loop is less sensitive to noise. If the delay is a time-varying function, $\alpha(t)$, then K should be selected to best compromise between the loop's ability to track $\alpha(t)$, and the loop's sensitivity to noise.

A feedback implementation based on this condition is shown in Figure 8.14. The matched-filter output is sampled on each side of the presumed peak. The difference in these samples approximates the derivative and provides the error signal that drives the controlled clock.

A baseband implementation of the feedback equation is shown in Figure 8.15. A passband implementation suggested by this equation is shown in Figure 8.16. The derivative of the matched-filter output is obtained directly by passing the received signal through a filter equal to the derivative of the matched filter $\dot{s}(-t)$.

Just as in the case of phase synchronization, the hyperbolic tangent can be approximated in either of two ways. For large signal-to-noise-ratios, the approximation

$$\tanh x \approx \operatorname{sgn} x$$

may be considered appropriate. Because a BPSK demodulator can be written as the function $\operatorname{sgn} x$, this approximation can be interpreted as replacing the nonlinearity

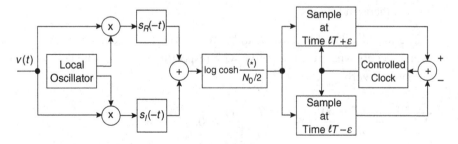

Figure 8.14. A sequential delay estimator for a BPSK waveform.

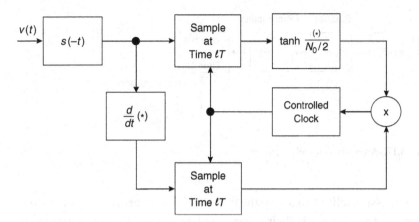

Figure 8.15. Another sequential delay estimator for a BPSK waveform.

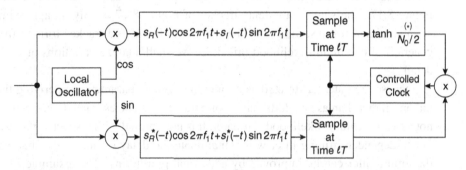

Figure 8.16. A sequential delay estimator at passband for a BPSK waveform.

shown in Figure 8.15 by a BPSK demodulator, thereby leading to a *decision-directed delay estimation*. Each BPSK data bit is estimated and then is used to invert or not invert the feedback signal according to the value of the data bit.

For small signal-to-noise ratios, the approximation of $\tanh x$ by x may be considered appropriate. This approximation is equivalent to replacing the feedback equation by

$$
\frac{d\alpha}{dt} = K\left[\frac{du_R(\ell T + \alpha)}{d\alpha} u_R(\ell T + \alpha)\right]
$$

$$
= \frac{1}{2} K \frac{du_R^2(t)}{dt}\bigg|_{t=\ell T+\alpha}.
$$

This approximation leads to the square-law delay estimator, which is given by

$$
\widehat{\alpha} = \operatorname{argmax}_\alpha \sum_{\ell=0}^{n-1} u_R^2(\ell T + \alpha).
$$

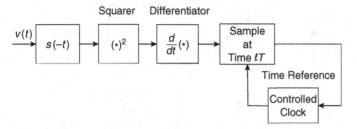

Figure 8.17. An equivalent delay tracker.

This expression tells us to divide the time axis into segments of length T, to align such segments of $u_R(t)$ and add their squares, and then to find the arg maximum of this sum as the estimate of α.

An equivalent delay tracker based on the square-law delay estimator is shown in Figure 8.17. The square-law delay tracker of Figure 8.17 is only an approximation, at small signal-to-noise ratios, to the maximum-likelihood tracker formed from the generalized maximum-likelihood principle. We shall analyze variations of it in some detail.

Because the data was treated as a random nuisance parameter in deriving the estimator – rather than as an additional set of parameters to be estimated – the estimator is not a true maximum-likelihood estimator. It is not optimal with respect to the data, and a data-dependent timing jitter will contaminate the delay estimate. We shall see that the timing jitter can be improved by additional prefiltering. This example illustrates that the generalized maximum-likelihood principle, though a valuable tool, does not always optimize with respect to the most important criteria.

The next task is to calculate the accuracy of the delay tracker both in the presence of noise and in the absence of noise. We shall study the timing jitter for the received BPSK waveform, given by

$$v(t) = \sum_{\ell=-\infty}^{\infty} a_\ell s(t - \ell T - \alpha) + n(t).$$

The output of the matched filter $s(-t)$ is

$$u(t) = \sum_{\ell=-\infty}^{\infty} a_\ell r(t - \ell T - \alpha) + n'(t).$$

The output $r(t)$ of the matched filter might be a Nyquist pulse to suit the needs of the data modulator, but the square-law delay tracker will see data-dependent timing jitter in the samples of the derivative of $u^2(t)$. This defect in the synchronization process can

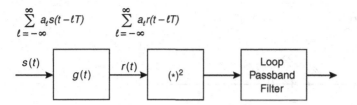

Figure 8.18. A simplified model of a delay tracker.

be seen by ignoring the noise term for the moment and writing

$$\frac{d}{dt}u^2(t) = \frac{1}{2} \sum_{\ell=-\infty}^{\infty} a_\ell r(t - \ell T - \alpha) \sum_{\ell=-\infty}^{\infty} a_\ell \frac{dr(t - \ell T - \alpha)}{dt}.$$

Then, at the sample point $t = \ell T + \alpha$, this becomes

$$\frac{d}{dt}u^2(t)\bigg|_{t=\ell'T+\alpha} = \frac{1}{2}a_{\ell'} \sum_{\ell=-\infty}^{\infty} a_\ell \frac{dr(t)}{dt}\bigg|_{t=(\ell'-\ell)T}$$

which need not equal zero because only that term of the sum with $\ell = \ell'$ is sure to have its derivative equal to zero.

The tracker is shown in its simplest representation in Figure 8.18. The delay-locked loop is depicted in Figure 8.18 simply as a narrow passband filter at frequency $1/T$. The desired output of the filter is a sinusoid. Alternate zero crossings (or peaks) of this desired sinusoid provide the data sample times. However, the output is not a pure sinusoid. It has phase jitter due to the random data modulating the input. We shall be concerned with this data-dependent phase jitter. The filter $g(t)$ has been the matched filter $s(-t)$, but for the delay tracker, we can use a different filter in order to reduce data-dependent timing jitter.

If the data symbols are independent and $E[a_\ell^2] = A^2$, the expected value of the squared output of filter $g(t)$ is

$$E[u^2(t)] = A^2 \sum_{\ell=-\infty}^{\infty} r^2(t - \ell T - \alpha) + \sigma_{n'}^2$$

where for now, $r(t) = g(t) * s(t)$. By using the relationship $r^2(t) \leftrightarrow R(f) * R(-f)$ and the Poisson sum formula, this can be expressed more conveniently in the frequency domain as

$$E[u^2(t)] = \frac{A^2}{T} \sum_{\ell=-\infty}^{\infty} \left[\int_{-\infty}^{\infty} R(f)R\left(\frac{\ell}{T} - f\right) df \right] e^{j2\pi \ell(t-\alpha)/T} + \sigma_{n'}^2$$

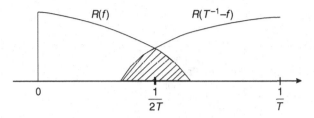

Figure 8.19. Illustrating the calculation of the tracking signal.

where now ℓ indexes translates on the frequency axis. The nature of the frequency integral appearing here is illustrated in Figure 8.19. For many pulses of interest, $R(f)$ and its translate overlap only for $\ell = -1, 0, 1$. This means that only the terms in the sum with $\ell = 0$ or ± 1 are nonzero. Therefore

$$E[u^2(t)] = U_0 + U_{1R} \cos 2\pi (t - \alpha)/T + U_{1I} \sin 2\pi (t - \alpha)/T + \sigma_{n'}^2$$

where

$$U_0 = \frac{A^2}{T} \int_{-\infty}^{\infty} R(f)R(-f)df$$

and

$$U_1 = \frac{A^2}{T} \int_{-\infty}^{\infty} R(f)R\left(\frac{1}{T} - f\right) df.$$

Indeed, for the case of a sinc pulse $r(t)$ with the first zero at T, the integral is nonzero only for $\ell = 0$, and so $E[u^2(t)]$ is a constant. This can be seen more directly in the time domain where $u^2(t) = A^2$. This means that, for sinc pulses, timing recovery is impossible by this method.

It is only the overlap in Figure 8.19 that provides a signal to drive the delay tracker. Consequently, there is no harm in filtering out, prior to the squarer, those regions of $R(f)$ that will not overlap.

The timing jitter is evaluated not from the expectation $E[u^2(t)]$ but from the actual squared signal $u^2(t)$. There will be no timing jitter if $u^2(t)$, in the absence of noise, is purely a (amplitude-modulated) cosine wave at frequency $1/T$. Thus we will choose the filter $g(t)$ to suppress the sine component. Suppose that the filter $g(t)$ is now chosen so that its output pulse, $r(t) = g(t) * s(t)$, has the form

$$r(t) = \sin(2\pi t/2T)p(t)$$

for some pulse $p(t)$. This filter output pulse $r(t)$ is an amplitude-modulated sinusoid that can be formed from the transmitted pulse by an appropriate filter $g(t)$, as shown

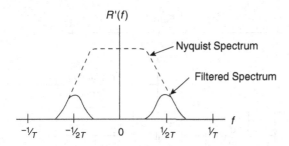

Figure 8.20. Spectrum of the filtered pulse.

in Figure 8.20. In any case, we have already seen that only the portion of the pulse spectrum in the vicinity of frequency $1/2T$ is used by the tracker. The new filter $g(t)$ will give the tracker signal a new shape that suppresses jitter. To see this, note that the output, in the absence of noise, is

$$u(t) = \sum_{\ell=-\infty}^{\infty} a_\ell \sin(2\pi(t - \ell T)/2T)p(t - \ell T)$$

$$= \sin \pi t/T \sum_{\ell=-\infty}^{\infty} (\pm a_\ell)p(t - \ell T).$$

The square of $u(t)$,

$$u^2(t) = (1 - \cos 2\pi t/T)\left[\sum_{\ell=-\infty}^{\infty} \pm a_\ell p(t - \ell T)\right]^2,$$

has a sinusoidal component at frequency $1/T$ that is amplitude modulated but has no phase modulation. Consequently, a time reference that is locked to this sinusoid will be free of data-dependent timing jitter.

Figure 8.21 shows a demodulator for a BPSK waveform that uses two receiver filters: one filter for data and the other filter, called the *bandedge timing recovery filter*, for the timing loop. The demodulator is free of data-dependent jitter, but not free of noise effects. By using a transmitted pulse such that the pulse out of the matched filter is a Nyquist pulse, there is no intersymbol interference at the sampling instants. By using a different filter $g(t)$ in the timing loop that is symmetric about the frequency $1/2T$, the time reference will predict the correct sampling instants without error caused by the signal. Of course, there will be sampling time jitter due to additive gaussian noise, which was not included in our analysis. The sensitivity to that noise can be made small by making the loop bandwidth small.

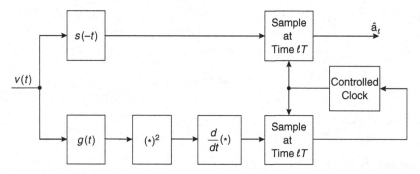

Figure 8.21. Demodulation of BPSK with a jitter-free clock.

8.8   Block synchronization

The task of symbol synchronization is to establish the proper sampling instants in the receiver. This task was studied in Section 8.7 and in this section, we regard symbol synchronization to be perfect and turn to block synchronization. The task of block synchronization is to establish a reference instant in time so that incoming symbols can be assigned their proper indices within a data block or data frame. If symbols are misframed, then subsequent operations will be incorrect and the entire communication link will go down.

Block synchronization may employ a special binary sequence as a marker; or a special pulse shape inserted periodically into the message waveform to mark the start of a new data block, as shown in Figure 8.22; or it may employ a synchronization-correcting code in which the data symbols are mapped into self-synchronizing codewords.

A binary marker for block synchronization is a special binary sequence that is embedded into the datastream to represent the start of a block, as shown in Figure 8.23. The appearance of this sequence designates the start of a block. If the random data contains this binary sequence, there may be a block synchronization error. In contrast, a pulse for block synchronization is a special waveform element, as shown in Figure 8.24, that is detected as a unit. A synchronization marker differs from a synchronization pulse in that it is superficially indistinguishable from the data. The bits of a marker may be first demodulated individually, then the bit sequence is recognized in the demodulated data. A synchronization pulse, on the other hand, is detected as a unit. A synchronization pulse may lead to a synchronization error only when the random data and the channel noise combine to form a received waveform segment that resembles the synchronization pulse.

The distinction we are making between a synchronization marker and a synchronization pulse is intended to underscore the variety of options from which one can choose in designing a block synchronizer. There is not a sharp distinction between the two

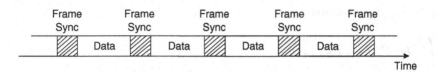

Figure 8.22. Embedding synchronization markers in the datastream.

Figure 8.23. Use of a sync marker.

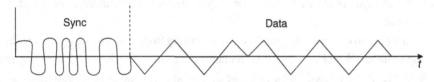

Figure 8.24. Use of a sync pulse.

concepts: the synchronization marker may be more suitable for a bandwidth-limited communication system, while the synchronization pulse may be more suitable for a power-limited communication system.

We define a block to have duration NT of which the first nT seconds contain the synchronization pulse or marker. The block structure defined for synchronization need not coincide with other block structures identified in the datastream for other purposes, such as channel coding. We shall study only the task of initial block synchronization using only a single block. However, the synchronization may be repeated every block, and so the proper synchronization can be anticipated in subsequent blocks once it is found. The block synchronizer then has the form of a delay-locked loop on the time scale of a block. Isolated synchronization errors can be rejected if they are clearly inconsistent with a synchronization reference propagated from earlier blocks.

The use of a special pulse for synchronization may lead to additional circuit complexity because there will be one matched filter for the sync pulse and one matched filter for each data pulse. Synchronization is easiest to analyze if the synchronization pulse is chosen to be orthogonal to every data pulse. This will not be acceptable if bandwidth is to be conserved, but might be attractive if bandwidth is plentiful, as for M-ary orthogonal signaling.

Let $s_m(t)$ for $m = 0, \ldots, M - 1$ be the data pulses used in the waveform. Let the synchronization pulse $b(t)$ be a pulse of duration nT and energy E_s satisfying the orthogonality condition

$$\int_{-\infty}^{\infty} b(t)s_m^*(t - \ell T)dt = 0$$

for each of the data pulses $s_m(t)$ that are used in the waveform, and for all ℓ. This means that, at the sampling instants, the output of a filter matched to $b(t)$ will have no output due to the data pulses. The output of the sync-matched filter equals E_s at every sync sampling instant and is zero at all data sampling instants.

The maximum-likelihood procedure is to sample the sync matched-filter output at each of the sampling instants, and then to choose the time at which the output is largest to estimate the block synchronization. It may be cumbersome, however, to store and sort the successive output values of the matched filter if the time interval between sync pulses is very large. Nevertheless, in contrast to the suboptimal procedure to be discussed next, there will always be a sync decision, and there is no fixed decision threshold.

A simpler procedure is to examine one sampling instant at a time, ignoring other sampling instants. Synchronization is declared each time the output of the sync matched filter rises above a fixed threshold. The detection of a sync pulse in gaussian noise then is similar to the task of demodulating an OOK pulse, except that the probability of a sync pulse being present at any sampling instant is $1/N$ whereas the probability of a pulse being present in a simple OOK waveform is $1/2$. Furthermore, the probabilities of error need not be equal. The threshold is set to compromise between the probability of sync detection and the probability of false synchronization. Synchronization may fail because of a missed sync pulse or because of a false sync detection. One or more false sync detections, combined with the true sync detection, will mean that the synchronization is ambiguous and unresolved. This is a *synchronization default*. Even worse, one may miss the true sync pulse and detect a false sync pulse. This is a *synchronization error*. At a higher level, it is possible to make inferences from a stream of such events because synchronization detections are expected periodically with period nT.

For signaling in the bandwidth-limited region, one prefers to avoid a sync pulse that is orthogonal to the data pulse because this requires additional bandwidth, and also because it requires an additional matched filter. Instead, one may use a special sequence of data bits for a synchronization marker. We shall investigate the use of a marker for the case of a BPSK waveform with an encoded bit sequence

$$(b_0, b_1, b_2, \ldots, b_{n-1}, d_0, d_1, \ldots, d_{N-1-n})$$

where $b_0, \ldots, b_{n-1}$ is the n-bit marker repeated at the beginning of each transmitted block, and $d_0, \ldots, d_{N-1-n}$ is the random data. In the absence of correct block synchronization, the marker may be delayed by any number of bit positions in the received data block. Possibly a beginning segment of the marker may appear at the end of one block

Blocklength	Maury–Styles Markers	Turyn Markers
7	1011000	1011000
8	10111000	
9	101110000	
10	1101111000	
11	10110111000	10110111000
12	110101100000	
13	1110101100000	1111100110101
14	11100110100000	11111001100101
15	111011001010000	111110011010110
16	1110101110010000	1110111000010110
17	11110011010100000	11001111101010010
18	111100110101000000	111110100101110011
19	1111100110010100000	1111000111011101101
20	11101101111000100000	11111011100010110100
21	111011101001011000000	111111010001011000110
22	1111001101101010000000	1111111100011011001010

Figure 8.25. Some binary markers.

and the remaining segment of the marker will then appear at the beginning of the next block. For this reason, we think of the N bits cyclically, and look for the n-bit marker appearing cyclically within an unframed block.

Figure 8.25 gives examples of binary markers known as the Maury–Styles markers and the Turyn markers. Those binary markers have been designed so that the Hamming distance[2] between a marker and each of its translates is large; the Hamming distance is computed only over the overlapping segment of the marker and its translate. The design criterion is selected so that it is unlikely for a translate of the marker, possibly with some error bits and filled out with random data bits, to cause a false synchronization. When the marker is embedded in data, however, there may be a false sync.

One can detect a synchronization marker by working either with demodulated data bits – called *postdetected data* – or by working directly with the received signal – called *predetected data*. When working with postdetected data bits, one searches an N-bit sequence for a known n-bit marker that can start on any of the N places. The probability is 2^{-n} that a randomly chosen sequence of n data bits is the marker sequence. This is the probability of a false synchronization at any sampling instant following n random data bits. To find the probability of at least one false occurrence of the sync

---

[2] The *Hamming distance* between two sequences of the same length is equal to the number of components in which the two sequences differ.

pattern in the block of N bits is a more difficult computation and depends on the specific pattern in the sync marker.

When E_b/N_0 is small, there will be bit errors in the detected data. Then the synchronization marker may be missed altogether, or may be found in the wrong place. At low E_b/N_0, it is better to use a maximum-likelihood synchronization detector that has full access to the predetection matched-filter outputs. The following theorem assumes that the matched-filter outputs (prior to the detection threshold) are partitioned into blocks of length N, with each block treated separately to find a cyclically shifted marker within it.

Theorem 8.8.1 *In the absence of intersymbol interference, a sufficient statistic for the detection of a sync marker in additive gaussian noise samples and cyclically embedded in a block of random BPSK data is*

$$\Lambda(\ell) = \sum_{i=0}^{n-1}\left[u_{((i+\ell))}b_i - \frac{N_0}{2}\log_e\cosh\frac{2A}{N_0}u_{((i+\ell))}\right]$$

for $\ell = 0,\ldots,N-1$, where u_i is the output of the matched filter $s^*(-t)$ at time iT, and $((i+\ell))$ denotes $(i+\ell)$ modulo n.

Proof It is sufficient to work with the output of the matched filter at the Nyquist sampling instants. The transmitted block

$$(b_0,b_1,b_2,\ldots,b_{n-1},d_0,d_1,\ldots,d_{N-1-n}),$$

is followed by the block

$$(b_0,b_1,b_2,\ldots,b_{n-1},d_0',d_1',\ldots,d_{N-1-n}').$$

The received block of matched-filter outputs is a vector random variable whose components are independent, gaussian random variables of variance σ^2 and mean given by the bit values. With a shift of ℓ places of the matched-filter samples, the probability density function of the block is

$$p(u|b,d,\ell) = \frac{1}{(\sqrt{2\pi}\sigma)^N}\prod_{i=0}^{n-1}e^{-(u_{i+\ell}-b_i)^2/2\sigma^2}\prod_{i=n}^{N-1}e^{-(u_{i+\ell}-d_{i-n})^2/2\sigma^2}.$$

Open the squares and collect the terms to write,

$$p(u|b,d,\ell) = B\prod_{i=0}^{n-1}e^{u_{i+\ell}b_i/\sigma^2}\prod_{i=n}^{N-1}e^{u_{i+\ell}d_{i-n}/\sigma^2}$$

where

$$B = \frac{1}{(\sqrt{2\pi}\sigma)^N} \prod_{i=0}^{N-1} e^{-(u_{i+\ell}^2/2\sigma^2)} \prod_{i=0}^{n-1} e^{-b_i^2/2\sigma^2} \prod_{i=n}^{N-1} e^{-d_{i-n}^2/2\sigma^2}.$$

The term denoted by B is actually independent of ℓ because the first product extends over all N indices. Therefore the term in brackets will have no effect on computing the maximum-likelihood and need not be inspected further. It will be dropped.

The data bits are random, independent, and equiprobable. Averaging over the random data gives

$$E[p(u|b,d,\ell)] = \prod_{i=0}^{n-1} e^{u_{i+\ell}b_i/\sigma^2} \prod_{i=n}^{N-1} E\left[e^{u_{i+\ell}d_{i-n}/\sigma^2}\right].$$

But

$$E[e^{u_{i+\ell}d_{i-n}/\sigma^2}] = \frac{1}{2}e^{u_{i+\ell}A/\sigma^2} + \frac{1}{2}e^{-u_{i+\ell}A/\sigma^2}$$
$$= \cosh(u_{i+\ell}A/\sigma^2).$$

Therefore the log-likelihood statistic is

$$\Lambda(\ell) = \log\left[\prod_{i=0}^{n-1} e^{u_{i+\ell}b_i/\sigma^2} \prod_{i=n}^{N-1} \cosh(u_{i+\ell}A/\sigma^2)\right]$$
$$= \log\left[\prod_{i=0}^{n-1} \frac{e^{u_{i+\ell}b_i/\sigma^2}}{\cosh(u_{i+\ell}A/\sigma^2)} \prod_{i=0}^{N-1} \cosh(u_{i+\ell}A/\sigma^2)\right].$$

The second product is now independent of ℓ because it ranges over all ℓ. It is constant and can be dropped. Then

$$\Lambda(\ell) = \log\prod_{i=0}^{n-1} \frac{e^{u_{i+\ell}b_i/\sigma^2}}{\cosh(u_{i+\ell}A/\sigma^2)}.$$

Expanding the logarithm leads to the log-likelihood statistic

$$\Lambda(\ell) = \sum_{i=0}^{n-1}\left[\frac{u_{i+\ell}b_i}{\sigma^2} - \log\cosh\frac{u_{i+\ell}A}{\sigma^2}\right].$$

Finally, replace σ^2 by $N_0/2$ and readjust the multiplying constant to complete the proof of the theorem. ∎

The maximum-likelihood block synchronizer will compute $\Lambda(\ell)$ for each ℓ from 0 to $N - 1$, and choose as the estimate $\widehat{\ell}$ that value of ℓ for which $\Lambda(\ell)$ is largest. If the sync marker appears in the data, then the maximum-likelihood block synchronizer cannot protect against the possibility of choosing the false marker. It will, however, reduce the possibility of an occasional badly corrupted bit, either in the marker or in the data, leading to a lost sync.

It is interesting and informative to examine approximations to the likelihood statistic for the case of a large signal-to-noise ratio, and for the case of a small signal-to-noise ratio. For a large signal-to-noise ratio, we can make the approximation that $\cosh x \sim \frac{1}{2} \exp|x|$, and $\log \cosh x \sim |x| - \log 2$. Then

$$\Lambda(\ell) \sim \sum_{i=0}^{n-1} \left[\frac{u_{i+\ell} b_i}{\sigma^2} - \frac{A|u_{i+\ell}|}{\sigma^2} + \log 2 \right].$$

Suppressing the constants, and noting that bits are detected by the rule $|u_i|/u_i = \widehat{b}_i$ this becomes

$$\Lambda(\ell) = \sum_{i=0}^{n-1} u_{i+\ell}[b_i - \widehat{b}_{i+\ell}]$$

where $\widehat{b}_i = \pm A$ is the regenerated value of the ith symbol. The sign of $\pm A$ is the sign of u_i. For any ℓ at which the demodulated sequence matches the marker, we have $\widehat{b}_{i+\ell} = b_i$ for $i = 0, \dots, n - 1$, so $\Lambda(\ell) = 0$. For other ℓ, $\Lambda(\ell)$ is negative because if $b_i = A$ and $\widehat{b}_{i+\ell} = -A$, then $u_{i+\ell}$ must be negative, while if $b_i = -A$ and $\widehat{b}_{i+\ell} = A$, then $u_{i+\ell}$ must be positive. Hence the approximation under consideration leads to the conclusion that the data sequence should be demodulated and searched for the sync pattern. If there are two or more copies of the sync pattern, their likelihoods under the approximation are identical, and the ambiguity cannot be resolved.

On the other hand, for a small signal-to-noise ratio, we use the approximation $\log \cosh x \sim x^2$ obtained from the first term of a Taylor series expansion. Then

$$\Lambda(\ell) \sim \sum_{i=0}^{n-1} \left[u_{i+\ell} b_i - \frac{2A^2}{N_0} u_{i+\ell}^2 \right].$$

This tells us that, to estimate ℓ, correlate the predetected data samples with the marker samples, offset the correlation by a term in the sample power for the segment of length n, and then select the largest. We can contrast this approximation with the earlier approximation if we write it as

$$\Lambda(\ell) = \sum_{i=0}^{n-1} u_{i+\ell}[b_i - \widetilde{b}_{i+\ell}]$$

where $\widetilde{b}_i = (2A^2/N_0)u_i$ is a soft output from the demodulator. Now the received signal $u_{i+\ell}$ is weighted by the soft error signal $b_i - \widetilde{b}_{i+\ell}$, and the sum forms the likelihood statistic.

8.9 Synchronization sequences

Sequences are used to construct pulses for block synchronization, as described in Section 8.8, and also for spread-spectrum communication, as described in Chapter 12. For the problem of block synchronization in some forms of wideband communications, we are interested in the properties of the pulse $s(t)$ when it appears in isolation. Then we speak of the aperiodic autocorrelation function, and in this section, we consider pulses with good aperiodic autocorrelation functions. For application to spread-spectrum communications, we will be interested in the properties of the infinitely long waveform obtained by periodically repeating $s(t)$. Then we speak of the periodic autocorrelation function.

A pulse for synchronization should have a strongly distinguishable characteristic. Typically, this means that it should have a strong and sharp matched-filter output. However, a narrow pulse will usually be unacceptable to the transmitter if its amplitude is much larger than the amplitude of the data waveform. To get a strong matched-filter output for a sync pulse with limited peak amplitude, one may need to use a synchronization pulse $s(t)$ that has both a large timewidth and a large bandwidth, perhaps one whose bandwidth is much larger than the reciprocal of its timewidth.

Other constraints may be imposed on synchronization pulses. A simple constraint is that $s(t)$ takes on only the values ± 1, and that transitions occur only at multiples of a fixed time interval T. Alternative constraints may be that $s(t) = \pm 1 \pm j$, or that $|s(t)| = 1$. Each such constraint leads to a class of synchronization pulses with a rich structure. These pulses have not been completely classified. We shall describe only one such class, the class of *Barker pulses*, as an example of such pulses for synchronization.

Definition 8.9.1 *A Barker sequence of blocklength n is a sequence of symbols c_j, for $j = 0, \ldots, n - 1$, taking values $+1$ and -1, with the properties that the discrete (aperiodic) autocorrelation function*

$$\phi(k) = \sum_{j=0}^{n-1} c_j c_{j+k}$$

satisfies $\phi(k) = n$ if $k = 0$ and $|\phi(k)| \leq 1$ if $k \neq 0$.

Thus when k is not zero, the only allowed values of $\phi(k)$ are -1, 0, and $+1$. The known Barker sequences are given in Figure 8.26. It is not known if there are other

n													
2	+	+											
2	−	+											
3	+	+	−										
4	+	+	−	+									
4	+	+	+	−									
5	+	+	+	−	+								
7	+	+	+	−	−	+	−						
11	+	+	+	−	−	−	+	−	−	+	−		
13	+	+	+	+	+	−	−	+	+	−	+	−	+

Figure 8.26. Known Barker sequences.

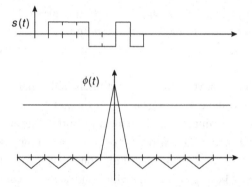

Figure 8.27. A Barker pulse and its autocorrelation function.

Barker sequences, but it is known that there are no others with n odd, and no others with n smaller than 12,100.

To form a pulse from the Barker sequence, choose a pulselet $s'(t)$. The Barker pulse is then

$$s(t) = \sum_{\ell=0}^{n-1} c_\ell s'(t - \ell T).$$

Figure 8.27 shows a Barker pulse using square pulselets together with its autocorrelation function.

It is not possible to design a binary sequence of finite length whose autocorrelation function has no sidelobes. However, it is possible to design a pair of binary sequences such that the sum of their two autocorrelation functions has no sidelobes. The sidelobes of the two autocorrelation functions must have opposite signs, and so they cancel. The mainlobes are both positive, and so they add positively. Such sequences can be used, for example, as the real and imaginary parts of a complex sequence. Then the real part of the autocorrelation function will have no sidelobes.

Definition 8.9.2 *A pair of finite sequences, of blocklength n, from the binary alphabet* $\{+1, -1\}$ *is called a complementary code (or a Golay complementary code) if the sum of their autocorrelation functions is zero for all nonzero integer shifts.*

There is essentially only one complementary code with a blocklength equal to two, namely the one with codewords $c_1 = (+1, +1)$ and $c_2 = (+1, -1)$. The autocorrelation sequences are $\phi_1 = (+1, +2, +1)$ and $\phi_2 = (-1, +2, -1)$. These sum to $(0, 4, 0)$. Either codeword c_1 or c_2 could be replaced by its negative or by its reciprocal without changing its autocorrelation sequence. New codes obtained by these slight changes are not essentially different from the given code.

There are no complementary codes with a blocklength equal to three. There is essentially only one complementary code with a blocklength equal to four. It has codewords $c_1 = (+1, +1, +1, -1)$ and $c_2 = (+1, +1, -1, +1)$. The autocorrelation sequences of c_1 and c_2 are

$$\phi_1 = (-1, 0, 1, 4, 1, 0, -1)$$

$$\phi_2 = (1, 0, -1, 4, -1, 0, 1).$$

These sum to $(0, 0, 0, 8, 0, 0, 0)$.

We can express the notion of a complementary code in the language of polynomials. A codeword with components c_i for $i = 0, \ldots, n-1$ can be represented by the codeword polynomial

$$c(x) = \sum_{i=0}^{n-1} c_i x^i.$$

The reciprocal polynomial of $c(x)$, denoted $\tilde{c}(x)$, is given by

$$\tilde{c}(x) = \sum_{i=0}^{n-1} c_{n-1-i} x^i$$

$$= x^{n-1} c(x^{-1}).$$

The two polynomials $c_1(x)$ and $c_2(x)$ of degree n form a complementary code if and only if

$$c_1(x)c_1(x^{-1}) + c_2(x)c_2(x^{-1}) = 2n.$$

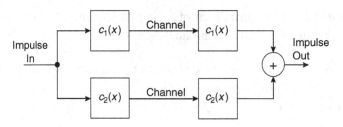

Figure 8.28. An application of complementary codes.

The right side is a discrete impulse of amplitude $2n$, the left side is a complicated way of writing the impulse. Each polynomial on the left can be represented by a finite-impulse-response filter, so this equation can be represented by a network of such filters, as shown in Figure 8.28. The coded signals can be passed through a pair of channels, each channel with a peak power constraint. The output of the complete circuit is an impulse whose amplitude is $2n$ times larger than the peak signal of either channel. In this way, the complementary code enables one to transmit a large impulse through a channel with a peak amplitude constraint. The advantage of using a complementary code is that there are no sidelobes in the output of the demodulator.

The two channels can be obtained by partitioning a single channel, as by frequency division or time division. Another possibility is to use the two polarization components of an electromagnetic signal. Of course, the most evident is to use the in-phase and quadrature components of a complex waveform.

Problems for Chapter 8

8.1. Sketch in detail a complete functional block diagram of a coherent receiver for offset QPSK, including carrier recovery and data clock synchronization. Can this receiver be used for MSK?

8.2. Two different methods are proposed for handling phase in a sixteen-ary orthogonal communication system. One method uses a noncoherent demodulator. The other method uses a coherent demodulator, but for carrier recovery, the waveform consists of alternating segments, 75 milliseconds of data symbols, followed by 25 milliseconds of unmodulated carrier, with the same average power as the modulated carrier. The average data rate of the two methods is the same. Which method should be preferred at a symbol error rate of 10^{-5}? Give a reason. Does the answer depend on error rate?

8.3. Based on the maximum-likelihood principle, develop and sketch a phase synchronization procedure for a binary OOK waveform.

8.4. A given digital communication system has a sync procedure that can track phase noise $\theta(t)$ to within an error signal $\theta_e(t)$. Model the error signal as bandlimited gaussian noise. Explain how, if the phase noise in the received signal is removed by eliminating its cause, the sync procedure can be used to increase the data rate. Does this imply that a waveform of maximum data rate for an additive noise channel cannot be self-synchronized? Repeat the argument for symbol synchronization.

8.5. A given digital communication system for a passband dual polarization channel uses four-dimensional modulation, consisting of QPSK on each polarization axis of an electromagnetic wave. Within the channel, there is an arbitrary phase shift on the carrier and an arbitrary polarization rotation. Use the maximum-likelihood principle to set up a method for recovering phase and polarization. Will the solution change if each polarization channel can have an independent phase shift on its carrier?

8.6. With data treated as a random nuisance parameter, derive the likelihood function for the phase of a binary FSK waveform. Sketch a coherent receiver designed around the maximum-likelihood estimator. Explain how the nonlinearity in the phase estimator may be approximated in a way that results in a decision-directed estimator.

8.7. The maximum-likelihood estimator developed in Problem 8.6 can be used bit by bit to estimate the phase on every received bit of a binary FSK waveform. Is the demodulator that results equivalent to, better than, or poorer than a non-coherent FSK demodulator? If these demodulators are not equivalent, explain the difference and decide whether this means that there is some weakness in the theory.

8.8. a. With data treated as a random nuisance parameter, develop and sketch a maximum-likelihood procedure for simultaneously estimating phase and symbol timing for a BPSK waveform.

      b. Repeat the development for a QPSK waveform.

8.9. By showing that the feedback signal is identical, prove that the Costas loop is mathematically equivalent to a phase-locked loop driven by the square of the received signal.

8.10. a. With data treated as a random nuisance parameter, develop a maximum-likelihood procedure for finding a marker, consisting of n QPSK symbols, in a QPSK block of length N. Compare the procedure to the solution for BPSK.

      b. Repeat for an eight-ary PSK waveform.

8.11. In a partially-coherent channel, the phase angle is not known, but it changes so slowly with respect to the bit rate that phase offset can be estimated from a long sequence of bits and can be corrected. However, there is a residual sign

ambiguity if the modulation is BPSK because the phase estimate has a 180°
ambiguity. Instead of using DPSK, it is proposed to use BPSK and to resolve
this ambiguity by inserting a sequence of four ones after every 100 channel bits,
and to use a data translation code to remove any sequence of four ones from
the channel bit stream. (An example of such a code inserts in the encoder an
artificial zero after every three consecutive ones and deletes in the decoder a
zero after every three consecutive ones.)

a. Explain how this resolves the ambiguity.

b. By how much, in the best case, is the data rate reduced?

c. Sketch a copy of Figure 3.10 and on the sketch plot a curve for the proposed
scheme. In the best case, what must be the change in E_b/N_0 at a bit error
rate of 10^{-5}?

8.12. The likelihood function with extraneous parameters averaged out is called
the generalized likelihood function. Maximizing the generalized likelihood
function is not the same as the maximum-likelihood principle. This can be
demonstrated by deriving the estimates of the amplitude of a BPSK waveform
from each point of view. Let $s(t)$ be a pulse such that

$$\int_{-\infty}^{\infty} s(t)s^*(t - \ell T)dt = 0 \qquad \text{for } \ell \neq 0.$$

The received BPSK waveform is

$$v(t) = \sum_{\ell=0}^{n-1} a_\ell s(t - \ell T) + n(t)$$

where $a_\ell = \pm A$ and A is an unknown parameter to be estimated.

a. Show that the simultaneous maximum-likelihood estimate of both data and
amplitude is

$$\widehat{a}_\ell = \text{sgn} [u_\ell]$$

$$\widehat{A} = \frac{1}{n}u_\ell \text{sgn} [u_\ell]$$

where u_ℓ is the ℓth output sample of the matched filter.

b. By averaging over all data sequences, develop a generalized log-likelihood
function

$$\Lambda(A) = \prod_{\ell=0}^{n-1} \left[\tfrac{1}{2}e^{2Au_\ell - A^2} + \tfrac{1}{2}e^{-2Au_\ell - A^2} \right].$$

What estimator for A maximizes this function? How does this estimator compare to the estimator of part a?

8.13. By reference to the noise analysis of the phase-locked loop, analyze the accuracy of the delay-locked loop used for symbol synchronization.

8.14. A passband sampler improperly samples with respect to a carrier frequency f_0' instead of with respect to the correct carrier frequency. Explain how a sufficiently good equalization procedure will automatically correct for small carrier offsets.

8.15. The first three Barker sequences of odd blocklength are

$$n = 3 : + + -; \quad n = 5 : + + + - +; \quad n = 7 : + + + - - + -.$$

Sketch the output of a matched filter for each of these Barker sequences under the following conditions:
a. The waveform consists of a single Barker sequence.
b. The waveform is periodic, consisting of a periodic repetition of the Barker sequence with period n.

8.16. Construct a list of all complementary codes of blocklength 4. How many pairs are there on the list? Show that essentially there is only one pair. We say that two codeword pairs are essentially the same if one can be turned into the other by the operations of replacing either codeword by its negative or of replacing either codeword by its reciprocal.

8.17. The two components of a real complementary code are used as the in-phase and quadrature components of an MSK pulse. Prove that all sidelobes in the matched-filter output for this MSK pulse are in phase quadrature to the main lobe.

Notes for Chapter 8

Techniques of automatic phase control go back to the early days of electronic systems and it is difficult to identify a specific starting point. Jaffe and Rechtin (1955) seem to be the first to explicitly identify the phase-locked loop as a technique to track a time-varying sinusoid in the presence of noise. The study of phase-locking in the presence of noise has been studied by Gruen (1953), by Viterbi (1963), and by many others. The application of the phase-locked loop to communication systems was discussed by Costas (1956). Bandedge timing recovery was discussed by Lyon (1975). The removal of data-dependent jitter from the timing recovery loop is due to Franks and Bubrouski (1974).

Marker design has been studied by Maury and Styles (1964), by Turyn (1968), and by others. Comprehensive surveys of phase and symbol synchronization have been written by Stiffler (1971), by Lindsey (1972), and by Franks (1980, 1983), and surveys of block synchronization have been written by Scholtz (1980).

The design of special pulse shapes and sequences for aperiodic synchronization has a rich literature starting with the work of Barker (1953), and including further work by Welti (1960), Turyn and Storer (1961), Frank (1963), Golomb and Scholtz (1965), and others. The maximum-likelihood strategy for finding such sequences embedded in a random datastream was studied by Massey (1972). The probability of synchronization error was studied by Nielsen (1973). Complementary codes were introduced by Golay (1949, 1961). Variations on this idea include the phase-modulated pulses studied by Frank (1980).

9 Codes for Digital Modulation

Rather than modulate one data symbol at a time into a channel waveform, it is possible to modulate the entire datastream as an interlocked unit into the channel waveform. The resulting waveform may exhibit symbol interdependence that is created intentionally to improve the performance of the demodulator. Although the symbol interdependence does have some similarity to intersymbol interference, in this situation it is designed deliberately to improve the minimum euclidean distance between sequences, and so to reduce the probability of demodulation error.

The methods developed in Chapter 4 for demodulating interdependent sequences led us to a positive view of intersymbol interdependence. This gives us the incentive to introduce intersymbol interdependence intentionally into a waveform to make sequences more distinguishable. The digital modulation codes that result are a form of data transmission code combined with the modulation waveform. The output of the data encoder is immediately in the form of an input to the waveform channel. The modulator only needs to apply the proper pulse shape to the symbols of the code sequence.

In this chapter, we shall study trellis-coded modulation waveforms, partial-response signaling waveforms, and continuous-phase modulation waveforms. Of these various methods, trellis-coded modulation is the more developed, and is in widespread use at the present time.

9.1 Partial-response signaling

The simplest coded-modulation waveforms are called partial-response signaling waveforms. These coded waveforms can be motivated by recalling the method of decision-feedback equalization. That method suffers from the occurrence of error propagation. A code can be used to eliminate, or reduce, the possibility of error propagation by moving the function of the decision feedback into a precoder prior to the modulator. By anticipating the intersymbol interference, its effect can be subtracted out before it happens provided the sampled channel coefficients g_ℓ are integers. This clever technique can be applicable without affecting the channel input alphabet.

A channel that might require great care to avoid intersymbol interference is the ideal rectangular baseband channel

$$H(f) = \begin{cases} 1 & |f| \leq W \\ 0 & |f| > W. \end{cases}$$

A pulse $s(t)$ will only completely fill the band of this channel if its transform satisfies

$$S(f) = \begin{cases} (2W)^{-1} & |f| \leq W \\ 0 & |f| > W. \end{cases}$$

The inverse Fourier transform of this $S(f)$ gives the pulse shape

$$s(t) = \frac{\sin 2\pi Wt}{2\pi Wt}$$

$$= \text{sinc } 2Wt.$$

The pulse $s(t)$ is a Nyquist pulse at symbol spacing $T = 1/2W$.

To use the ideal bandlimited channel to transmit data that is represented in the form of a stream of real numbers, simply change the stream of real numbers into a stream of pulse-amplitude modulated sinc pulses, given by

$$c(t) = \sum_{\ell=-\infty}^{\infty} a_\ell \text{sinc } 2W(t - \ell T).$$

If $a_\ell = \pm 1$, this is binary phase-shift keying using sinc pulses. The choice of a sinc pulse as the pulse shape ensures that the bit rate of the BPSK waveform is as high as possible for the given ideal bandlimited channel. If one tried to signal with BPSK at a faster bit rate, the channel response would surely create intersymbol interference.

Of course, waveform design with sinc pulses to make full use of an ideal bandlimited channel, though mathematically simple, is impossible to implement. Not only is the sinc pulse unrealizable, but the ideal bandlimited channel is unrealizable also. In this section, we shall come to a remarkable reformulation of this signaling waveform by the observation that the unrealizability of the waveform and the unrealizability of the channel can be made to cancel each other, leaving a scheme that is realizable in both respects.

The method of partial-response signaling can be viewed as a way to obtain high data rate by allowing symbols to overlap intentionally. Partial-response signaling creates intentional intersymbol interference. Usually the constraint is that the intersymbol interference coefficients take on integer values. The partial-response waveform can be demodulated by a decision-feedback demodulator or by a maximum-likelihood sequence demodulator using the Viterbi algorithm.

A practical alternative to a BPSK waveform with sinc pulses is the simplest partial-response waveform, which is known as the *duobinary (partial-response) waveform.* Define the *duobinary pulse* as

$$\text{duob}(t) = \frac{\sin \pi t}{\pi t(1 - t)}.$$

A sketch of the duobinary pulse is shown in Figure 9.1. It is not a Nyquist pulse because duob(1) is equal to one. For other nonzero values of ℓ, duob(ℓ) = 0. We can eliminate the need to generate the pulse $s(t)$ from the modulator if we choose a channel with the impulse response $s(t)$. The Fourier transform of duob(t) is

$$S(f) = \begin{cases} 2e^{-j\pi fT} \cos \pi fT & |f| \leq 1/2T \\ 0 & |f| > 1/2T. \end{cases}$$

The complex exponential corresponds to a delay of $T/2$ and can be absorbed into the timing of the modulator and demodulator. Thus with the timing adjusted and the transmitted pulses replaced by impulses, the channel can have the transfer function $H(f) = \cos \pi fT$ for $|f| \leq 1/2T$, and otherwise $H(f) = 0$. Not only is this lowpass channel a practical channel in contrast to the ideal rectangular lowpass channel, but it makes the modulator trivial, as is shown in Figure 9.2.

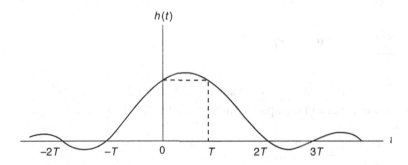

Figure 9.1. The duobinary pulse.

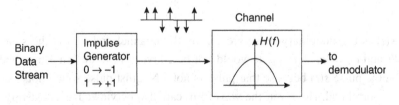

Figure 9.2. Equivalent modulation scheme.

The duobinary pulse can be viewed as a pair of pulses by observing that

$$s(t) = \text{sinc}(t) + \text{sinc}(t - 1)$$

$$= \frac{\sin \pi t}{\pi t (1 - t)}$$

$$= \text{duob}(t).$$

This has the Fourier transform

$$S(f) = \text{rect}(f) + e^{-j2\pi f} \text{rect}(f)$$

$$= 2e^{-j\pi f} \cos \pi f \, \text{rect}(f).$$

Accordingly, the duobinary partial-response signaling waveform is defined as binary antipodal signaling using the duobinary pulse as the modulation pulse. Then

$$c(t) = \sum_{\ell=-\infty}^{\infty} a_\ell \text{duob} \, 2W(t - \ell T)$$

$$= \sum_{\ell=-\infty}^{\infty} (a_\ell + a_{\ell-1}) \text{sinc} \, 2W(t - \ell T).$$

Thus, the waveform $c(t)$, although formed as BPSK with the duobinary pulse, can be interpreted as

$$c(t) = \sum_{\ell=-\infty}^{\infty} c_\ell \text{sinc} \, 2W(t - \ell T)$$

where $c_\ell = a_\ell + a_{\ell+1}$, and so takes values $-2, 0$, and $+2$.

How should the channel output be filtered? The received signal is

$$v(t) = \sum_{\ell=-\infty}^{\infty} a_\ell s(t - \ell T) + n(t)$$

$$= \sum_{\ell=-\infty}^{\infty} c_\ell p(t - \ell T) + n(t)$$

where $s(t)$ is the duobinary pulse created by the channel and $p(t)$ is the sync pulse sinc $2Wt$ and $c_\ell = a_\ell + a_{\ell+1}$. It would not be convenient to design a matched filter based on the pulse $s(t)$ because that pulse is not a Nyquist pulse so the noise samples would be correlated. However, the waveform can also be viewed as consisting of the coded data c_ℓ modulating sinc pulses at the Nyquist rate. Theorem 7.4.2 tells us that, for this situation, the outputs of the matched filter $p(-t)$ form a sufficient statistic. The matched filter is the sinc pulse

$$p(-t) = \text{sinc}(t/T).$$

The output is a Nyquist pulse, so the output noise samples are uncorrelated if the input noise is white.

The conversion of the data symbols a_ℓ into the channel code symbols can be regarded as an encoding operation. The binary datastream a_ℓ, which is a stream of $+1$ and -1 symbols, can be represented by the polynomial

$$a(x) = \sum_{\ell=-\infty}^{\infty} a_\ell x^\ell.$$

Conceptually, this datastream can be regarded as encoded by passing it through a real, finite-impulse-response filter represented by the polynomial $g(x) = 1+x$. The encoded datastream c_ℓ, which is represented by the polynomial

$$c(x) = \sum_{\ell=-\infty}^{\infty} c_\ell x^\ell,$$

is defined by the polynomial product

$$c(x) = g(x)a(x)$$
$$= (1+x)a(x).$$

Figure 9.3 shows this form of the conceptual encoder for duobinary signaling. The waveform is formed by first passing the datastream, represented by a series of impulses, through a discrete-time finite-impulse-response filter with generator polynomial $g(x)$, and then passing the encoded output through an ideal lowpass filter that will turn

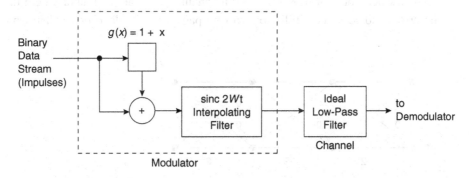

Figure 9.3. Conceptual modulation scheme.

impulses into sinc pulses. But this is equivalent to passing the data through the single filter that is equal to the cascade of a finite-impulse-response filter and the ideal low-pass filter. The single filter has an impulse response equal to the convolution of the impulse responses of its two subfilters, and this is the duobinary pulse. Similarly, in the frequency domain, the cosine pulse spectrum is the sum of two rectangular spectra as modified by the translation theorem.

The demodulator samples the matched-filter output $v(t)$ at the sampling instants ℓT. The intersymbol interference is easy to see by inspection of the pulse shape in Figure 9.5. A single pulse will produce two nonzero samples, both equal to one; all remaining samples of the intersymbol interference are equal to zero. In effect, the samples are noisy copies of the output of the encoding polynomial $g(x) = 1 + x$. The sampled output of the matched filter is

$$u_\ell = \sum_{\ell=-\infty}^{\infty} \left[\sum_{\ell'=-\infty}^{\mu} g_{\ell'} a_{\ell-\ell'} + n_\ell \right]$$

$$= \sum_{\ell=-\infty}^{\infty} c_\ell + n_\ell$$

where $c_\ell = a_\ell + a_{\ell-1}$ for duobinary signaling. For binary signaling, a_ℓ takes values in the set $\{-1, 1\}$, so c_ℓ takes values in the set $\{-2, 0, 2\}$. The possible sequences of received samples c_ℓ for the duobinary waveform are compactly displayed by the trellis in Figure 9.4.

The data symbols can be recovered from the samples by means of a Viterbi demodulator or by a decision-feedback demodulator. Figure 9.5 shows the waveform used with a decision-feedback demodulator. A decision-feedback demodulator is subject to error propagation; a single demodulation error will produce incorrect feedback, which can lead to more errors. In this way, one demodulation error may propagate into many subsequent errors.

A better method is to use a precoder. A precoder can be thought of as the decision-feedback demodulator moved across the channel into the modulator where there are no errors and so no possibility of error propagation. The decision-feedback precoder

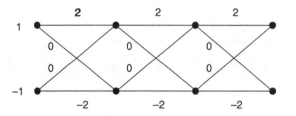

Figure 9.4. Trellis for duobinary partial-response.

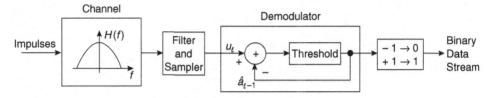

Figure 9.5. Decision-feedback demodulator.

anticipates what will happen in the channel and does the inverse to the data before it reaches the channel. However, the decision feedback deals with real numbers and real arithmetic, whereas the precoder deals with binary data and modulo-two arithmetic. For this reason, it is not completely obvious that the decision feedback can be moved into the modulator.

The structure of the precoder will be easy to develop if we first replace our demodulator by one that uses digital logic in the form of a postcoder. This means that we want to postpone the decision feedback to a point at which all variables are binary. Specifically, the threshold will be disentangled from the decision feedback.

The threshold test is

$$
\widehat{a}_\ell = \begin{cases} 1 & \text{if } u_\ell - \widehat{a}_{\ell-1} \geq 0 \\ -1 & \text{if } u_\ell - \widehat{a}_{\ell-1} < 0. \end{cases}
$$

We can rewrite this as

If $\widehat{a}_{\ell-1} = 1$,

$$
\widehat{a}_\ell = \begin{cases} 1 & \text{if } u_\ell \geq 1 \\ -1 & \text{if } u_\ell < 1. \end{cases}
$$

If $\widehat{a}_{\ell-1} = -1$,

$$
\widehat{a}_\ell = \begin{cases} 1 & \text{if } u_\ell \geq -1 \\ -1 & \text{if } u_\ell < -1. \end{cases}
$$

We want to decouple the threshold tests from the decision feedback, which means that the threshold circuit is not privy to $\widehat{a}_{\ell-1}$. It is sufficient, then, to quantize u_ℓ into the three intervals $u_\ell \geq 1$, $-1 < u_\ell < 1$, and $u_\ell \leq -1$. Because we want a binary-valued implementation, we will use instead a suboptimal two-level quantization by rewriting the case $-1 < u_\ell < 1$ as $|u_\ell| < 1$, and merging the other two cases, $u_\ell \geq 1$ and $u_\ell \leq -1$ as $|u_\ell| \geq 1$.

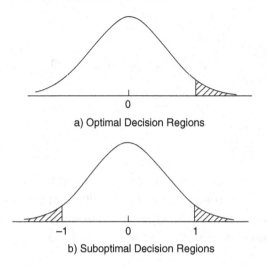

a) Optimal Decision Regions

b) Suboptimal Decision Regions

Figure 9.6. Error probabilities for two threshold rules.

The decision rule now is replaced by the suboptimal rule

If $\widehat{a}_{\ell-1} = 1$,

$$\widehat{a}_\ell = \begin{cases} 1 & \text{if } |u_\ell| \geq 1 \\ -1 & \text{if } |u_\ell| < 1. \end{cases}$$

If $\widehat{a}_{\ell-1} = -1$,

$$\widehat{a}_\ell = \begin{cases} 1 & \text{if } |u_\ell| < 1 \\ -1 & \text{if } |u_\ell| \geq 1. \end{cases}$$

The suboptimal decision rule will have degraded performance because the decision regions have been altered. Figure 9.6 compares the probability of error of the optimal ternary decision regions with the probability of error of the suboptimal binary decision regions. For example, for the case in which $a_\ell = -1$ and $a_{\ell-1} = 1$, the mean of u_ℓ is zero. Then, for the optimal decision rule, an error will occur only if u_ℓ is greater than one. For the suboptimal decision rule, an error will also occur if u_ℓ is less than minus one. By inspection of Figure 9.6, it becomes clear that for the suboptimal decision regions, the probability of error is twice as large as for the optimal decision rule. In return for accepting this larger error probability, we can now design the decision feedback circuit as a binary postcoder in series with the threshold, as shown in Figure 9.7.

We now have reached the point where the design of a precoder is obvious. Indeed, for the example of duobinary partial-response, we can simply move the postcoder into the modulator where it becomes the required precoder. The duobinary partial-response modulation with a precoder is shown in Figure 9.8.

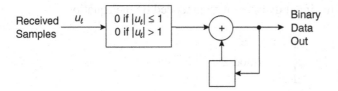

Figure 9.7. Alternative decision-feedback logic.

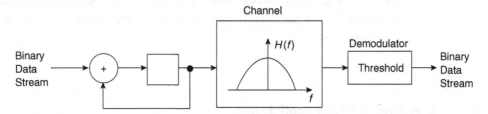

Figure 9.8. Duobinary partial-response modulation with precoding.

Our detailed study of the duobinary partial-response waveform typifies other forms of partial-response signaling, and these are also important in applications. Each choice of the generator polynomial $g(x)$ gives another partial-response waveform. Several choices for the generator polynomial of partial-response signaling are in common use, and the simpler of them have conventional names. The most widely used partial-response waveforms are defined as follows:

$$\text{duobinary} \quad g(x) = 1 + x$$
$$\text{dicode} \quad g(x) = 1 - x$$
$$\text{modified duobinary} \quad g(x) = 1 - x^2.$$

Notice that the two latter examples have negative signs, which leads to a considerable difference both in the pulse shape and in the spectrum. The signaling pulse $s(t)$ will have a Fourier transform $S(f)$ that is zero for $f = 0$. This is an important consideration for channels, such as magnetic recording channels, that cannot pass zero frequency.

The output of the dicode channel is the arithmetic difference, as real numbers, of the past two channel input bits. The channel output alphabet is $\{-2, 0, 2\}$ as for the duobinary waveform but, in this case, the output is 0 if the last two input symbols are the same, and otherwise it is either -2 or 2. It is 2 if the current input symbol is 1 and the previous channel input symbol is -1.

Most partial-response waveforms can be used with a precoder. The formulation of the precoder is generalized as follows. Suppose that g_k is an integer for all k and that

g_0 is not even. Let the input datastream be precoded by the equation

$$b_i = d_i - \frac{1}{g_0} \left[\sum_{i=1}^{n-1} g_j b_{i-j} \right] \qquad (\text{mod } 2).$$

Except for the modulo-two operation, this amounts to moving the feedback operation from the demodulator to the modulator. The precoding is the inverse of the encoding. By doing this inverse operation before the encoding, we eliminate the need to do it in the demodulator. It is trivial to verify that the output of the encoder is identical to the input of the precoder.

9.2   Continuous-phase modulation

A *phase-modulated signaling waveform* is a waveform of the form

$$c(t) = \cos(2\pi f_0 t + \theta(t)).$$

A *continuous-phase signaling waveform* is a phase-modulated waveform for which $\theta(t)$ is a continuous function of time. We are interested in continuous-phase waveforms for digital communications. The special feature of a continuous-phase waveform is that its amplitude is constant. This makes it tolerant of severe nonlinearities in the transmission system. A continuous-phase waveform is a generalization of the minimum-shift keying waveform, which was studied in Section 5.7.

We shall require that the datastream is modulated into the phase $\theta(t)$ in such a way that the data can be conveniently demodulated directly from the in-phase and quadrature components of $c(t)$. This requirement is imposed because phase demodulation – recovering an unstructured phase waveform $\theta(t)$ from the waveform $c(t)$ – is a nonlinear operation that can behave poorly in noise and so should be avoided. The maximum-likelihood demodulator tells us how to recover the data sequence when given the received signal $v(t) = c(t) + n(t)$. We want the maximum-likelihood demodulator to be a matched filter followed by a (complex) sequence demodulator. Consequently, we want to choose the method of modulating data into $\theta(t)$ in such a way that the maximum-likelihood demodulator is easy to implement – in particular, so that the Viterbi algorithm can be used to process matched-filter output samples.

We begin the discussion with the study of a demodulator for MSK that will recover the FSK phase sequence directly by using a maximum-likelihood demodulator. This is an alternative to using a matched filter on the in-phase and quadrature half-cosine pulses.

When viewed as a phase-modulated waveform at complex baseband, the MSK waveform is

$$c(t) = e^{j\theta(t)}$$

where

$$\theta(t) = \theta_\ell \pm 2\pi \frac{t - \ell T_b}{4T_b} \qquad \ell T_b \leq t < (\ell + 1)T_b.$$

These phase trajectories were illustrated in Figure 5.22. This is a form of FSK modulation with the frequencies $f_0 \pm 1/4T_b$ and the starting phase of each pulse adjusted to make the phase continuous. The phase of this waveform changes by $\pm \pi/2$ as ℓ changes to $\ell + 1$. For odd ℓ, $\theta_\ell = \pm \pi/2$, and for even ℓ, $\theta_\ell = 0$ or π. Although we could proceed using this definition of $c(t)$, the trellis for this description of the MSK waveform and the Viterbi demodulator will be easier to discuss if we first multiply $c(t)$ by $e^{j2\pi t/4T_b}$ so that the phase changes by 0 or $+\pi$ for every change in ℓ. This is equivalent to offsetting the definition of the carrier frequency of the passband representation so that the two FSK frequencies are now f_0 and $f_0 + 1/2T_b$. Define

$$c'(t) = c(t)e^{j2\pi t/4T_b}$$

so that the phase changes either by 0 or by $+\pi$ as ℓ changes to $\ell + 1$. Thus

$$c'(t) = e^{j\theta'(t)}$$

where, on the interval $\ell T_b < t \leq (\ell + 1)T_b$,

$$\theta'(t) = \begin{cases} \theta_\ell' & \text{if the } \ell\text{th data bit is a zero} \\ \theta_\ell' + 2\pi \frac{t - \ell T_b}{2T_b} & \text{if the } \ell\text{th data bit is a one.} \end{cases}$$

These modified phase trajectories are illustrated in Figure 9.9. The advantage of the phase trajectories depicted in Figure 9.9 as compared to those in Figure 5.22 is that the structure is now invariant to a translation of the time index. The modified trellis is the same at even time indices as at odd time indices.

We shall develop a coherent demodulator for this representation of the MSK waveform. Normally, when we speak of a coherent demodulator, we mean that the reference phase is known modulo 2π. In this case, it is actually enough that the reference phase is known modulo $\pi/2$.

The FSK description of the MSK waveform now is

$$c(t) = \sum_{\ell=-\infty}^{\infty} [a_\ell e^{j\theta_\ell} s_0(t - \ell T_b) + \bar{a}_\ell e^{j\theta_\ell} s_1(t - \ell T_b)]$$

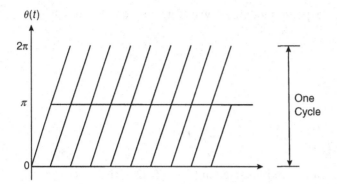

Figure 9.9. Modified phase trajectories for MSK.

where $\theta_\ell = 0$ or π,

$$(a_\ell, \bar{a}_\ell) = \begin{cases} (A, 0) & \text{if the } \ell\text{th data bit is a zero} \\ (0, A) & \text{if the } \ell\text{th data bit is a one,} \end{cases}$$

and

$$s_0(t) = \text{rect}\left(\frac{t}{T_b} - \frac{1}{2}\right)$$

$$s_1(t) = \text{rect}\left(\frac{t}{T_b} - \frac{1}{2}\right) e^{j\pi t/T_b}.$$

The maximum-likelihood demodulator consists of first forming the decision statistics used for coherent FSK, then using the Viterbi algorithm to deduce the data sequence. The decision statistics are formed by passing the received waveform through the pair of matched filters, $s_0^*(-t)$ and $s_1^*(-t)$, and sampling the real parts of the outputs at time ℓT_b to give $u_{0\ell} = u_0(\ell T_{\hat{b}})$ and $u_{1\ell} = u_1(\ell T_{\hat{b}})$. Because the pulses for different ℓ do not overlap, the matched filter can be implemented with integrate and dump correlators.

The pulses $s_0(t)$ and $s_1(t)$ are not orthogonal because

$$\int_0^{T_b} s_0(t) s_1^*(t) dt = \frac{-2j}{\pi/T_b}.$$

However, this term is purely imaginary, which, for our present purposes, is as good as orthogonal because we deal only with the real part of the matched-filter outputs. Further, although the noise samples caused by $n(t)$ passing through the two matched filters are not uncorrelated, the real parts of the noise samples are uncorrelated.

State

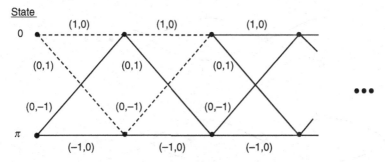

Figure 9.10. Trellis for the MSK waveform.

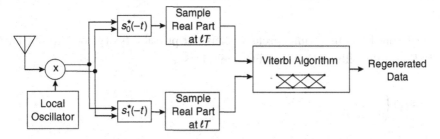

Figure 9.11. FSK demodulator for an MSK waveform.

Now we are ready to draw the trellis shown in Figure 9.10. There are two states corresponding to the two values, 0 and π, that θ_ℓ can take. There are two paths out of each node corresponding to the two possible values of the data bit. A data bit zero causes no change in state. A data bit one causes a change in state. Each branch is labeled with the expected value of the pair of decision statistics: $(Ae^{j\theta_\ell}, 0)$ for a data bit zero, and $(0, -Ae^{j\theta_\ell})$ for a data bit one, with A set equal to one. The Viterbi algorithm is used to search this trellis for the path that best agrees with the sequence of matched-filter outputs.

Figure 9.11 shows the maximum-likelihood demodulator for MSK that we have developed. The performance of the demodulator is easily determined by finding the minimum distance of the set of sequences described by the trellis. The minimum distance is set by the two paths highlighted in the trellis, from which we conclude that $d_{min} = 4$. In fact, every path has one neighboring path at distance $d_{min} = 2$. This is the same minimum distance as BPSK (whose trellis is shown in Figure 9.12). Consequently, the probability of an error event for the demodulator of Figure 9.11 is given by the approximation

$$p_e \approx Q\left(\sqrt{\frac{2E_b}{N_0}}\right).$$

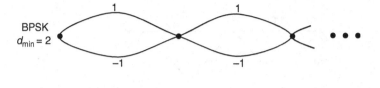

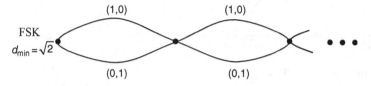

Figure 9.12. Degenerate trellises for BPSK and coherent FSK.

On the one hand, this is unexpected because the probability of error of a coherent binary FSK (whose trellis is shown in Figure 9.11) is

$$p_e \approx Q\left(\sqrt{\frac{E_b}{N_0}}\right).$$

Remarkably, by using a continuous-phase FSK waveform rather than a more elementary FSK waveform, a 3 dB energy advantage has been obtained. On the other hand, however, we might have anticipated this because we know that an MSK waveform can be demodulated by demodulating its in-phase and quadrature modulation components separately, and each of these demodulators will have the probability of error of BPSK.

Many other continuous-phase waveforms can be designed. Each may be regarded as a generalization of the MSK waveform. The general case at passband is described by

$$c(t) = \cos(2\pi f_0 t + \theta(t)),$$

and at complex baseband

$$c(t) = e^{j\theta(t)}$$

where the phase modulation $\theta(t)$ is given by

$$\theta(t) = 2\pi h \int_{-\infty}^{t} \sum_{\ell=0}^{\infty} a_\ell s(\xi - \ell T) d\xi.$$

The phase modulation may also be specified by giving its derivative

$$\dot{\theta}(t) = 2\pi h \sum_{\ell=0}^{\infty} a_\ell s(t - \ell T).$$

The parameter h is a constant called the *modulation index*. The modulation index h is usually chosen to be a rational number

$$h = \frac{K}{N}$$

because otherwise $\theta(t)$ would take on an infinite number of states and our demodulation methods would not apply. The integers K and N should be chosen to be coprime. The data symbols a_ℓ take values in an M-ary real-valued and uniformly-spaced signal constellation. When M is even, we may choose the symmetric signal constellation

$$a_\ell \in \{\pm 1, \pm 3, \ldots, \pm M/2\}.$$

The pulse shape $s(t)$ is chosen based on its effect on the spectrum of $c(t)$ and to simplify the modulator and demodulator. We shall consider only the simplest pulse shape

$$s(t) = \text{rect}\left(\frac{t}{T} - \frac{1}{2}\right)$$

for the phase modulation. In this case,

$$\dot{\theta}(t) = 2\pi h a_\ell \quad \ell T \leq t < (\ell+1)T.$$

Consequently, the continuous-phase waveforms are also M-ary FSK waveforms. Such waveforms are called *continuous-phase* FSK (CPFSK) waveforms. During each symbol time, an M-ary data symbol is modulated onto one of M frequencies, $f_0 + h a_\ell$, forming one of M sinusoids not necessarily orthogonal. The sinusoids of a CPFSK waveform fill out the T second interval, and the phase of each sinusoid is adjusted so that the phase is continuous at the symbol boundaries. The waveform is

$$c(t) = e^{j\theta(t)}$$

where

$$\theta(t) = \frac{2\pi h}{T} a_\ell \left(\frac{t}{T} - \ell - \frac{1}{2}\right) + \sum_{i=0}^{\ell} 2\pi h a_i + \theta_0 \quad \ell T \leq t \leq (\ell+1)T,$$

and θ_0 is the initial phase angle, which is known if the demodulator is coherent and unknown if the demodulator is noncoherent.

Because of the imposed phase continuity, computation of the power density spectrum of a CPFSK waveform can be quite difficult, but generally, the spectrum of $c(t)$ will fall off much more quickly with f than will the spectrum of $s(t)$.

Binary CPFSK with modulation index $h = 1/2$ is MSK, which we have already studied in detail. There we found it convenient to redefine the signal constellation by redefining the carrier frequency. Similarly, by redefining the carrier frequency of CPFSK, we can choose to view the signal constellation as

$$a_\ell \in \{0, 1, 2, \ldots, M - 1\},$$

which has the same form for odd M as for even M.

To complete the discussion of continuous-phase modulation, we must describe the demodulator. In principle it is straightforward to formulate a demodulator. Simply inspect the likelihood function to construct the appropriate bank of matched filters and the appropriate trellis, then apply the Viterbi demodulator. When the additive noise is gaussian, maximizing the likelihood function is equivalent to minimizing the euclidean distance

$$d(c(t), v(t)) = \int_{-\infty}^{\infty} |v(t) - e^{j\theta(t)}|^2 dt$$

over all $e^{j\theta(t)}$ that are legitimate waveforms that can be produced by the modulator. Expanding the square, we see that minimizing the euclidean distance for a block of length L is equivalent to maximizing the correlation

$$\rho(a_0, a_1, \ldots, a_L) = \mathrm{Re}\left[\int_{-\infty}^{\infty} v(t)e^{-j\theta(t)} dt\right]$$

$$= \mathrm{Re}\left[\sum_{\ell=0}^{L} e^{-j\theta_\ell} \int_{\ell T}^{(\ell+1)T} v(t)e^{-j2\pi h a_\ell (t - \ell T)/T}\right]$$

where

$$\theta_\ell = \theta(\ell T) = 2\pi h \sum_{i=0}^{\ell} a_i$$

and t has been offset by $1/2$.

For the example of MSK, shown in Figure 9.10, it was possible to replace the distance $d(v(t), c(t))$ by a euclidean distance between sequences for labeling and searching the trellis. For the general case of CPFSK, the likelihood statistic does not take the form of a euclidean-distance computation in sequence space[1]. It is, however, still in the form

---

[1] The form of a euclidean distance could be obtained by passing $v(t)$ through an infinite bank of orthogonal filters to form the output samples equal to the Fourier coefficients of a Fourier expansion of $v(t)$ in one symbol interval. Then each branch of the trellis would be labeled with an infinite vector of Fourier coefficients, and the distance computation would be the computation of the euclidean distance between infinite sequences on each branch.

of a sum

$$p(a_0, a_1, \ldots, a_L) = p(a_0, a_1, \ldots, a_{L-1}) + \text{Re}\left[e^{-j\theta_L} \int_{LT}^{(L+1)T} v(t) e^{-j2\pi h a_L(t-LT)/T} dt \right].$$

Therefore the Viterbi algorithm still applies, using the second term on the right as the branch distance.

For each data sequence, the correlation $p(a_0, a_1, \ldots, a_L)$ can be recursively computed from the sequence of output samples of the bank of M filters with impulse responses

$$g_m(t) = e^{j2\pi hmt/T} \quad m = 0, \ldots, M - 1.$$

Let $u_m(t)$ be the output of filter $g_m(t)$ excited by $v(t)$. Then the set of

$$\{u_m(\ell T): \quad m = 0, \ldots, M - 1; \ell = 0, 1, 2, \ldots, \}$$

is a sufficient statistic for demodulation because it is sufficient for reconstructing $p(a_0, a_1, \ldots, a_L)$, and so for reconstructing $d(c(t), v(t))$ for each possible codeword. However, the computation of $p(a_0, a_1, \ldots, a_L)$ from the statistics $u_m(\ell T)$ does not have the form of a euclidean-distance computation. Rather, we have the recursion

$$p(a_0, a_1, \ldots, a_\ell) = p(a_0, a_1, \ldots, a_{\ell-1}) + \text{Re}[e^{-j\theta_\ell} u_m(\ell T)]$$

where $a_\ell = m_\ell$. This is exactly the kind of additive structure needed to apply the Viterbi algorithm.

The structure of the demodulator is illustrated by a binary CPFSK waveform that is similar to MSK except that the phase changes by $\pm 45°$ rather than by $\pm 90°$. For convenience, we will redefine the frequency reference so that the phase changes by $0°$ for a data bit zero and by $90°$ for a data bit one. The trellis is shown in Figure 9.13. The highlighted pair of paths is separated by the minimum distance

$$d_{\min}^2 = \int_0^T |1 - e^{j(\pi/2)(t/T)}|^2 dt + \int_0^T |e^{j(\pi/2)(t/T)} - j|^2 dt$$

$$= 4T \left(1 - \frac{2}{\pi} \right)$$

$$= 4 \left(1 - \frac{2}{\pi} \right) E_b.$$

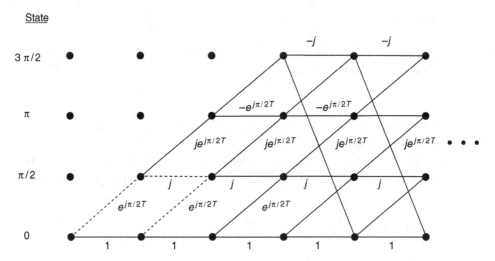

Figure 9.13. Trellis for a binary CPFSK waveform.

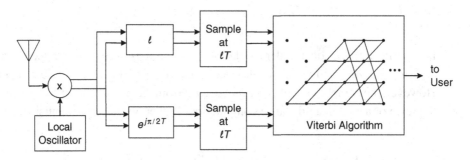

Figure 9.14. Demodulator for a binary CPFSK waveform.

Because $d_{\min} = 2E_b$ for MSK, this form of binary CPFSK has a smaller minimum distance and a larger bit error rate than MSK. Indeed, based on the discussion of Section 4.3, we have the approximation

$$p_e \approx Q\left(\frac{d_{\min}}{2\sigma}\right)$$

$$\approx Q\left(\sqrt{\frac{(1 - 2/\pi)E_b}{\sigma^2}}\right).$$

Because of its poorer minimum distance, this waveform will not be preferred to MSK very often, but it does have a different spectrum, and it does provide a simple example of a demodulator for CPFSK. The demodulator is shown in Figure 9.14. One matched filter is provided for each pulse, in this case,

$$s_0(t) = \text{rect}\left(\frac{t}{T} - \frac{1}{2}\right)$$

$$s_1(t) = e^{j(\pi/2)(t/T)}\text{rect}\left(\frac{t}{T} - \frac{1}{2}\right).$$

The outputs of the matched filters are sampled and, for each path of the trellis, the appropriate sample is multiplied by the appropriate phase term $e^{j\theta_\ell}$. The Viterbi algorithm then searches the trellis for the maximum-likelihood path.

9.3 Trellis codes for digital modulation

An M-ary signal constellation in the complex plane can be used to transmit $k = \log_2 M$ bits per signaling interval, as was discussed in Section 2.3. Figure 9.15 shows an example of a complex (or two-dimensional) sixteen-ary signal constellation and a scatter diagram of received samples at some fixed signal-to-noise ratio. The scatter diagram shows a large number of matched-filter output samples to the same scale as the input signal constellation. Each point in the scatter diagram is one of the points of the signal constellation perturbed by noise. In this example, the noise is severe, and it is obvious that there will be many demodulation errors. To reduce the error rate, we must make changes in the waveform. The timid approach is simply to replace the sixteen-ary signal constellation by an eight-ary or a four-ary signal constellation. This reduces the probability of error but also reduces the data rate. A much better approach is to use a code to increase the minimum euclidean sequence distance d_{min}. The minimum euclidean distance, or the *free distance*, is the smallest euclidean distance between any

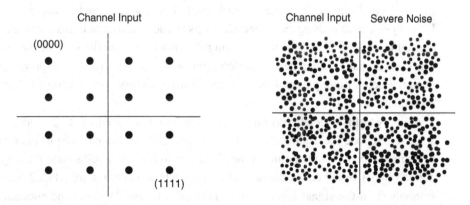

Figure 9.15. Effect of severe channel noise on a sixteen-ary signal constellation.

two codewords of the code. In Section 4.5, we studied in some detail the approximation

$$p_e \approx N_{d_{\min}} Q\left(\frac{d_{\min}}{2\sigma}\right).$$

This section is based on the observation that the probability of error p_e can be reduced by increasing the minimum sequence distance provided the number of nearest neighbors $N_{d_{\min}}$ is not too large. A trellis code is so designed to increase the minimum sequence distance. In comparison to an uncoded signaling waveform of the same data rate, a trellis code uses a larger signal constellation, but does not use all of the possible sequences.

These motivating comments can be repeated: if an uncoded four-ary or an eight-ary signaling waveform has too high a probability of error, increase the number of points in the signal constellation and use a trellis code on this larger signal constellation such that the original data rate is unchanged, but the bit error rate is less. By using a larger signal constellation and a trellis code, we shall see that one can reduce the required E_b/N_0 at a fixed probability of error.

The capacity curves that are shown later in Figure 11.8 of Chapter 11 provide a way to understand the advantages of coding. If we compare, for example, the capacity of eight-ary PSK and sixteen-ary PSK shown in Figure 11.8, we see that to transmit at a rate of three bits per symbol with a low probability of symbol error will take considerably more energy with eight-ary PSK than with sixteen-ary PSK. Therefore better performance is achieved by using the larger signal constellation and some form of coding. For this application, trellis codes have been preferred to block codes, in large part because the special distance structure of the signal constellation can be treated in a simple way by the Viterbi algorithm.

A trellis code that is defined on a finite set of points of the complex plane is called an *Ungerboeck code*, and the resulting waveform $c(t)$ is called a *trellis-coded modulation* signaling waveform. The Ungerboeck codes form a class of trellis codes with much of the structure of convolutional codes, which are discussed in Chapter 10. These codes are not convolutional codes because the codewords do not satisfy the necessary linearity condition at the code symbol level. For a given fixed k and n, an (n, k) Ungerboeck code is designed to encode k bits in each dataframe into a complex signal constellation with 2^n points, the encoding depending also on the state of the encoder. The code alphabet is the signal constellation, which is a discrete set of points chosen from the field of complex numbers. A codeword is a sequence of points of this signal constellation.

A binary trellis code of constraint length v is based on a trellis with 2^v nodes at each stage. Each node has 2^k branches leaving it, and 2^k branches entering it. Each branch is labeled with one of the points of the signal constellation. A trellis code is completely defined by its signal constellation and a trellis whose branches are labeled with code symbols from the signal constellation. However, to use the code, the encoding rule must also be given. Each of the 2^k possible dataframes must be assigned to one of the branches leaving a node. The data value of the incoming dataframe then determines which branch is selected. The label on that branch determines the output symbol of the encoder.

The complex representation of the transmitted signal is

$$c(t) = \sum_{\ell=-\infty}^{\infty} c_\ell s(t - \ell T)$$

where $s(t)$ is such that $s(t) * s^*(-t)$ is a Nyquist pulse and the c_ℓ are generated by the encoder for the Ungerboeck code. We now have arrived at a waveform for digital communication wherein the data is indeed deeply buried. The waveform $c(t)$ is such that only when suitably mixed to complex baseband and passed through a matched filter will its expected waveform samples pass through the points of the complex signal constellation. The user data residing in the history of those points is determined according to the dictates of the Ungerboeck code.

We shall study larger examples of Ungerboeck codes for QAM signal constellations in Section 9.5. In this section, we will construct two simple $(3, 2)$ codes with constraint length 2 for the eight-ary-PSK signal constellation; the first example is the one that initially suggests itself, but the second example is actually a much better code. To design the code, we draw a trellis and label the trellis with channel symbols. Because the constraint length $\nu = 2$ is chosen, the trellis consists of four states. Because $k = 2$, each state has four branches entering and four branches leaving. Two such trellises are shown in Figure 9.16. We shall design codes for each of them.

To design a code, we must label the branches of the trellis with the symbols of the eight-ary PSK signal constellation. This is done so as to make the euclidean free distance of the code as large as possible. Figure 9.17 shows one way to label the branches of the first trellis of Figure 9.16 that has large euclidean free distance. By inspection, we see that the path labeled 001 011 000 ... is at the minimum squared euclidean distance from the all-zero codeword. Specifically, by inspection of Figure 9.17, the following two squared euclidean distances are found:

$$d^2(000, 001) = d_0^2$$
$$d^2(000, 011) = d_1^2.$$

Therefore, the sequence distance between $(000, 000)$ and $(001, 011)$ is $d_0^2 + d_1^2$. The squared free euclidean distance is defined as the squared euclidean distance between any two paths of the trellis for which the distance is smallest, and no distance between two paths is smaller. Therefore the free squared euclidean distance of the trellis code is $d_0^2 + d_1^2$.

To judge the performance of the code, we will compare it to simple uncoded QPSK, which also transmits two data bits per symbol. Figure 9.18 shows QPSK as a degenerate

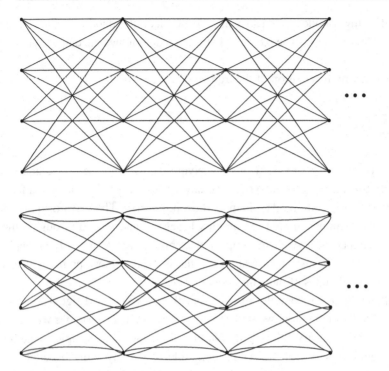

Figure 9.16. Two trellises with $k = 2$ and $\nu = 2$.

trellis code of the same energy per data bit. The free squared euclidean distance of the uncoded QPSK waveform is d_1^2. In comparing the two codes, we see that the asymptotic coding gain is $G = 10\log_{10}[(d_0^2 + d_1^2)/d_1^2]$ or 1.1 dB. This gain is only a small improvement, but it is surprisingly good for such a simple code. However, we shall see in Figure 9.21 that there is another trellis code for this signal constellation that yields a 3 dB improvement and is not more complicated, so our first example is unlikely to be used in practice.

The trellis and signal constellation in Figure 9.17 completely define the code, but to use the code, we also need to assign four two-bit datawords uniquely to each of the four paths leaving each node. We may do this in any convenient way. One way is to assign, at each node, the four data patterns 00, 01, 10, 11 to the four branches in order from the topmost to the bottommost. Then the encoder can be a simple look-up table. The two data bits and the two state bits are used as a four-bit address, at which address are stored five bits. Two of the five bits specify the new state, and three of the five bits specify the constellation point. A table look-up encoder is shown in Figure 9.19. This kind of encoder requires $2^{\nu+k}$ words of memory and will be impractical if $\nu + k$ is large.

An equivalent encoder can be designed in the form of a shift register if we assign input bits to branches carefully. With the same assignment – the four data patterns 00,

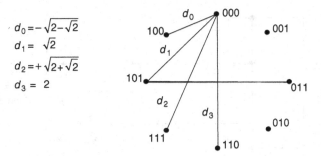

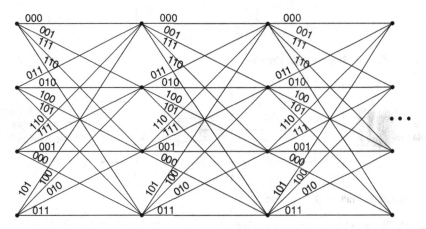

Figure 9.17. A simple trellis code for eight-ary PSK.

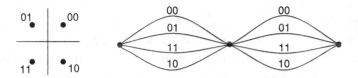

Figure 9.18. QPSK described as a degenerate trellis code.

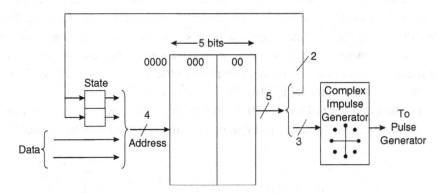

Figure 9.19. Table look-up encoder for a trellis code.

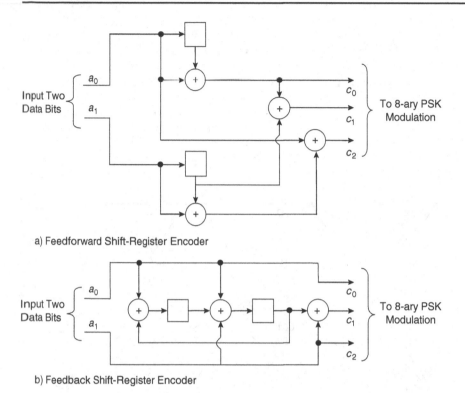

a) Feedforward Shift-Register Encoder

b) Feedback Shift-Register Encoder

Figure 9.20. Encoders for a simple trellis code.

01, 10, 11 assigned to the four branches in order from the topmost to the bottommost –
we notice that the two data bits are equal to the next state bits. Therefore we can
configure an encoder by using two bits of memory and associated logic, as shown as
the feedforward encoder in Figure 9.20.

There are other ways of assigning data bits to the branches of the trellis. The assignment can be made in a way that makes the implementation easy. We notice that at
each node of the trellis in Figure 9.17, the last two bits of the code label are different.
We will assign the data bits to agree with these two bits in expectation of obtaining a
simple implementation. This leads to the encoder with feedback shown in Figure 9.20.
It is easy to verify that this encoder will produce the codewords shown on the trellis of
Figure 9.17 if the states are labeled 00, 01, 10, 11 from top to bottom.

The first encoder in Figure 9.20 can be described by a matrix of polynomials. By
following the incoming data bits through the shift-register circuit, we deduce that

$$[c_0(x) \quad c_1(x) \quad c_2(x)] = [a_0(x) \quad a_1(x)] \begin{bmatrix} 1+x & 1+x & 1 \\ 0 & x & 1+x \end{bmatrix}.$$

The polynomial matrix

$$G(x) = \begin{bmatrix} 1+x & 1+x & 1 \\ 0 & x & 1+x \end{bmatrix}$$

is called the *generator matrix* of the trellis code.

To design an encoder in which the data bits appear explicitly within the code bits, we must transform the generator matrix so that all the columns of an identity matrix appear as columns in the generator matrix. This is called the *systematic form* of the generator matrix. Hence we will form an identity submatrix in the matrix $G(x)$ by elementary row operations. We will divide $G(x)$ by $1+x$ because a feedback shift-register corresponding to $1/(1+x)$ is preferred to one corresponding to $1/x$. Consequently, the matrix becomes

$$\begin{bmatrix} 1 & 1 & \dfrac{1}{1+x} \\ 0 & \dfrac{x}{1+x} & 1 \end{bmatrix}.$$

We will form an identity matrix in the first and third columns. To the first row, add $1/(1 + x)$ times the second row, using modulo-two polynomial arithmetic, to obtain the generator matrix in systematic form as

$$G'(x) = \begin{bmatrix} 1 & 1+\dfrac{x}{1+x^2} & 0 \\ 0 & \dfrac{x}{1+x} & 1 \end{bmatrix}$$

$$= \begin{bmatrix} 1 & 1+\dfrac{x}{1+x^2} & 0 \\ 0 & \dfrac{x+x^2}{1+x^2} & 1 \end{bmatrix}.$$

This systematic generator matrix leads to the *systematic encoder* with feedback, shown in Figure 9.20b. It is easy to verify that the systematic encoder also produces the codewords shown on the trellis of Figure 9.17 – this time with the states labeled 00, 11, 10, 01 from top to bottom.

The generator matrix is a k by n matrix of polynomials. Another kind of polynomial matrix describing the trellis code, called a *check matrix*, is any $n - k$ by n matrix of polynomials, denoted $H(x)$, with the property that $H(x)G(x) = 0$. To compute $H(x)$ from $G(x)$, divide by an appropriate polynomial and permute columns to put $G(x)$ in systematic form as the new generator matrix $G(x) = [I \quad P(x)]$ where I is a k by k identity matrix, and $P(x)$ is a k by $n - k$ matrix of rational forms of polynomials. Then $H(x) = [-P(x)^T \quad I]$ is a check matrix in systematic form. It can be put into a form corresponding to the original nonsystematic form of $G(x)$ by inverting permutations and clearing the denominators.

For example, the systematic generator matrix for the trellis code just given can be rewritten as

$$G'(x) = \begin{bmatrix} 1 & 0 & 1 + \dfrac{x}{1+x^2} \\ 0 & 1 & \dfrac{x+x^2}{1+x^2} \end{bmatrix}$$

by transposing two bit positions, so

$$H'(x) = \begin{bmatrix} 1 + \dfrac{x}{1+x^2} & \dfrac{x+x^2}{1+x^2} & 1 \end{bmatrix}.$$

Now multiply by $1 + x^2$ and retranspose the bit positions to obtain

$$H(x) = [1 + x + x^2 \quad 1 + x^2 \quad x + x^2].$$

It is easy to verify that $G(x)H(x) = 0$ for the original generator matrix $G(x)$.

There is another useful trellis having four nodes and four branches per node, as shown in Figure 9.16. It may be surprising that this trellis can define a better code than the one in Figure 9.17. Clearly, the free distance cannot be larger than the distance between points assigned to "parallel" branches beginning and ending on the same node. To maximize the free distance, each pair of branches should be labeled with pairs of points from the signal constellation that are 180° apart. This means that the free euclidean distance is not larger than d_3. The task then is to assign labels so that the free euclidean distance is not smaller than d_3. It turns out that, by trial and error, points of the signal constellation can be assigned to the trellis branches so that d_3 is, indeed, the free distance of the code. A labeled trellis is shown in Figure 9.21, and an encoder is shown in Figure 9.22. The generator matrix is

$$G(x) = \begin{bmatrix} 1 & 0 & 0 \\ 0 & 1+x^2 & x \end{bmatrix}.$$

This encoder is both feedforward and systematic, except that odd and even data bits are transmitted in different frames. If desired, a simple delay in either the encoder or decoder will recover the timing between odd and even bits. The gain of the code is

$$\text{Gain} = 10 \log_{10}(d_3^2/d_1^2) = 3 \text{ dB}.$$

The performance of this better four-state code for eight-ary PSK is shown in Figure 9.23 and is contrasted to QPSK. For an uncoded QPSK, $d_{min} = \sqrt{2}A$ and there are two

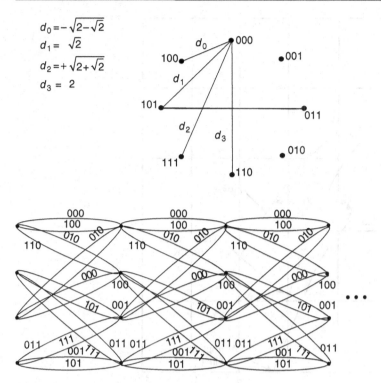

$$d_0 = -\sqrt{2-\sqrt{2}}$$
$$d_1 = \sqrt{2}$$
$$d_2 = +\sqrt{2+\sqrt{2}}$$
$$d_3 = 2$$

Figure 9.21. A better trellis code for eight-ary PSK.

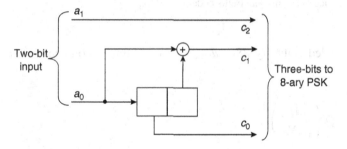

Figure 9.22. Encoder for the better trellis code.

nearest neighbors. Thus our approximation to the probability of an error event for QPSK is

$$p_e \approx 2Q\left(\frac{d_{\min}}{2\sigma}\right)$$
$$= 2Q\left(\sqrt{\frac{2E_b}{N_0}}\right).$$

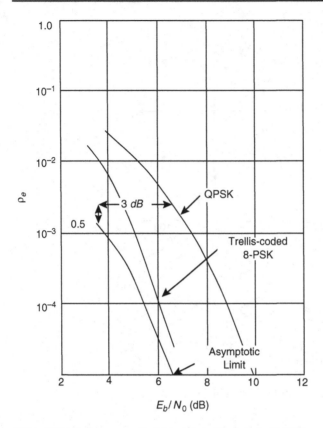

Figure 9.23. Performance of a four-state trellis code for 8-PSK.

For the trellis-coded eight-ary PSK, $d_{min} = 2A$ and there is one nearest neighbor. Thus

$$p_e \approx Q\left(\frac{2A}{2\sigma}\right) = Q\left(\sqrt{\frac{4E_b}{N_0}}\right).$$

Consequently, the asymptotic bound in Figure 9.23 can be obtained from the curve for QPSK by sliding that curve to the left by 3 dB and down by the factor $\frac{1}{2}$. Figure 9.23 also shows a more accurate curve for p_e showing poorer performance, which was obtained by simulation and includes error events due to neighbors other than the nearest neighbor. A study of Figure 9.23 is a good way to judge the accuracy of the free euclidean distance as a figure of merit.

It is not possible to achieve a better $(3, 2)$ trellis code for an eight-ary PSK with constraint length 2. To improve the asymptotic coding gain, it is necessary to use a code with a larger constraint length – that is, to increase the number of states in the trellis. Figure 9.24 shows an Ungerboeck code with a constraint length 3 and a 3.6 dB coding

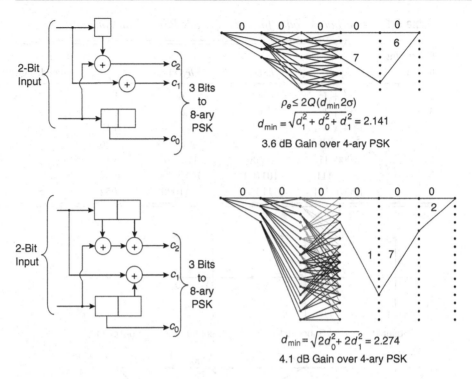

$$\rho_e \leq 2Q(d_{min}2\sigma)$$

$$d_{min} = \sqrt{d_1^2 + d_0^2 + d_1^2} = 2.141$$

3.6 dB Gain over 4-ary PSK

$$d_{min} = \sqrt{2d_0^2 + 2d_1^2} = 2.274$$

4.1 dB Gain over 4-ary PSK

Figure 9.24. More trellis encoders for the eight-ary PSK channel.

gain, and another Ungerboeck code with constraint length 4 and a 4.1 dB coding gain. These trellises do not contain the curious double paths like those seen in Figure 9.21.

The trellises shown in Figure 9.24 are already becoming too large to draw conveniently. For larger constraint lengths, we rarely draw the trellis, though we may discuss it conceptually. A code with a large constraint length is described by giving its k by n polynomial generator matrix or its $(n - k)$ by n polynomial check matrix. For the $(k + 1, k)$ codes, which are the codes of interest, the polynomial check matrix is easier to give.

Table 9.1 is a table of polynomial check matrices for good Ungerboeck codes for the eight-ary PSK signal constellations with the polynomials for the last three codes expressed in terms of the binary coefficients of the monomials. Any of these codes can be used as a transparent plug-in replacement for the popular uncoded (four-ary) QPSK modulator. The data rate is still two bits per symbol. There is no change in the symbol rate, so the coded system has the same bandwidth as the uncoded system and transmits the same number of data bits per symbol; hence the user of the system is unaware of the code's presence. However, the system now can run at a lower signal-to-noise ratio; the code with constraint length 9 has a gain of 5.7 dB.

The encoder that was shown in Figure 9.22 is so simple that we should be able to find a simplified explanation for it and for the parallel paths of its trellis. Indeed, we can if

Table 9.1. *Ungerboeck codes for the eight-ary PSK signal constellation*

Constraint length	$h_1(x)$	$h_2(x)$	$h_3(x)$	d_{min}^2/d_0^2	Gain over uncoded QPSK
2	x^2+1	x	0	4.00	$\sim$3.0 dB
3	x^3+1	x	x^2	4.59	3.6
4	x^4+x+1	x^2	x^3+x^2+x	5.17	4.1
5	x^5+x^2+1	x^3+x^2+x	$x^4+x^3+x^2$	5.76	4.6
6	1000011	11000	110110	6.34	5.0
7	10111111	101100	1010010	6.59	5.2
8	100011101	111010	1011000	7.52	5.7

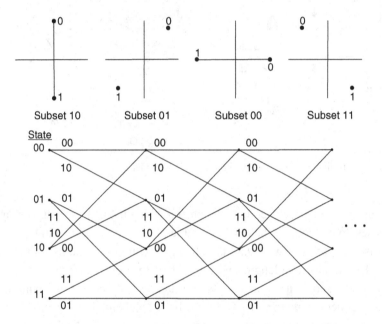

Figure 9.25. Another view of a trellis code for eight-ary PSK.

we adopt the view that the symbols of the code are subsets of the signal constellation instead of points of the signal constellation. This is a somewhat abstract view that can be understood by studying Figure 9.21. Close examination of the branch labels on the trellis of Figure 9.21 will show that the labels can be decomposed into two parts. The second and third bits label the pair of branches, and the first bit labels the individual branch within a pair. Therefore we can choose to regard the code as a mapping into a subset of the signal constellation rather than a point. Figure 9.25 shows how the description of the trellis is recast into this framework. Now the underlying code has $k = 1$ and $v = 2$. The symbols, however, are pairs of points of the signal constellation, and the extra data bit is encoded into the choice from the two points of the pair.

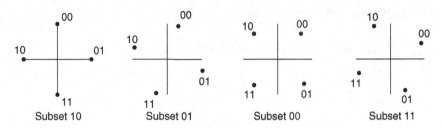

Figure 9.26. A partition of the sixteen-ary PSK constellation.

This idea of regarding the trellis code as a mapping from a subset of the data bits into a subset of the signal constellation followed by the selection of one point of the subconstellation is actually quite useful. For example, we can reuse the trellis for the eight-ary PSK signal constellation, shown in Figure 9.25, to create a rate three-fourths code for a sixteen-ary PSK signal constellation. Simply double the total number of points, assigning eight points to each of two subconstellations. Each symbol of the code is actually a pair of points, one from each subconstellation, chosen to be far apart. The output of the eight-ary PSK encoder is used to select one of eight such two-point symbols. The extra data bit is used to select one point from the two-point symbol.

As a second example, Figure 9.26 shows an appropriate subdivision of a sixteen-ary PSK constellation into four subconstellations, each with four points. With these subconstellations replacing the subconstellations in Figure 9.25, the design of the new code is complete. The encoder is exactly the encoder of Figure 9.22 except that there is one new data bit a_2, and one new code bit c_3 with $c_3 = a_2$.

The free distance of the code for a sixteen-ary PSK code and the free distance of the code for an eight-ary PSK code are not established by the same pairs of paths. To compute the free distance of the code for the sixteen-ary PSK signal constellation, it is convenient initially to write down the distances between each pair of subconstellations, which are defined as the smallest distance between any point in one subconstellation and a point in the other subconstellation. Referring to Figure 9.26, we have

$$d(10, 01) = d_3$$

$$d(10, 00) = d_2$$

$$d(10, 11) = d_3$$

$$d(01, 00) = d_3$$

$$d(01, 11) = d_2$$

$$d(00, 11) = d_3.$$

Referring to the trellis in Figure 9.21, we can find that the free distance is established by the path that stays in state 00, and the path that goes from state 00 to state 01, then to state 10, and then back again to state 00. Thus

$$d_{\min}^2 = d^2(10,00) + d^2(01,00) + d^2(10,00)$$
$$= 2d_2^2 + d_3^2$$
$$= 2.259A^2.$$

The gain with respect to uncoded eight-ary PSK is $G = 10\log_{10}(d_{\min}^2/d_1^2) = 3.54$ dB. However, for this code, $N_{d_{\min}} = 4^3$, so the asymptotic coding gain will not be fully effective. It will be diminished by the large number of nearest neighbors.

9.4   Lattices and lattice cosets

The construction of many of the Ungerboeck trellis codes, as illustrated at the end of the last section, can be placed in a formal mathematical setting by introducing the notions of a lattice and a lattice coset. The Ungerboeck codes on a square signal constellation will be developed in this way in Section 9.5. While we could develop such codes in an informal way, as was done for the eight-ary PSK signal constellation in the previous section, we shall use the formal setup because it gives new insights into the structure of the codes. Figure 9.27 shows a 32-ary signal constellation broken down into eight subsets of four points each, with a somewhat regular structure so that, in each subset, the points are far apart. Such a partition has a structure that underlies the design of good codes.

A two-dimensional lattice Λ is a periodic arrangement of points in the plane. Specifically, let A be a fixed nonsingular two by two matrix. The lattice is the set of all points in the plane that can be written as the two-component vector $s = Am$, where $m = (m_1, m_2)^T$ is a two-dimensional vector with integer components. The columns of A are called the generators of the lattice. The most important example is the two-dimensional square lattice whose points are at the intersections of an infinitely large

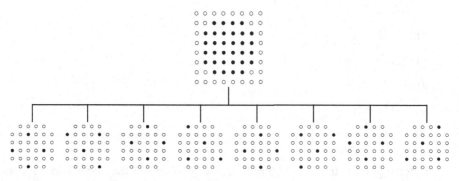

Figure 9.27. Partitioning a 32-ary signal constellation.

checkerboard. More generally, an N-dimensional lattice is a periodically arranged set of points in N-dimensional space, formally defined as follows:

Definition 9.4.1 *Let A be a fixed, nonsingular, real N by N matrix. An N-dimensional lattice is the set of all N-tuples that can be written* $s = Am$ *where* $m = (m_1, \ldots, m_N)^T$ *is a vector of dimension N with integer components.*

The matrix A, in the general case, is called the *generator matrix* of the lattice, and its columns are called *generators*. The *dual lattice* is the lattice with generator matrix $(A^{-1})^T$. The *minimum distance* of the lattice is defined as the smallest distance between any two of its points. The quantity det A is the *volume* of one cell of the lattice. This cell may be taken as the parallelepiped cell formed by the generators of A or, alternatively, as the "Voronoi region", consisting of those points that are closer to the origin than to any other point of the lattice. The Voronoi regions form a partition of the space with the same number of cells per unit volume as the partition formed by the parallelepiped cells.

The integers Z are themselves a one-dimensional lattice with generator 1. The set of all integer-valued N-tuples is the N-dimensional *square lattice* Z^N, whose generator matrix is the N by N identity matrix. The lattices Z^N have a minimum distance equal to one. Figure 9.28 shows the two-dimensional lattice Z^2.

The two-dimensional hexagonal lattice is shown in Figure 9.29. The two-dimensional hexagonal lattice, denoted A_2, has the generator matrix

$$A = \begin{bmatrix} 0 & \sqrt{3} \\ 2 & 1 \end{bmatrix}.$$

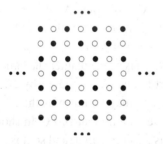

Figure 9.28. The lattice Z^2 and sublattice RZ^2 (black dots).

Figure 9.29. The hexagonal lattice.

The lattice A_2 is the densest of any lattice in two dimensions with the same minimum distance. Many of the frequently used two-dimensional signal constellations are subsets of either $\mathbf{Z}^2$ or A_2.

As the dimension of the space increases, the number of interesting lattices increases rapidly. Many of the most important lattices have been given conventional names. For example, the *Schlafli lattice* D_4 consists of all points of the four-dimensional lattice $\mathbf{Z}^4$ whose coordinates add to an even integer. Good four-dimensional signal constellations can be constructed by working with the Schlafli lattice D_4, choosing a subset of points with a good structure. More generally, D_N consists of all points of $\mathbf{Z}^N$ whose coordinates add to an even integer. Also worth mention in passing is the *Gosset lattice* E_8, which is the densest known eight-dimensional lattice, and the ultradense *Leech lattice* Λ_{24}, which is the densest known 24-dimensional lattice.

A *sublattice* Λ' of a given lattice Λ is another lattice, all of whose points are in the given lattice Λ. This can be so if and only if the generators of the sublattice are elements of the given lattice. If in Figure 9.28, the origin is at the center dot, then the set of black dots forms a sublattice. The sublattice, in fact, is seen to be a magnified and rotated version of the original lattice. A *coset* of the sublattice Λ' is the set of all N-tuples obtained by summing all sublattice points Λ' with some fixed lattice point c, for instance, one that is in the lattice but not in the sublattice. The coset is the set $\{\lambda + c : \lambda \in \Lambda'\}$. Each coset is simply a translation of the sublattice Λ'. In Figure 9.28 there are two cosets: the set of white dots and the set of black dots that is the same set as the sublattice Λ'. All cosets are of the same type and have the same minimum squared distance as the sublattice itself. In general, given the sublattice Λ', a lattice has an m-way partition into m cosets of the sublattice, including the sublattice itself. All of the cosets are implied as soon as the sublattice Λ' is specified. The set of cosets is specified by the notation Λ/Λ' because we think of Λ' as dividing Λ into the collection of cosets. The set of all cosets of a lattice is called a *partition* of the lattice. A partition is called a binary partition if the number of cosets is a power of 2.

For example, the N-dimensional square lattice $\mathbf{Z}^N$ has as a sublattice the lattice of all even integer N-tuples denoted $2\mathbf{Z}^N$, and $\mathbf{Z}^N$ is partitioned into 2^N cosets by $2\mathbf{Z}^N$, each with minimum squared distance 4. The coset leaders may be taken as the 2^N binary combinations of the N generators that comprise the identity matrix $\mathbf{I}_N$. In shorthand, this partition is written $\mathbf{Z}^N/2\mathbf{Z}^N$. The partition has 2^N cosets in it, and so it is a binary partition.

9.5   Trellis codes on lattice cosets

Now we are ready to return to the study of trellis codes. For a signal constellation that is a subset of a lattice, one can construct a trellis code in terms of an algebraic

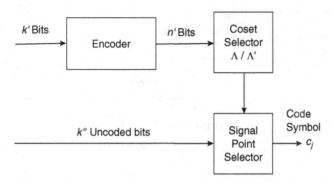

Figure 9.30. General structure of a trellis encoder on a lattice coset.

construction on a lattice. The lattice Λ is partitioned into $2^{n'}$ cosets, and each coset is labeled with n' bits. A set of $2^{k''}$ points near the origin is chosen from each coset to form a subconstellation, and the full signal constellation is the union of these subconstellations. Normally, one would choose the subconstellations to be translates of each other, but this is not a requirement.

Next, choose a suitable (n', k') polynomial generator matrix. We take the symbols of the code to be cosets – that is, elements of Λ/Λ' for some lattice Λ and sublattice Λ' – and the symbols of the transmitted waveform to be elements of these cosets. A set of $2^{k''}$ points near the origin forms a subconstellation within the coset.

Figure 9.30 shows the general structure of a trellis code on a lattice coset. The lattice is partitioned into $2^{n'}$ cosets, each labeled by n' bits. An (n', k') code generates an n'-bit symbol that selects one of the $2^{n'}$ cosets. One of the $2^{k''}$ points of the subconstellation of the encoded coset is then selected by the remaining k'' uncoded data bits. In each frame, a total of $k = k' + k''$ bits is modulated into the sequence of code symbols from the $2^{n'+k''}$-point signal constellation. Usually, $n' = k' + 1$ so that, in each frame, k data bits are mapped into a 2^{k+1}-point signal constellation.

Figure 9.31 shows a simple example of encoding five data bits into 64-point signal constellations. Either one of the two variations of the signal constellation shown can be used. The lattice $\mathbf{Z}^2$ is partitioned into four cosets that are denoted $\{\mathcal{A}, \mathcal{B}, \mathcal{C}, \mathcal{D}\}$. The output of the $(2,1)$ encoder selects one of the four cosets. Four uncoded bits then select one of the points from that coset. This means that sixteen of the points in each coset have been singled out to be assigned a four-bit binary index. The specification of these subconstellations is independent of the design of the encoder. Either the square signal constellation or the cross signal constellation (or some other signal constellation) may be selected to specify the set of points chosen from the cosets. The number of data bits k can be readily enlarged or reduced by enlarging or reducing the number of uncoded data bits k''. This means simply that the size of the subconstellation is enlarged or reduced, thereby enlarging or reducing the size of the signal constellation.

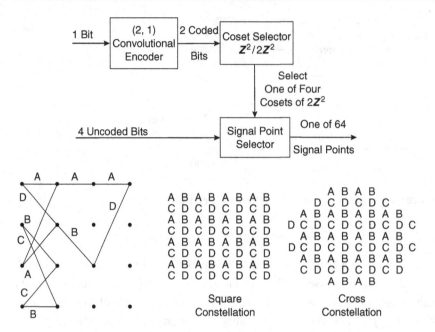

Figure 9.31. A $v = 2$ Ungerboeck code with two alternative signal constellations.

To find the asymptotic coding gain of one of the resulting codes is largely a matter of finding the free distance of the code. We shall find that the free distance of the trellis code is due either to the euclidean free distance of the underlying code, whose code alphabet consists of cosets, or is due to the euclidean distance between points within a coset. The latter, which we will call $d_{\text{min}}^{(1)}$, dictates the distance when the coded bits are the same and the uncoded bits are different. The former, which we will call $d_{\text{min}}^{(2)}$, dictates the distance when the coded bits are different and the uncoded bits are arbitrary. The free distance of the trellis code is

$$d_{\text{min}} = \min(d_{\text{min}}^{(1)}, d_{\text{min}}^{(2)}).$$

The smallest squared euclidean distance between two points in the same coset of our example is 2^2. The smallest distances (in units of the lattice spacing d_0) between points in different cosets is given by

$$d^2(\mathcal{A}, \mathcal{B}) = 1$$
$$d^2(\mathcal{A}, \mathcal{C}) = 1$$

$$d^2(\mathcal{A}, \mathcal{D}) = 1$$

$$d^2(\mathcal{B}, \mathcal{C}) = 1$$

$$d^2(\mathcal{B}, \mathcal{D}) = 1$$

$$d^2(\mathcal{C}, \mathcal{D}) = 1.$$

Therefore the square of the free euclidean distance of the code is found from Figure 9.31 to be

$$(d_{\min}^{(1)})^2 = d^2(\mathcal{A}, \mathcal{D}) + d^2(\mathcal{A}, \mathcal{B}) + d^2(\mathcal{A}, \mathcal{D})$$
$$= 5.$$

Consequently,

$$d_{\min}^2 = \min(5, 4)$$
$$= 4.$$

The structure described, as portrayed in Figure 9.31, essentially decouples the choice of the encoding and the coset selection from the choice of the signal constellation. By using the language of lattice cosets, the treatment is streamlined, although not essentially different from the development in Section 9.3. However, the extensive topic of systematically choosing the map from the output of the encoder into the collection of cosets so as to maximize the minimum distance has not yet been discussed. This task consists of labeling the branches of the trellis with cosets as illustrated in Figure 9.31. It can be done by hand, by trial and error, if the trellis is not too large.

Figure 9.32 shows an example of a (3, 2) trellis code with eight states. This code has a free distance $d_{\min} = \sqrt{5}d_0$, and no path has more than four nearest neighbors at distance $d_{\min}$, though some have fewer. Thus

$$p_e \approx 4Q\left(\frac{\sqrt{5}d_0}{2\sigma}\right).$$

Each node has four branches leaving it, and these must be labeled with the four two-bit patterns. Every labeling gives the same code, but different labelings give different encoders. The labeling is chosen to make the encoder easy to design.

Table 9.2 shows the check polynomials for some good Ungerboeck codes for the sixteen-ary QAM signal constellation. Each of the last three entries has the same asymptotic coding gain so the larger constraint length does not improve the asymptotic coding

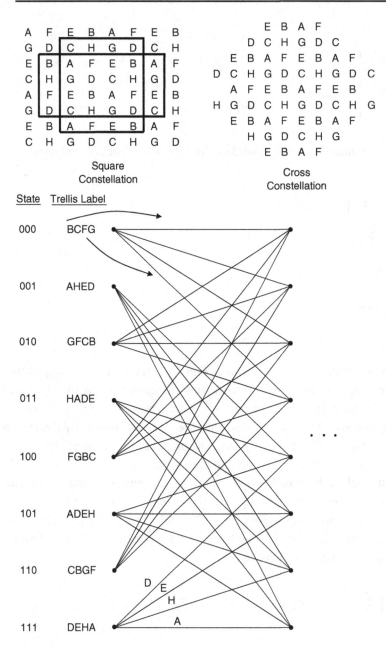

Figure 9.32. A (3, 2) trellis code with eight states.

gain. It does, however, reduce the number of nearest neighbors from 344 to 44 to 4, as the constraint length changes from 7 to 8 to 9. By appending some uncoded bits, each of these codes can be used for a larger QAM signal constellation but, of course, the coding gain must then be recomputed.

Table 9.2. *Ungerboeck codes for the sixteen-ary QAM signal constellation*

Constraint length	$h_1(x)$	$h_2(x)$	$h_3(x)$	d_{min}^2/d_0^2	Gain over uncoded 8-PSK (dB)
2	$x^2 + 1$	x	0	4.0	~4.36
3	$x^3 + 1$	x	x^2	5.0	5.33
4	$x^4 + x + 1$	x^2	$x^3 + x^2 + x$	6.0	6.12
5	$x^5 + 1$	$x^2 + x$	x^3	6.0	6.12
6	$x^6 + 1$	$x^3 + x^2 + x$	$x^5 + x^4 + x^2$	7.0	6.79
7	$x^7 + x + 1$	$x^3 + x^2$	$x^5 + x$	8.0	7.37
8	$x^8 + 1$	101110	$x^7 + x^6 + x^2$	8.0	7.37
9	$x^9 + 1$	11100110	$x^8 + x^6 + x^3$	8.0	7.37

9.6   Differential trellis codes

A differential method for digital modulation maps the datastream into a continuous-time waveform in such a way that the datastream can be recovered despite some form of ambiguity in the received waveform, usually a phase ambiguity. The simplest example is differential PSK. This modulation technique protects against a 180° phase ambiguity in the received waveform. Other differential modulation codes are more sophisticated, and protect against multiple ambiguities.

The differential trellis codes discussed in this section are a generalization of differential QPSK. They are used for passband channels in which the carrier phase is known only to within a multiple of 90°. These codes can simplify the carrier acquisition problem because, after a temporary interruption in carrier phase synchronization, it is enough for the carrier phase to relock in any one of four phase positions. The phase of the received waveform may be offset by 0°, 90°, 180°, or 270° from the phase of the transmitted waveform. Nevertheless, for each of these four possibilities, the demodulated datastream is the same. The technique is to design a code such that every 90° rotation of a codeword is another codeword. Then by differential encoding, the end-to-end transmission is made transparent to 90° offsets of carrier phase.

The idea of differential trellis codes will be easier to understand if we first examine differential modulation in the absence of coding. We shall describe differential sixteen-ary signaling. Figure 9.33 shows a sixteen-ary square signal constellation that we will visualize as four subconstellations, each with four points. Each subconstellation is invariant under a rotation by any multiple of 90°. Each of its points is given an identical two-bit label characterizing that subconstellation.

Two bits of data are encoded into a point of the signal constellation by using the indicated labels. Two more bits are used to select the quadrant. The quadrant is selected

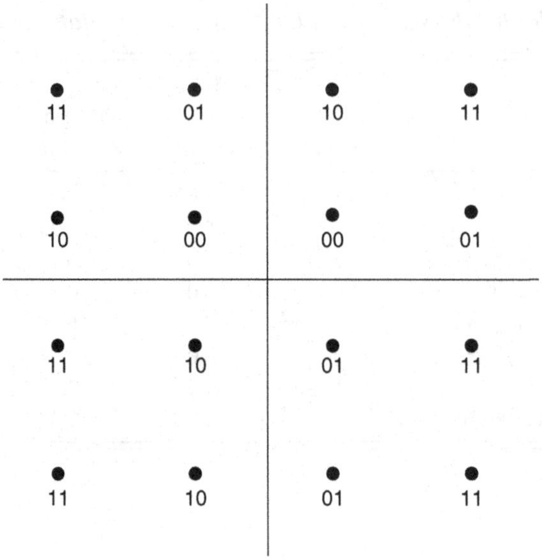

Figure 9.33. A symmetrically labeled signal constellation.

differentially by the rule

$$00 \rightarrow \Delta\phi = 0°$$
$$10 \rightarrow \Delta\phi = 90°$$
$$11 \rightarrow \Delta\phi = 180°$$
$$01 \rightarrow \Delta\phi = 270°.$$

A modulator for the differential sixteen-ary signal constellation is shown in Figure 9.34. Two of the data bits define a point in the subconstellation, and two of the data bits define a phase rotation as one of the four multiples of 90°. The demodulator is easily designed to undo what the modulator has done, first by detecting the proper points in the signal constellation, then mapping the sequence of labels into the sequence of data.

The primary requirement of differential encoding is that the signal constellation be labeled in a special way. A secondary requirement that the data be expressed in a special differential form is easily accomplished by proper precoding and postcoding. Consequently, for suitable trellis codes, the method of differential coding can be combined with the method of trellis coding. Some trellis codes are invariant under a 90° phase rotation; others are not. If a code does have this property, it should be exploited by the design of the encoder and decoder to make the system transparent to 90° rotations in the channel. Therefore the encoder and decoder should employ some form of differential encoding.

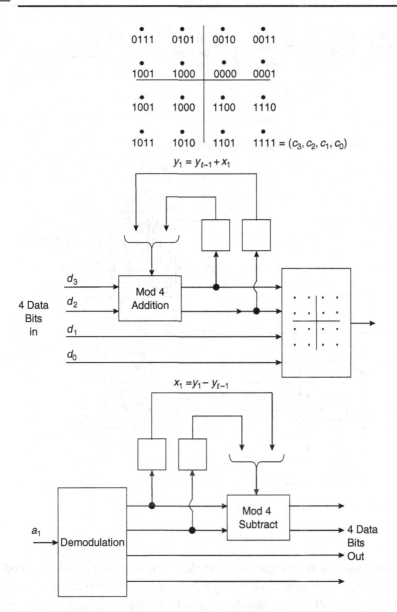

Figure 9.34. A differential modulator for a sixteen-ary constellation.

Some trellis codes on $\mathbf{Z}^2$ have a structure that is immediately compatible with differential encoding. In particular, trellis codes with two or more uncoded bits may have this property. Recall the partition of sixteen-ary PSK that was shown in Figure 9.26. Differential encoding can be readily used with that partition by changing the way in which the two uncoded data bits are mapped into points of the coset. We regard these two bits as an integer modulo-four and, instead of encoding this integer directly, encode the

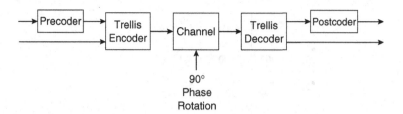

Figure 9.35. Construction of a differential trellis code.

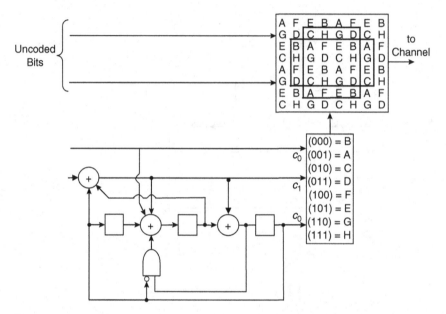

Figure 9.36. A nonlinear encoder for a rotationally invariant trellis code.

modulo-four sum of it and the previous encoded value. This amounts to a precoding of
the uncoded data bits, which is undone by a postcoder after demodulation. Figure 9.35
shows this kind of differential encoding. Essentially, the same idea can be used with
the partition of an eight-ary PSK signal constellation shown in Figure 9.25, to give an
eight-ary PSK code invariant under 180° offsets of carrier phase.

Other trellis codes are not invariant under a 90° phase rotation. If the best (n, k)
code of a given constraint length is not invariant under a 90° phase rotation, then
one may choose to use a code with less coding gain in order to obtain the property
of rotational invariance. Even if the trellis code is invariant under 90° rotations, it
may not be possible to find a linear encoder that is transparent to rotations and also
can be decomposed into a precoder and a linear encoder such that the encoder has
only linear feedback or feedforward shift registers. A transparent encoder must then be

nonlinear, using nonlinear elements such as complex rotation. This makes the task of finding the minimum distance harder.

Figure 9.36 shows a nonlinear encoder for the trellis code of Figure 9.32 that is transparent to 90° phase rotations. This encoder is referred to as nonlinear because, as a consequence of the *and* gates in the encoder, there is no longer a linearity property between the input and output of the logic circuit under modulo-two addition.

9.7   Four-dimensional trellis codes

Nothing in the definition of a trellis code is peculiar to a two-dimensional signal constellation. Trellis codes can be constructed for signal constellations of any dimension; we shall consider binary trellis codes in four dimensions. A four-dimensional symbol can be sent over a two-dimensional channel such as a complex baseband channel, by sending two components of the four-dimensional symbol at a time. The reason for using a four-dimensional signal constellation is to make the minimum distance larger without increasing the computational complexity, although the conceptual complexity is increased. The potential gains, however, are modest. Using four-dimensional lattice cosets in the construction of these codes makes it easy to construct the four-dimensional trellis codes.

Choose any four-dimensional lattice that can be partitioned into $2^{n'}$ cosets, in each of which $2^{k''}$ points are selected as the points of a signal subconstellation. Choose any (n', k') binary polynomial generator matrix to form a code. To encode k bits, set $k = k' + k''$ and encode k' bits each frame using the (n', k') encoder. The n' bits out of the encoder each frame select a subconstellation, and the k'' uncoded bits select a point from that subconstellation. In this way in each frame, k bits are represented by $n' + k''$ bits. Commonly, $n' = k' + 1$ so that, in each frame, k data bits are mapped into a 2^{k+1}-point four-dimensional signal constellation.

There are two reasons why the four-dimensional code may be superior to a two-dimensional code. First, points can be packed more densely into four dimensions than into two dimensions. Second, if k is even, to send $k/2$ points using a two-dimensional trellis code would require a $2^{k/2+1}$-point signal constellation, while to send k bits would require, in effect, a $(2^{k/2+1})^2 = 2^{k+2}$ four-dimensional signal constellation. Designing the code in four dimensions makes a four-dimensional signal constellation possible with fewer points, thereby improving the minimum distance of the signal constellation (though not necessarily of the code).

We will construct a trellis code for the four-dimensional square trellis Z^4. This example is not too difficult and yet provides a practical code. Consider the points of Z^4 as pairs of points from Z^2, that is, $Z^4 = Z^2 \times Z^2$, and Z^2 is partitioned, as shown in Figure 9.37, into subsets $\mathcal{A}, \mathcal{B}, \mathcal{C}$, and $\mathcal{D}$, each of which has minimum distance 4.

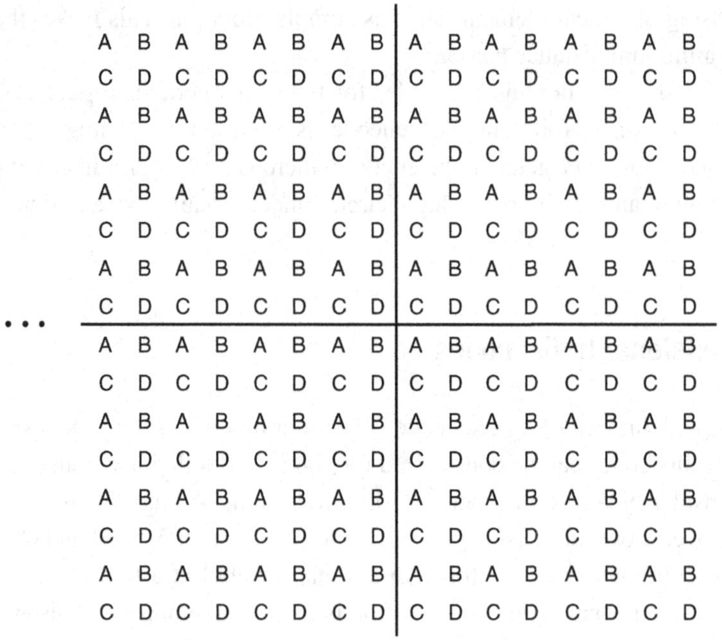

```
A  B  A  B  A  B  A  B │ A  B  A  B  A  B  A  B
C  D  C  D  C  D  C  D │ C  D  C  D  C  D  C  D
A  B  A  B  A  B  A  B │ A  B  A  B  A  B  A  B
C  D  C  D  C  D  C  D │ C  D  C  D  C  D  C  D
A  B  A  B  A  B  A  B │ A  B  A  B  A  B  A  B
C  D  C  D  C  D  C  D │ C  D  C  D  C  D  C  D
A  B  A  B  A  B  A  B │ A  B  A  B  A  B  A  B
C  D  C  D  C  D  C  D │ C  D  C  D  C  D  C  D
─────────────────────────────────────────────
A  B  A  B  A  B  A  B │ A  B  A  B  A  B  A  B
C  D  C  D  C  D  C  D │ C  D  C  D  C  D  C  D
A  B  A  B  A  B  A  B │ A  B  A  B  A  B  A  B
C  D  C  D  C  D  C  D │ C  D  C  D  C  D  C  D
A  B  A  B  A  B  A  B │ A  B  A  B  A  B  A  B
C  D  C  D  C  D  C  D │ C  D  C  D  C  D  C  D
A  B  A  B  A  B  A  B │ A  B  A  B  A  B  A  B
C  D  C  D  C  D  C  D │ C  D  C  D  C  D  C  D
```

Figure 9.37. A partition of $\mathbf{Z}^2$.

Eight subsets of $\mathbf{Z}^4$ are defined as follows:

$$\mathcal{E}_0 = \{C, C\} \cup \{B, B\} \qquad \mathcal{E}_4 = \{C, A\} \cup \{B, D\}$$
$$\mathcal{E}_1 = \{A, A\} \cup \{D, D\} \qquad \mathcal{E}_5 = \{A, B\} \cup \{D, C\}$$
$$\mathcal{E}_2 = \{C, B\} \cup \{B, C\} \qquad \mathcal{E}_6 = \{C, D\} \cup \{B, A\}$$
$$\mathcal{E}_3 = \{A, D\} \cup \{D, A\} \qquad \mathcal{E}_7 = \{A, C\} \cup \{D, B\}$$

where, for example, $\{A, B\}$ denotes the set of points $\mathbf{Z}^4$, combining one point from two-dimensional subset A and one from two-dimensional subset B. It takes three bits to specify the set $\mathcal{E}_\ell$ and twelve bits to specify an element of $\mathcal{E}_\ell$. The minimum distance of each two-dimensional pair is again four because, in any two elements of $\{A, B\}$, either the first component or the second component must be different. Further, the eight four-dimensional subsets are defined in such a way that the minimum distance between them is again four. This is because, for example, the first component of $\{A, B\} \cup \{C, D\}$ must lie in $A \cup C$, which has minimum squared euclidean distance 2 and the second must lie in $B \cup D$, which again has minimum squared euclidean distance 2.

Because of the relationship

$$d_{\min} = \min(d_{\min}^{(1)}, d_{\min}^{(2)}),$$

we see that it is sufficient to choose a generator matrix for a $(3, 2)$ code encoding two data bits into the eight subsets $\mathcal{E}_\ell, \ell = 1, \ldots, 8$. An additional twelve uncoded bits select an element of $\mathcal{E}_\ell$ with $d_{\min}^{(2)}$ at least 4. Then we will have a $(15, 14)$ four-dimensional trellis code with $d_{\min} = 4$.

Problems for Chapter 9

9.1. The "modified duobinary" partial-response signaling waveform has the generator polynomial $g(x) = 1 - x^2$.
   a. Find the impulse response.
   b. Find the channel transfer function that will provide the encoding.
   c. Show that even and odd output samples are independent and can be deinterleaved and demodulated separately.
   d. Describe the output sequences on a trellis.
   e. Design a precoder that will make the outputs independent.

9.2. Precoding for partial-response signaling fails if the generator polynomial $g(x)$ has its low-order coefficient g_0 equal to zero modulo 2. For example, $g(x) = 2 + x - x^2$ cannot be precoded.
   a. Explain why this is so.
   b. Prove that the data can be "partially precoded" by using

$$b_{j-\ell} = \frac{1}{g_\ell} \left[d_j - \sum_{i=\ell+1}^{n-1} g_i b_{j-i} \right] \quad (\text{mod } 2)$$

   where g_ℓ is the lowest index coefficient of $g(x)$ that is not zero modulo 2. What is the channel output? Are the symbols of the received datastream independent?

9.3. Partial-response signaling can also be used when the input to the channel takes more than two amplitudes.
   a. Describe a duobinary system with a four-level input alphabet. Give a decision-feedback demodulator including the detection rule.
   b. Can a precoder be used in place of the decision-feedback demodulator? If so, describe it.
   c. Can one choose a duobinary system for an eight-ary PSK signal constellation?
   d. If so, can a precoder be used?

9.4. A partial-response waveform can be designed by using any Nyquist pulse in place of the sinc pulse.
   a. What is the equivalent channel for a duobinary partial-response waveform based on the Nyquist pulse

$$p(t) = \frac{\sin \pi t/T}{\pi t/T} \cdot \frac{\cos a\pi t/T}{1 - 4a^2t^2/T^2}$$

   and the generator polynomial $g(x) = x + 1$?

b. What is the matched filter for this waveform? Sketch a decision-feedback demodulator. Sketch an alternative implementation that uses a precoder.

9.5.   a. By examining the trellis for duobinary partial-response signaling, show that there are some paths that have an infinite number of neighbors at a distance equal to the minimum euclidean distance. How do the error events differ? What does this mean in the application of the Viterbi demodulator to this waveform?

b. Sketch a labeled trellis for a duobinary signaling waveform used with a precoder.

9.6.   A telephone channel with echo (idealized) is an ideal passband channel from 300 Hz to 2700 Hz with an echo of 1/1200 sec and an echo amplitude of 0.1 of the desired signal. That is, the true impulse response $h'(t)$ is

$$h'(t) = h(t) + 0.1h\left(1 - \frac{1}{1200}\right)$$

where $h(t)$ is the impulse response without the echo. The transmitted signal is a sixteen-ary PSK using a sinc pulse to fill the bandwidth.

a. What is the data rate?

b. Describe how to use the Viterbi algorithm to cancel the effects of the echo. How many states are in the trellis? Is it possible to use interleaving to simplify the Viterbi receiver?

c. We now desire to add a modified duobinary partial-response technique so that $h(t)$ can be replaced by a more practical $h(t)$, but the echo is still present. How many states will there be in the Viterbi algorithm?

d. Can (partial-response) precoding be used with a sixteen-ary PSK? Why?

9.7.   Differential QPSK is a modulation scheme that works in the presence of a fixed unknown phase offset. Each pair of bits is modulated into a phase change as follows:

$$00 \rightarrow \Delta\phi = 0°$$
$$01 \rightarrow \Delta\phi = 90°$$
$$11 \rightarrow \Delta\phi = 180°$$
$$10 \rightarrow \Delta\phi = 270°.$$

a. If noninterfering sinc pulses are used to modulate the QPSK points, what is the bandwidth of the signal as a function of the data rate?

b. Give a demodulator for a differential QPSK as a variation of a demodulator for QPSK. What is the probability of error as a function of E_b/N_0?

c. Describe a duobinary partial-response modulator, including a precoder, for this passband waveform. Include all major modulator functions. What is the channel transfer function?

9.8. A $(2, 1)$ trellis code of constraint length 1 for the real signal constellation $\{-3, -1, 1, 3\}$ has generator polynomials $g_1(x) = x + 1$ and $g_2(x) = 1$. The output of the encoder is two bits and is used to specify a point in the signal constellation.

a. Sketch an encoder. Is it systematic?

b. Sketch a trellis.

c. Label the trellis with the points of the signal constellation to make the free euclidean distance large. (Choose the map from pairs of encoder output bits to signal constellation points.)

d. What is the free euclidean distance?

e. Compare the performance with BPSK.

9.9. A $(2, 1)$ trellis code of constraint length 1 for the *real* signal constellation $\{-3, -1, 1, 3\}$ has generator polynomials $g_1(x) = x^3 + x + 1$ and $g_2(x) = 1$. The output of the encoder is two bits and is used to specify a point in the signal constellation.

a. Sketch an encoder.

b. Sketch a trellis.

c. Label the trellis with the points of the signal constellation to make the euclidean free distance large. (Choose the map from pairs of encoder output bits to points of the signal constellation.)

d. What is the euclidean free distance?

e. Compare the performance with BPSK.

9.10. The points of an eight-ary PSK signal constellation are sequentially labeled with three-bit numbers in the sequence (000, 111, 110, 101, 100, 011, 010, 001, 000). An Ungerboeck code of constraint length 4 and rate $\frac{2}{3}$ for this signal constellation has check polynomials

$$h_1(x) = x^4 + x + 1 \quad h_2(x) = x^2 \quad h_3(x) = x^3 + x^2 + x.$$

a. Design a nonsystematic encoder without feedback.

b. Design a systematic encoder with feedback.

c. The free distance is $d_{min} = \sqrt{2d_0^2 + 2d_1^2}$ where d_0 and d_1 are the smallest and next smallest euclidean distances between two points in the signal constellation. Find two input data patterns that produce codewords separated by the free distance.

d. What is the asymptotic coding gain?

9.11. Prove that the following is an equivalent definition of a lattice. A lattice is a set of points in R^n such that if x and y are in the set, then so is $ax + by$ for any integers a and b.

9.12. a. Show that no $(2, 1)$ trellis code can have a 90° phase invariance when its four two-bit codeword frames are mapped onto the four points of the QPSK signal constellation.

b. Show that some $(4, 2)$ trellis codes do have a 90° phase invariance. Explain how to design an encoder and decoder so that the 90° phase offsets in the channel are transparent to the user.

c. Show that no trellis code can have a 45° phase invariance when used with the eight-ary PSK signal constellation.

9.13. A rate $\frac{2}{3}$, constraint length 2, Ungerboeck code for an eight-ary PSK signal constellation with generator polynomials $(1, 1 + x^2, x)$ has an asymptotic coding gain of 3 dB. Describe how this code can be used as the basis for an Ungerboeck code for a 2^m-ary PSK with a rate of $m/(m + 1)$ and a constraint length of 2. What is the asymptotic coding gain for large m? How does $N_{d_{min}}$ depend on m?

9.14. As outlined below, show that, when the spectral bit rate density is large, using a signal constellation without coding uses about eight times (9 dB) as much energy as does a more sophisticated waveform.

a. Show that the real 2^r-point signal constellation $\{\pm 1, \pm 3, \ldots, \pm 2^r - 1\}$ has an average energy $E_c = (4^r - 1)/3$. Show that the 2^{2r}-point square signal constellation that uses this real signal constellation on the real and imaginary axes has an average energy $E_c = 2(4^r - 1)/3$.

b. Show that the probability of symbol error for the real signal constellation is approximated by

$$p_e \approx Q\left(\sqrt{\frac{6^r E_b}{4^r N_0}}\right).$$

Show that this same approximation holds for the complex square signal constellation.

c. Given the value

$$Q\left(\sqrt{22.4}\right) = 10^{-6},$$

show that achieving a symbol error rate of 10^{-6} requires an energy satisfying

$$\frac{E_b}{N_0} \approx \left(\frac{2^r}{6^r}\right) 22.4.$$

Compare this to the Shannon capacity bound

$$\frac{E_b}{N_0} > \frac{2^r - 1}{r}.$$

Notes for Chapter 9

The MSK waveform was generalized to continuous-phase modulation by Pelchat, Davis, and Luntz (1971), and by de Buda (1972). The performance of suboptimal methods of demodulation of CPFSK was studied by Osborne and Luntz (1974), and by Schonhoff (1976). Guided by the ideas of partial-response signaling for the baseband channel and by the CPFSK signaling waveform, Miyakawa, Harashima, and Tanaka (1975), and Anderson and Taylor (1978), developed continuous-phase digital modulation waveforms. This line of work was continued by Aulin, Rydbeck, and Sundberg (1981). Rimoldi (1988) developed an alternative formulation of continuous-phase modulation.

Ungerboeck (1977, 1982) was the first to recognize that convolutional codes designed with a large free Hamming distance need not be the best codes when the decoder is based on euclidean distance in the complex plane. For these applications, he introduced the use of trellis codes that are designed with a large free euclidean distance. These codes rapidly found wide application, and many tried their hands at alternative descriptions, including Calderbank and Mazo (1984), and Forney and his coauthors (1984). This elaboration of the Ungerboeck codes included the introduction of multidimensional trellis codes by Forney and his coauthors (1984). The best rotationally invariant trellis codes were discovered by Wei (1984), who recognized that nonlinear elements were necessary in the encoder to obtain the best such codes.

The use of lattices to construct block codes started with unpublished work by Lang, and included further work by de Buda (1975), Blake (1971), and Leech and Sloane (1970). Lattices were studied in a classical paper by Voronoi (1908). The best tables of lattices have been published by Sloane (1981), and by Conway and Sloane (1987).

Forney and his coauthors (1984) recognized the role of lattices in describing the Ungerboeck codes, and made an explicit distinction between the design of codes and the design of constellations. Calderbank and Sloane (1987) discussed the role of lattice partitions in constructing codes. Forney (1988) provided an elegant formal development of trellis codes as a construction within lattice cosets. Trellis codes for four-dimensional and eight-dimensional signal constellations were studied by Wilson, Sleeper, and Srinath (1984); by Calderbank and Sloane (1985, 1986); and by Wei (1987). Forney explained the advantages of obtaining a nonequiprobable probability distribution on lattice points, and Calderbank and Ozarow (1990) discussed ways to do this.

The use of partial-response signaling to increase data rate was introduced by Lender (1964) and developed by Kretzmer (1966) under the terminology of *partial-response classes*, as well as by others. Kabal and Pasupathy (1975) surveyed and unified the many contributions. The idea of precoding the decision-feedback within the transmitter to eliminate error propagation can be found in the work of Tomlinson (1971) and also Miyakawa and Harashima (1969). Partial-response signaling plays an essential role in most digital magnetic recording systems, as was suggested by Kobayashi and Tang (1970). The applicability of the Viterbi algorithm to partial-response signaling was observed by Omura (1970) and Kobayashi (1971).

10 Codes for Data Transmission

The modulator and demodulator make a waveform channel into a discrete communication channel. Because of channel noise, the discrete communication channel is a noisy communication channel; there may be errors or other forms of lost data. A *data transmission code* is a code that makes a noisy discrete channel into a reliable channel. Despite noise or errors that may exist in the channel output, the output of the decoder for a good data-transmission code is virtually error-free. In this chapter, we shall study some practical codes for data transmission. These codes are designed for noisy channels that have no constraints on the sequence of transmitted symbols. Then a data transmission code can be used to make the noisy unconstrained channel into a reliable channel.

For the kinds of discrete channels formed by the demodulators of Chapter 3, the output is simply a regenerated stream of channel input symbols, some of which may be in error. Such channels are called *hard-decision* channels. The data transmission code is then called an error-control code or an error-correcting code. More generally, however, the demodulator may be designed to qualify its output in some way. Viewed from modulator input to demodulator output, we may find a channel output that is less specific than a hard-decision channel, perhaps including erasures or other forms of tentative data such as likelihood data on the set of possible output symbols. Then it is not possible to speak of an error or of an error-correcting code. Thus, for the general case, the codes are called *data transmission codes*.

10.1 Block codes for data transmission

A block code for data transmission is a code that protects a discrete-time sequence of symbols from noise and random errors. To every k data symbols in the alphabet of the code, a t-error-correcting code forms a codeword by appending $n - k$ new symbols called *check symbols*. The codeword now has blocklength n and is constructed so that the original k data symbols can be recovered even when up to t codeword symbols are corrupted by additive channel noise. The relationship between the number of check

0	0	0	0	0	0	0
1	1	1	0	0	0	1
1	1	0	0	0	1	0
0	0	1	0	0	1	1
1	0	1	0	1	0	0
0	1	0	0	1	0	1
0	1	1	0	1	1	0
1	0	0	0	1	1	1
0	1	1	1	0	0	0
1	0	0	1	0	0	1
1	0	1	1	0	1	0
0	1	0	1	0	1	1
1	1	0	1	1	0	0
0	0	1	1	1	0	1
0	0	0	1	1	1	0
1	1	1	1	1	1	1

Figure 10.1. The $(7, 4)$ Hamming code.

symbols, $n - k$, and the number of symbols t that can be corrected can be quite complicated, in general. We shall only study several popular and often-used examples of block codes.

There are a great many kinds of block codes known for data transmission. We shall study those bit-organized codes known as (n, k) Hamming codes and those byte-organized codes known as (n, k) Reed–Solomon codes.

The simplest Hamming code is the $(7, 4)$ binary Hamming code shown in Figure 10.1. There are sixteen binary codewords. A four-bit dataword is represented by a seven-bit codeword. By inspection of Figure 10.1 we can see that every codeword differs from every other codeword in at least three places. We express this in the language of geometry, saying that every codeword is at *Hamming distance* at least three from every other codeword. If a codeword is sent through a channel and the channel makes a single error, then the senseword will differ from the correct codeword in one place, and will differ from every other codeword in at least two places. The decoder will recover the correct codeword if it decides that the codeword closest to the senseword in Hamming distance was the codeword transmitted. However, if the channel makes more than one error, then the decoder will be wrong. Thus, the $(7, 4)$ Hamming code can correct one bit error but not more. It is called a single-error-correcting code.

The seven bit positions of the $(7, 4)$ Hamming code can be rearranged into a permuted order without changing the distance structure. Similarly, the sixteen codewords can be

0 0 0 0	0 0 0 0 0 0 0
0 0 0 1	1 1 1 0 0 0 1
0 0 1 0	1 1 0 0 0 1 0
0 0 1 1	0 0 1 0 0 1 1
0 1 0 0	1 0 1 0 1 0 0
0 1 0 1	0 1 0 0 1 0 1
0 1 1 0	0 1 1 0 1 1 0
0 1 1 1	1 0 0 0 1 1 1
1 0 0 0	0 1 1 1 0 0 0
1 0 0 1	1 0 0 1 0 0 1
1 0 1 0	1 0 1 1 0 1 0
1 0 1 1	0 1 0 1 0 1 1
1 1 0 0	1 1 0 1 1 0 0
1 1 0 1	0 0 1 1 1 0 1
1 1 1 0	0 0 0 1 1 1 0
1 1 1 1	1 1 1 1 1 1 1

Figure 10.2. Systematic encoding of $(7, 4)$ Hamming code.

rearranged into a permuted order without changing the distance structure. Any such change creates what is called an equivalent $(7, 4)$ Hamming code. In general, two codes are called equivalent if they differ only in some trivial way, such as a permutation of bit positions.

The mapping between four-bit datawords and seven-bit codewords is arbitrary but usually the *systematic form* of encoding shown in Figure 10.2 is used. There the first four bits of the codeword are equal to the four data bits. In general, a systematic encoding rule is one in which the data bits are included unchanged as part of the codeword. The remaining bits are the check bits.

In general, for each positive integer m, there is a binary $(2^m - 1, 2^m - 1 - m)$ Hamming code that can correct a single error. The construction of the codes can be expressed in matrix form using the following definition of addition and multiplication of bits:

+	0	1		·	0	1
0	0	1		0	0	0
1	1	0		1	0	1

The operations are the same as the *exclusive-or* operation and the *and* operation, but we will call them *addition* and *multiplication* so that we can use a matrix formalism to define the code. The two-element set $\{0, 1\}$ together with this definition of addition and multiplication is a number system called a *finite field* or a *Galois field*, and is denoted by the label $GF(2)$. In Section 10.2 we shall discuss the finite field known as $GF(2^m)$, which has 2^m elements.

To construct a $(2^m - 1, 2^m - 1 - m)$ Hamming code, we first write down a matrix H, known as a *check matrix* and consisting of all of the $2^m - 1$ nonzero m-bit numbers as columns. If the code is to be systematic, then the first m columns should contain, in the form of an identity matrix, those m-bit numbers with a single one. For example, when $m = 3$,

$$H = \begin{bmatrix} 1 & 0 & 0 & 0 & 1 & 1 & 1 \\ 0 & 1 & 0 & 1 & 0 & 1 & 1 \\ 0 & 0 & 1 & 1 & 1 & 0 & 1 \end{bmatrix}.$$

We partition this as

$$H = [I \vdots P]$$

where I is the m by m identity matrix and P is an m by $(2^m - 1 - m)$ matrix containing all other nonzero m-bit binary numbers as columns.

Next define the *generator matrix*

$$G = [P^T \vdots I]$$

where I is now a $(2^m - 1 - m)$ by $(2^m - 1 - m)$ identity matrix.

For our running example with $m = 3$,

$$G = \begin{bmatrix} 0 & 1 & 1 & 1 & 0 & 0 & 0 \\ 1 & 0 & 1 & 0 & 1 & 0 & 0 \\ 1 & 1 & 0 & 0 & 0 & 1 & 0 \\ 1 & 1 & 1 & 0 & 0 & 0 & 1 \end{bmatrix}.$$

The codewords are given as the row vectors c formed by the vector-matrix product

$$c = aG$$

with operations in the field $GF(2)$, where the row vector a is a $2^m - 1 - m$ bit dataword and the row vector c is a $2^m - 1$ bit codeword. Because G has a $(2^m - 1 - m)$ by $(2^m - 1 - m)$ block equal to the identity, the generator matrix will construct the code in systematic form.

The codeword is transmitted and some bit errors are made. The senseword v with components $v_i, i = 0, \ldots, n-1$, is received with errors. The senseword has components

$$v_i = c_i + e_i \quad i = 0, \ldots, n - 1.$$

If there is a single bit error, it must be corrected, and by assumption, e_i is nonzero for at most one value of i. (If there are two or more bit errors, correction is neither required nor possible.) Then we can compute the matrix-vector product

$$s^T = Hv^T$$
$$= HG^T a^T + He^T$$
$$= He^T$$

in the field $GF(2)$. Now observe that

$$HG^T = [I \ P] \begin{bmatrix} P \\ I \end{bmatrix}$$
$$= P + P$$
$$= 0.$$

The last line follows because $1 + 1 = 0$ and $0 + 0 = 0$ under the exclusive-or definition of addition in $GF(2)$. Therefore

$$s^T = He^T.$$

The vector s, called the *syndrome*, is equal to zero if there is no error. If there is one error the syndrome is equal to the corresponding column of H. Every column of H is distinct so it is trivial to find the column of H matching s and then to invert that bit of v to obtain c.

The Hamming codes are quite simple, and their application is limited. In contrast, the Reed–Solomon codes, described next, are more complicated and are more widely used in communication systems. A Reed–Solomon code, for many practical reasons, is normally constructed in a number system called a Galois field, which will be studied in Section 10.2.

We shall first study Reed–Solomon codes constructed in the complex number system because these codes are easier to understand in the complex number system. The complex number system is a very familiar example of an arithmetic structure known as an algebraic field. The complex field is not used in practice for the construction of Reed–Solomon codes because of issues of computational precision that arise. In the next section we shall see how a Galois field can be used as an alternative number system in place of the complex field so that the issue of precision is avoided. By defining Reed–Solomon codes in the complex field first, we are able to separate our study of the principle behind the code from our study of the Galois fields that are used to bypass the precision problems of the complex field.

A Reed–Solomon code will be defined using the language of the discrete Fourier transform. Let c be a vector of blocklength n over the complex field with discrete

Fourier transform C given by

$$C_j = \sum_{i=0}^{n-1} \omega^{ij} c_i \quad i = 0, \dots, n-1$$

where ω is an nth root of unity in the complex field.[1] Specifically,

$$\omega = e^{-j2\pi/n}.$$

Definition 10.1.1 *The t-error-correcting Reed–Solomon code of blocklength n is the set of all vectors c whose spectrum satisfies $C_j = 0$ for $j = n - 2t, \dots, n - 1$. This set is described briefly as an $(n, n - 2t)$ Reed–Solomon code over the complex field.*

One way to find the Reed–Solomon codewords is to encode in the frequency domain. This means setting C_j equal to zero for $j = n - 2t, \dots, n - 1$, and setting the remaining $n - 2t$ components of the transform equal to the $n - 2t$ data symbols given by $a_0, \dots, a_{n-2}$. That is,

$$C_j = \begin{cases} a_j & j = 0, \dots, n - 2t - 1 \\ 0 & j = n - 2t, \dots, n - 1. \end{cases}$$

An inverse Fourier transform produces the codeword c. The number of data symbols encoded equals $n - 2t$ and there are $2t$ extra check symbols in the codeword to correct t errors. A Reed–Solomon code always uses two check symbols for every error to be corrected.

Using the Fourier transform is not the only way to encode the $n - 2t$ data symbols into the codewords – others may yield a simpler implementation – but the frequency-domain encoder is the most instructive because it exhibits very explicitly the notion that the codewords are all those words with the same set of $2t$ zeros in the transform domain.

An alternative encoder in the time domain works as follows. The $n - 2t$ data symbols are expressed as a polynomial

$$a(x) = a_{n-2t-1} x^{n-2t-1} + a_{n-2t-2} x^{n-2t-2} + \cdots + a_1 x + a_0$$

where $a_0, a_1, \dots, a_{n-2t-1}$ are the $n - 2t$ data symbols. Then the n codeword symbols are given by the coefficients of the polynomial product

$$c(x) = g(x) a(x)$$

---

[1] The letter j is used both for $\sqrt{-1}$ and as an integer-valued index throughout this section. This should not cause any confusion.

where $g(x)$ is a fixed polynomial called the *generator polynomial*. The generator poly-nomial is the unique monic (leading coefficient equals 1) polynomial of degree $2t$ that has zeros at $\omega^{n-2t}, \omega^{n-2t+1}, \ldots, \omega^{n-1}$. It can be obtained by multiplying out the expression

$$g(x) = (x - \omega^{n-2t})(x - \omega^{n-2t+1}) \cdots (x - \omega^{n-1}).$$

We can verify as follows that this time-domain encoder does indeed produce a Reed–Solomon code. The Fourier transform of the codeword is formally the same as evaluating the polynomial $c(x)$ at ω^j. That is,

$$C_j = \sum_{i=0}^{n-1} c_i \omega^{ij} = c(\omega^j) = g(\omega^j)a(\omega^j).$$

By the definition of $g(x)$, $g(\omega^j)$ equals zero for $j = n - 2t, \ldots, n - 1$. Consequently $C_j = 0$ for $j = n-2t, \ldots, n-1$. Therefore the encoding in the time domain does produce legitimate Reed–Solomon codewords. The set of codewords produced by the time-domain encoder is the same as the set of codewords produced by the frequency-domain encoder, but the mapping between datawords and codewords is different.

Both methods of encoding discussed so far have the property that the symbols of the dataword do not appear explicitly in the codeword. Encoders with this property are called *nonsystematic encoders*. Another method of encoding, known as *systematic encoding*, leaves the data symbols unchanged and contained in the first $n - 2t$ compo-nents of the codeword. Multiplication of $a(x)$ by x^{2t} will move the components of $a(x)$ left $2t$ places. Thus, we can write an encoding rule as

$$c(x) = x^{2t}a(x) + r(x)$$

where $r(x)$ is a polynomial of degree less than $2t$ appended to make the spectrum be a legitimate codeword spectrum. The spectrum will be right if $c(x)$ is a multiple of $g(x)$ and this will be so if $r(x)$ is chosen as the negative of the remainder when $x^{2t}a(x)$ is divided by $g(x)$. Thus, because it gives a multiple of $g(x)$,

$$c(x) = x^{2t}a(x) - R_{g(x)}[x^{2t}a(x)]$$

defines a systematic form of encoder, where the operator $R_{g(x)}$ takes the remainder under division by $g(x)$. This definition of $c(x)$ is indeed a codeword because it has zero remainder when divided by $g(x)$. Thus,

$$R_{g(x)}[c(x)] = R_{g(x)}[x^{2t}a(x)] - R_{g(x)}[R_{g(x)}[x^{2t}a(x)]]$$
$$= 0$$

because remaindering can be distributed across addition.

A decoder for a Reed–Solomon code does not depend on how the codewords are used to store information except for the final step of reading the data symbols out of the codeword after error correction is complete.

To prove that an $(n, n - 2t)$ Reed–Solomon code can correct t symbol errors, it is enough to prove that every pair of codewords in the code differ from each other in at least $2t + 1$ places. This is because, if this is true, then changing any t components of any codeword will produce a word that is different from the correct codeword in t components and is different from every other codeword in at least $t + 1$ components. If at most t errors occur, then choosing the codeword that differs from the noisy senseword in the fewest number of components will recover the correct codeword. If each symbol of the senseword is more likely to be correct than to be in error, then choosing the codeword that differs from the senseword in the fewest places will recover the most likely codeword and will minimize the probability of decoding error.

Theorem 10.1.2 *An $(n, n - 2t)$ Reed–Solomon code can correct t symbol errors.*

Proof By definition of the code,

$$C_j = 0 \quad j = n - 2t, n - 2t + 1, \ldots, n - 1.$$

By linearity of the Fourier transform, the difference in two codewords (computed componentwise) then must also have a spectrum that is zero for $j = n - 2t, \ldots, n - 1$ and so itself is a codeword. We only need to prove that no codeword has fewer than $2t + 1$ nonzero components unless it is zero in every component. Let

$$C(y) = \sum_{j=0}^{n-2t-1} C_j y^j.$$

This is a polynomial of degree at most $n - 2t - 1$, so by the fundamental theorem of algebra it has at most $n - 2t - 1$ zeros. Therefore,

$$c_i = \frac{1}{n} \sum_{j=0}^{n-1} \omega^{-ij} C_j$$

$$= \frac{1}{n} C(\omega^{-i})$$

can be zero in at most $n - 2t - 1$ places, and so it is nonzero in at least $2t + 1$ places. This completes the proof of the theorem. ∎

10.2   Codes constructed in a finite field

Reed–Solomon codes in the complex number system have symbols that are arbitrary complex numbers and hence may require an infinite number of bits to represent exactly. This leads to two concerns. One must consider how the code symbols are to be transmitted through the channel, and one must consider the effects of numerical precision in the encoder and decoder. Because of limited wordlength, questions of precision arise in the computations of the encoder and decoder. In a large code, in which the number of correctable errors t is large, the precision difficulties encountered in inverting large matrices in the decoder may cause the decoder calculations to collapse.

In addition to the question of computational precision, one must also consider the question of what an error is, and this depends on how the code symbols are transmitted. This question might be severe if analog pulse amplitude modulation were used in which the components of the codeword are converted to analog signals for passage through the channel, and then redigitized. There may be an error in the low-order bit of every component of the codeword. This sort of error must be distinguished from the t "major" errors that the code is designed to correct.

Rather than tackle these two problems head on, it has been found to be cleaner and more elegant to simply step around the problems by using a different arithmetic system in place of the complex field, an arithmetic system with only a finite number of elements. As long as the data consist merely of bit packages, it does not matter what arithmetic system is used by the encoder and decoder. We are free to invent any definition of addition and multiplication, however artificial, as long as the theory of Reed–Solomon codes is still valid in that number system.

Fortunately, there is a suitable system of arithmetic known as a *Galois field*. In general, an *algebraic field* is a set of elements, called *numbers* or *field elements*, and a definition of two operations, called "addition" and "multiplication" such that the formal properties satisfied by the real arithmetic system are satisfied. This means that there are two field elements 0 and 1 that satisfy $a + 0 = a$ and $a \cdot 1 = a$; the inverse operations of subtraction and division are implicit in the definitions of addition and subtraction; and the rules of algebra known as commutativity, associativity, and distributivity apply.

For each positive integer m, there is a Galois field called $GF(2^m)$ that has 2^m elements in it. The set of elements of $GF(2^m)$ can be represented as the set of m-bit binary numbers (m-bit bytes). Thus, $GF(256)$ consists of the 256 distinct eight-bit bytes, and $GF(16)$ consists of the sixteen distinct hexadecimal symbols represented as four-bit bytes. To complete the description of the Galois field $GF(2^m)$, we need to define addition and multiplication. Addition is the easiest. It is defined as bit-by-bit modulo-two addition. This is the same as bit-by-bit "exclusive-or". For example, two of the elements of

$GF(16)$ are 1101 and 1110. Their addition is

$$(1101) + (1110) = (0011).$$

Multiplication is more complicated to define. The multiplication in $GF(2^m)$ must be consistent with addition in the sense that the distributive law

$$(a + b)c = ac + bc$$

holds for any a, b, and c in $GF(2^m)$, where addition is a bit-by-bit exclusive-or. This suggests that multiplication should have the structure of a shift and exclusive-or, rather than the conventional structure of shift and add.

We may try to define multiplication in accordance with the following product:

```
              1  1  1  0
              1  1  0  1
           _____
              1  1  1  0
        0  0  0  0
     1  1  1  0
  1  1  1  0
 _____
  1  0  0  0  1  1  0
```

where the addition is $GF(2)$ addition (exclusive or). However, the arithmetic system must be closed under multiplication. The product of two elements of $GF(2^m)$ must produce another element of $GF(2^m)$; the wordlength must not increase beyond m bits. The product of two m-bit numbers must produce an m-bit number. If a shift and exclusive-or structure is the right definition for the multiplier, then we also need a rule for folding back the overflow bits into the m bits of the product. The trick is to define the overflow rule so that division is meaningful. This requires that $b = c$ whenever $ab = ac$ and a is nonzero. The overflow rule will be constructed in terms of polynomial division.

Let $p(x)$ be an *irreducible polynomial* over $GF(2)$ of degree m. This means that $p(x)$ can have only coefficients equal to zero or one, and that $p(x)$ *cannot* be factored into the product of two smaller nontrivial polynomials over $GF(2)$. Factoring $p(x)$ means writing

$$p(x) = p^{(1)}(x)p^{(2)}(x)$$

where polynomial multiplication uses modulo-two arithmetic on the coefficients.

Multiplication of two elements a and b in $GF(2^m)$ to produce the element $c = ab$ is defined as a polynomial multiplication modulo the irreducible polynomial $p(x)$.

Let a and b be numbers in $GF(2^m)$. These are m-bit binary numbers with the binary representations

$$a = (a_0, \ldots, a_{m-1})$$
$$b = (b_0, \ldots, b_{m-1}).$$

They also have the polynomial representations

$$a(x) = \sum_{i=0}^{m-1} a_i x^i$$
$$b(x) = \sum_{i=0}^{m-1} b_i x^i$$

where a_i and b_i are the ith bits of a and b, respectively. Then the product is defined as the sequence of coefficients of the polynomial

$$c(x) = a(x)b(x) \quad (\text{mod } p(x)).$$

Because the coefficients are added and multiplied by the bit operations of $GF(2)$, the polynomial product is equivalent to the shift and exclusive-or operations mentioned before. The modulo $p(x)$ operation specifies the rule for folding overflow bits back into the m bits of the field element. With this definition of multiplication and the earlier definition of addition, the description of the Galois field $GF(2^m)$ is complete.

Because the polynomial $p(x)$ is an irreducible polynomial, division will always exist, although we do not prove this fact in this book. Existence of the division operation means that the equation $ax = b$ has only one solution denoted b/a, provided a is not zero.

As an example of a Galois field, we will construct $GF(2^4)$. The elements are the set of four-bit bytes

$$GF(2^4) = \{0000, 0001, 0010, \ldots, 1111\}$$

and addition of two elements is bit-by-bit modulo-two addition. To define multiplication, we use the polynomial

$$p(x) = x^4 + x + 1.$$

To verify that this polynomial is irreducible, we can observe that $p(x)$ must have either a first-degree factor or a second-degree factor if it is reducible, then check that $x, x+1$, $x^2 + 1, x^2 + x + 1$ are not factors. There are no other possibilities.

Thus, to multiply, say, 0101 by 1011, we represent these by the polynomials $a(x) = x^2 + 1$ and $b(x) = x^3 + x + 1$ and write

$$c(x) = (x^2 + 1)(x^3 + x + 1) \qquad (\text{mod } p(x))$$
$$= x^5 + x^2 + x + 1 \qquad (\text{mod } x^4 + x + 1).$$

The modulo $p(x)$ operation consists of division by $p(x)$, keeping only the remainder polynomial. Carrying out the division for the sample calculation gives

$$c(x) = 1$$

which is the polynomial representation for binary 0001, so we have the product in $GF(16)$

$$(0101)(1011) = (0001).$$

Because the product happens to be equal to 1, this example also tells us how to divide in $GF(16)$. We see that

$$(0101)^{-1} = (1011)$$

because $(0101)^{-1}$ is defined to be the field element for which $(0101)^{-1}(0101) = 1$. Likewise,

$$(1011)^{-1} = (0101).$$

To divide by 0101 we multiply by 1011, while to divide by 1011 we multiply by 0101. For example

$$\frac{(0110)}{(1011)} = (0110)(0101)$$
$$= (1101)$$

where the product is calculated as already described.

Division will be possible for every field element a, if a^{-1} exists for every a. The inverse a^{-1} is the value of b that solves

$$ab = 1.$$

This equation always has a solution (except when $a = 0$) if $p(x)$ is chosen to have no polynomial factors. This is why $p(x)$ was chosen in the definition to be an irreducible polynomial.

Although the multiplication and division rules of a Galois field may be unfamiliar, logic circuits or computer subroutines to implement them are straightforward. One could even build a programmable computer with Galois field arithmetic as primitive instructions.

Algebraic manipulations in the Galois field $GF(2^m)$ behave very much like manipulation in the fields more usually encountered in engineering problems such as the real field or the complex field, although we will not take the trouble to verify this. The conventional algebraic properties of associativity, commutativity, and distributivity all hold. Methods of solving linear systems of equations are valid, including matrix algebra, determinants, and so forth. There is even a discrete Fourier transform in the finite field $GF(2^m)$ and it has all the familiar properties of the discrete Fourier transform. The Fourier transform is particularly important to our purposes because it is the basis of the definition of the Reed–Solomon code.

Let ω be an element of order n in $GF(2^m)$. That is, $\omega^n = 1$, and no smaller power of ω equals 1. (Such an ω exists only if n divides $2^m - 1$; consequently n cannot be even.) Then

$$V_j = \sum_{i=0}^{n-1} \omega^{ij} v_i \quad j = 0, \ldots, n-1$$

$$v_i = \sum_{j=0}^{n-1} \omega^{-ij} V_j \quad i = 0, \ldots, n-1$$

are the equations of the discrete Fourier transform and the inverse Fourier transform. The equations look quite familiar, but the product and sums they express are products and sums in the Galois field $GF(2^m)$. The only exception is in the exponent of ω. Integers in the exponent are conventional integers: ω^r means ω multiplied by itself $r - 1$ times. The factor n^{-1} that normally appears in the inverse Fourier transform is always equal to 1 in a Fourier transform in $GF(2^m)$ and so is omitted. This point is a consequence of the fact that $1 + 1 = 0$ in such fields, and n is always equal to the sum of an odd number of ones.

The proof that the equation for the inverse Fourier transform does indeed produce v_i is exactly the same as the proof in the complex number system. To verify the inverse Fourier transform, first notice that

$$(1 + \omega + \omega^2 + \cdots + \omega^{n-1})(1 - \omega) = 1 - \omega^n = 0$$

because ω has order n. But this means that the first term on the left must equal zero because $1 - \omega$ does not equal zero. The same argument applies to ω^{ij} for any distinct i

and j because ω^{i-j} has order that divides n. Consequently

$$\sum_{k=0}^{n-1} \omega^{(i-j)k} = 0$$

unless $i - j = 0$, in which case the sum is equal to 1. Finally

$$\sum_{k=0}^{n-1} \omega^{-jk} \left[\sum_{i=0}^{n-1} \omega^{ik} v_i \right] = \sum_{i=0}^{n-1} v_i \sum_{k=0}^{n-1} \omega^{(i-j)k} = v_i$$

which verifies the inverse Fourier transform.

As an example of a Fourier transform, in $GF(16)$ the element (1000), represented by the polynomial x^3, has order 5. Hence, we have a five-point Fourier transform

$$V_j = \sum_{i=0}^{4} \omega^{ij} v_i \quad j = 0, \ldots, 4$$

with $\omega = 1000$. If $v = (0001, 0010, 0011, 0100, 0101)$, then

$$
\begin{bmatrix} V_0 \\ V_1 \\ V_2 \\ V_3 \\ V_4 \end{bmatrix} =
\begin{bmatrix}
1 & 1 & 1 & 1 & 1 \\
1 & \omega & \omega^2 & \omega^3 & \omega^4 \\
1 & \omega^2 & \omega^4 & \omega^1 & \omega^3 \\
1 & \omega^3 & \omega^1 & \omega^4 & \omega^2 \\
1 & \omega^4 & \omega^3 & \omega^2 & \omega^1
\end{bmatrix}
\begin{bmatrix} 0001 \\ 0010 \\ 0011 \\ 0100 \\ 0101 \end{bmatrix}.
$$

We can carry out these calculations with the aid of a multiplication table for $GF(16)$. Alternatively, we can express the computation in the polynomial representation with ω represented by the polynomial x^3. The equation then becomes

$$
\begin{bmatrix} V_0(x) \\ V_1(x) \\ V_2(x) \\ V_3(x) \\ V_4(x) \end{bmatrix} =
\begin{bmatrix}
1 & 1 & 1 & 1 & 1 \\
1 & x^3 & x^6 & x^9 & x^{12} \\
1 & x^6 & x^{12} & x^3 & x^9 \\
1 & x^9 & x^3 & x^{12} & x^6 \\
1 & x^{12} & x^9 & x^6 & x^3
\end{bmatrix}
\begin{bmatrix} 1 \\ x \\ x+1 \\ x^2 \\ x^2+1 \end{bmatrix} \quad (\mathrm{mod}\ x^4 + x + 1)
$$

$$
= \begin{bmatrix} 1 \\ x^3 + x^2 + 1 \\ x^3 + x \\ x^3 + x^2 + 1 \\ x^3 + x \end{bmatrix}.
$$

Consequently,

$$\begin{bmatrix} V_0 \\ V_1 \\ V_2 \\ V_3 \\ V_4 \end{bmatrix} = \begin{bmatrix} 0001 \\ 1101 \\ 1010 \\ 1101 \\ 1010 \end{bmatrix}$$

by the rules of arithmetic in $GF(16)$.

There is one way in which the topic of Fourier transforms in $GF(2^m)$ differs from those in the complex field. In the complex field, there is a discrete Fourier transform of every blocklength because $\omega = e^{-j2\pi/n}$ exists in the complex field for every n. In $GF(2^m)$, only a few blocklengths have Fourier transforms because an element ω of order n does not exist for every n. Specifically, there is a Fourier transform of blocklength n in $GF(2^m)$ if and only if n divides $2^m - 1$.

Now we are ready to look at Reed–Solomon codes in $GF(2^m)$. Nothing in the definition of Reed–Solomon codes is unique to the complex field. We can use the same definition in $GF(2^m)$. Definition 10.1.1 applies, but with the understanding that the computations are in the field $GF(2^m)$. To define the Reed–Solomon code of blocklength n, one needs a Fourier transform of blocklength n. Thus, we can construct a Reed–Solomon code of blocklength n in $GF(2^m)$ if and only if n divides $2^m - 1$ because a Fourier transform of blocklength n exists if and only if n divides $2^m - 1$.

For example, in $GF(2^3)$, we can choose $n = 7$. If we construct $GF(2^3)$ using the irreducible polynomial $p(x) = x^3 + x + 1$, then the element $\omega = (010)$ has order 7. To verify this, we can compute the powers of ω

$$\omega = (010)$$

$$\omega^2 = (100)$$

$$\omega^3 = (011)$$

$$\omega^4 = (110)$$

$$\omega^5 = (111)$$

$$\omega^6 = (101)$$

$$\omega^7 = (001).$$

Then, using this ω as the kernel of the Fourier transform, we can construct the $(7, 7 - 2t)$ Reed–Solomon code in $GF(2^3)$ for any value of t provided $7 - 2t$ is nonnegative. In particular, we can construct a $(7, 5)$ single-error-correcting Reed–Solomon code in

$GF(2^3)$. The generator polynomial is

$$g(x) = (x - \omega)(x - \omega^2)$$
$$= x^2 + (\omega + \omega^2)x + \omega^3$$
$$= x^2 + \omega^4 x + \omega^3$$
$$= (001)x^2 + (110)x + (011)$$

which can be used to encode the data into the set of codewords. This generator polynomial is not the same as the polynomial

$$g(x) = (x - \omega^5)(x - \omega^6)$$

called for by Definition 10.1.1. However, by the translation properties of the Fourier transform, it gives an equivalent, but different, Reed–Solomon code.

Figure 10.3 shows some of the codewords in the $(7, 5)$ Reed–Solomon code using an octal notation for the seven components of the codewords. The Fourier transform of each

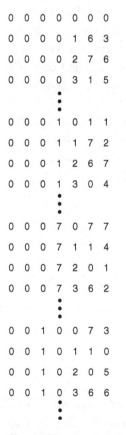

Figure 10.3. The $(7, 5)$ Reed–Solomon code.

codeword has components C_1 and C_2 equal to zero. In all, there are 8^5 (about 32,000) codewords in the $(7, 5)$ Reed–Solomon code, so it is pointless and impractical to list them all even for this small code. In a larger Reed–Solomon code such as a $(255, 223)$ Reed–Solomon code over GF(256), the number of codewords is astronomical. The encoder and decoder, however, have no need for an explicit list of codewords. Specific codewords are easily computed as needed.

10.3   Decoding of block codes

A practical Reed–Solomon decoder cannot exhaustively compare the senseword directly to every codeword to determine which codeword is at minimum Hamming distance from the senseword. An efficient computational procedure to find the correct codeword is needed. Specifically, a computational procedure is needed to find the error word e with the smallest number of nonzero components whose Fourier transform has $2t$ components E_j for $j = n - 2t, \ldots, n - 1$ that agree with that set of $2t$ components of the senseword. The codeword is then found by subtracting this error word from the senseword. The procedure that we shall derive is predicated on the assumption that there is such an error word e with at most t nonzero components because this is the error-correcting capability of the code. (If there are more than t errors, the decoder is not able, nor expected, to correct the senseword.)

The codeword c is transmitted and some symbol errors are made. The senseword v is received and, if there are not more than t errors, the senseword must be corrected. Thus, the channel errors are represented by the vector e, which is assumed to be nonzero in not more than t places. The senseword v is written componentwise as

$$v_i = c_i + e_i \quad i = 0, \ldots, n - 1.$$

The decoder must process the senseword v so as to remove the error word e; the data symbols are then recovered from c. The senseword v is a noisy codeword and has a Fourier transform

$$V_j = \sum_{i=0}^{n-1} \omega^{ij} v_i \quad j = 0, \ldots, n - 1$$

with components $V_j = C_j + E_j$ for $j = 0, \ldots, n - 1$. But, by construction of a Reed–Solomon code,

$$C_j = 0 \quad j = n - 2t, \ldots, n - 1.$$

Hence

$$V_j = E_j \quad j = n - 2t, \ldots, n - 1.$$

This block of $2t$ components of V gives us a window through which we can look at $2t$ of the n components of E, the transform of the error pattern e. The decoder must find all n components of E given that segment consisting of $2t$ consecutive components of E, and the additional information that at most t components of the time-domain error pattern e are nonzero. Once E is known, the computation is trivial because the relationship $C_j = V_j - E_j$ recovers C. From C one can compute c. The data symbols are then recovered easily in various ways depending on the method of encoding.

To find a procedure to so compute E, we will use properties of the Fourier transform. Suppose for the moment that there exists a polynomial $\Lambda(x)$ of degree at most t and with $\Lambda_0 = 1$ such that

$$\Lambda(x)E(x) = 0 \quad (\text{mod } x^n - 1)$$

where

$$E(x) = \sum_{j=0}^{n-1} E_j x^j.$$

The polynomial product is equivalent to a cyclic convolution. Using the defining property of $\Lambda(x)$ we can rewrite the coefficients of the polynomial product as

$$\Lambda_0 E_j + \sum_{k=1}^{t} \Lambda_k E_{j-k} = 0 \quad j = 0, \ldots, n - 1$$

or, because $\Lambda_0 = 1$,

$$E_j = -\sum_{k=1}^{t} \Lambda_k E_{j-k} \quad j = 0, \ldots, n - 1.$$

This equation may be recognized as a description of the kind of filter known as an *autoregressive filter* with t taps. It defines component E_j in terms of the preceding t components $E_{j-1}, \ldots, E_{j-t}$. But we know $2t$ components of E so by setting in turn $j = n - t, \ldots, n - 1$, we can write down the following set of $n - 1$ equations

$$
\begin{aligned}
E_{n-t} &= -\Lambda_1 E_{n-t-1} & -\Lambda_2 E_{n-t-2} & - \cdots & -\Lambda_t E_{n-2t} \\
E_{n-t+1} &= -\Lambda_1 E_{n-t} & -\Lambda_2 E_{n-t-1} & - \cdots & -\Lambda_t E_{n-2t+1} \\
&\ \ \vdots & \ \ \vdots & & \\
E_{n-1} &= -\Lambda_1 E_{n-2} & -\Lambda_2 E_{n-3} & - \cdots & -\Lambda_t E_{n-t-1}.
\end{aligned}
$$

There are t equations here, linear in the unknown components Λ_k of Λ and involving only known components of E. Hence, provided there is a solution, we can find Λ by solving this system of linear equations. The solution will give the polynomial $\Lambda(x)$ that was assumed earlier. As long as the linear system of equations has a solution, the introduction of the polynomial $\Lambda(x)$ is meaningful. Once $\Lambda(x)$ is known, all other values of E can be obtained by recursive computation using the equation

$$E_j = -\sum_{k=1}^{t} \Lambda_k E_{j-k} \quad j = 0, \ldots, n - 2t - 1$$

recalling that the indices are modulo n.

The development of the decoding procedure is now complete provided that the system of linear equations can be inverted. We must verify that at least one solution to the system of equations will always exist if there are at most t errors. We know that there cannot be more than one solution by our geometrical reasoning earlier. Suppose that there are $\nu \leq t$ nonzero errors at locations with indices i_ℓ for $\ell = 1, \ldots, \nu$. Define the polynomial $\Lambda(x)$ by

$$\Lambda(x) = \prod_{\ell=1}^{\nu} (1 - x\omega^{i_\ell})$$

which has degree at most t and $\Lambda_0 = 1$. We shall show that this polynomial, which is known as the *error-locator polynomial*, is the polynomial that we have assumed exists. The zeros of the error-locator polynomial "locate" the errors, and a decoder that first finds the error-locator polynomial is called a *locator decoder*. The vector Λ of length n whose components Λ_j are coefficients of the polynomial $\Lambda(x)$ (padded with $n - \nu$ zeros) has an inverse Fourier transform

$$\lambda_i = \frac{1}{n} \sum_{j=0}^{n-1} \Lambda_j \omega^{-ij}.$$

This can be obtained from $\Lambda(x)$ by evaluating $\Lambda(x)$ at $x = \omega^{-i}$. That is,

$$\lambda_i = \frac{1}{n} \Lambda(\omega^{-i}).$$

Therefore

$$\lambda_i = \frac{1}{n} \prod_{\ell=1}^{\nu} (1 - \omega^{-i}\omega^{i_\ell})$$

which is zero if and only if $i = i_\ell$ for some ℓ, where the i_ℓ for $\ell = 1, \ldots, \nu$ index the nonzero components of e. That is, in the time domain, $\lambda_i e_i = 0$ for all i. Therefore, because a product in the time domain corresponds to a cyclic convolution in the frequency domain, we see that the convolution in the frequency domain is equal to zero

$$\Lambda * E = 0.$$

Hence, a polynomial $\Lambda(x)$ solving the cyclic convolution equation $\Lambda(x)E(x) = 0$ does exist because the error-locator polynomial is such a polynomial. Thus the decoding procedure is sound.

The decoding procedure that we have now developed requires solution of the matrix-vector equation

$$\begin{bmatrix} E_{n-t-1} & E_{n-t-2} & \cdots & E_{n-2t} \\ E_{n-t} & E_{n-t-1} & \cdots & E_{n-2t+1} \\ \vdots & & & \\ E_{n-2} & E_{n-3} & \cdots & E_{n-t-1} \end{bmatrix} \begin{bmatrix} \Lambda_1 \\ \Lambda_2 \\ \vdots \\ \Lambda_t \end{bmatrix} = - \begin{bmatrix} E_{n-t} \\ E_{n-t+1} \\ \vdots \\ E_{n-1} \end{bmatrix}.$$

A straightforward way to solve the equation is to compute the matrix inverse. The matrix, however, is a special kind of matrix known as a Toeplitz matrix, because the elements in any subdiagonal are equal. The computational problem is one of inverting a Toeplitz system of equations. This kind of computational problem arises frequently in the subject of signal processing, as in the decoding of Reed–Solomon codes, in the design of autoregressive filters, and in spectral analysis. To invert a Toeplitz system of equations, special fast algorithms – which we shall not study – are commonly used because they are computationally more efficient than computing the matrix inverse by a general method.

10.4 Performance of block codes

A decoder for a block code with minimum distance $2t + 1$ can correct all error patterns containing t or fewer symbol errors, and there will be at least one pattern with $t+1$ errors that cannot be corrected properly. There may be some error patterns with more than t errors that could be corrected properly, but this is rarely done in practice. A decoder that decodes only up to a fixed number of errors, say t errors, is called a *bounded-distance decoder*. When there are more than t errors, the decoder will sometimes miscorrect the codeword, and sometimes flag the codeword as uncorrectable. The user, depending on this application, may or may not consider a flagged erroneous message (a decoding failure) as less serious than an unflagged erroneous message (a decoding error).

The decoding task can be described geometrically. Regard each codeword to have a sphere of radius t drawn around it. Each sphere encompasses all of the sensewords within distance t of that codeword. They will be decoded into that codeword. Between the many decoding spheres lie many other sensewords that do not lie within distance t

of any codeword and so are not decoded. Correspondingly, the computations of locator decoding for correcting t errors in a Reed–Solomon codeword involve matrix operations as discussed in Section 10.2, but when there are more than t errors the computations lose the clean structure of matrix operations and are impractical. When more than t errors occur, the senseword will often lie between the decoding spheres and then the decoder can declare that it has an uncorrectable message. Occasionally, however, the error pattern is such that the senseword will actually lie within the decoding sphere of an incorrect codeword. Then the decoder makes a decoding error. Hence, the decoder output can be either the correct message, an incorrect message, or a decoding default (an erased message). An incorrect message will contain incorrect symbols, but also may contain many correct symbols, though such a message is rarely of value to the recipient.

We shall study the performance of codes when used with a discrete memoryless channel described by a probability of symbol error at the channel output. When used with a discrete channel, the performance of the code is expressed in terms of the probability of an error at the output of the decoder as a function of the probability of symbol error of the channel. We may be interested in the probability of block decoding error, the probability of symbol decoding error, or the probability of decoding default. Because not all symbols need be wrong in a wrong message, the probability of symbol error will usually be smaller than the probability of block error. Because all symbols are usually rejected when a message is found to be undecidable, the probability of symbol decoding failure is usually the same as the probability of message decoding failure.

There are three regions into which the senseword can fall. The probability of correct decoding is the probability that the senseword lies in the decoding sphere about the transmitted codeword. The probability of incorrect decoding is the probability that the senseword lies in any decoding sphere about any other codeword. The probability of decoding default is the probability that the senseword lies in the space between spheres. The sum of these three probabilities equals one, so formulas for only two of them are needed. In some applications, decoder defaults are treated differently from incorrect outputs, so the probability of decoding error is equal to the probability of incorrect decoding. In other applications there is no distinction and the probability of error is equal to the sum of the probability of decoding error and the probability of failure.

The discrete channels that we consider are the M-ary symmetric channels that make independent symbol errors with probability p_e in each component. The probability that v errors lie within a particular subset consisting of v symbols and the other $n - v$ symbols are error-free is $(1 - p_e)^{n-v} p_e^v$. Conditional on a particular transmitted codeword, a particular senseword with v errors has probability $(1 - p_e)^{n-v}(p_e/(M - 1))^v$ of occurring. The decoder will decode every senseword to the closest codeword provided that it is within distance t of that codeword. Otherwise there is a decoding default.

For a linear code such as a Hamming code or a Reed–Solomon code, we can analyze the performance conditional on the all-zero word being transmitted. This is because the linear structure of the code ensures that the geometric configuration of the codewords will look the same as viewed by any codeword. Every codeword will have the same conditional probability of error. Therefore, the probability of error conditional on the all-zero word being transmitted is equal to the unconditional probability of error.

The computation of the probability of decoding error is quite tedious. This can be seen by referring to Figure 10.4. The probabilities of all words lying in all spheres other than the central sphere must be summed; the probabilities of points lying between spheres must not be included in the sum. To count up only the points within the spheres is a complicated combinatorial procedure, which we do not give in this book. The probability of correct decoding, on the other hand, is easy to compute.

Theorem 10.4.1 *A decoder for a t-error-correcting code used with a memoryless channel has a probability of correct block decoding given by*

$$1 - p_{em} = \sum_{\ell=0}^{t} \binom{n}{\ell} p_e^\ell (1 - p_e)^{n-\ell}.$$

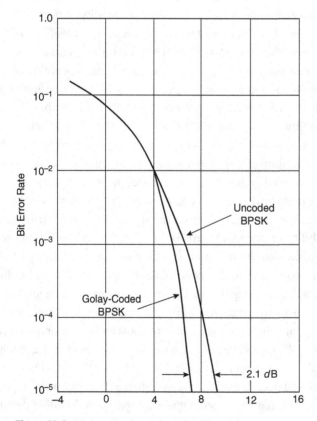

Figure 10.4. Illustrating the notion of coding gain.

Proof   There are $\binom{n}{\ell}$ ways that the ℓ places with errors can be selected; each occurs with probability $p_e^{\ell}(1 - p_e)^{n-\ell}$. The theorem follows.   ∎

Although the theorem holds for any channel alphabet size, only the probability of channel error enters the formula. It does not matter how this probability of symbol error is divided among the individual symbols.

We may study the performance of the code when it is used on an additive gaussian noise channel. Then we may express the performance of the code in terms of the probability of symbol error at the output of the channel as a function of E_b/N_0 of the channel waveform. Convention requires that E_b is the average energy per *information* bit. This is n/k larger than the energy per channel bit because the energy in the check symbols is divided among the information bits in this calculation.

When a decoder is used with a demodulator for a gaussian noise channel, the decoder is called a *hard-decision decoder* if the output of the demodulator is simply a demodulated symbol. Sometimes the demodulator passes other data to the decoder such as the digitized output of the matched filters. Then the decoder is called a *soft-decision decoder*.

When used in conjunction with a demodulator for an additive gaussian noise channel, the performance of the code is described by a quantity known as the *coding gain*. The coding gain is defined as the difference between the E_b/N_0 required to achieve a performance specification, usually probability of error, with the code and without the code. Figure 10.4 illustrates the coding gain with BPSK as the uncoded reference. The binary code used for this illustration is a code known as the Golay (23, 12) triple-error-correcting code. The figure shows that at a bit-error-rate of 10^{-5}, the Golay code has a coding gain of 2.1 dB.

It is natural to ask if the coding gain can be increased indefinitely by devising better codes. As shown in Chapter 11, we cannot signal at any E_b/N_0 smaller than -1.6 dB, so there is a very definite negative answer to this question. At a bit error rate of 10^{-5}, the maximum coding gain relative to BPSK of any coding system whatsoever is 11.2 dB. If, moreover, a hard-decision BPSK demodulator is used, then a soft-decision decoder is inappropriate, and 2 dB of coding gain is not available. Then the maximum coding gain is 9.2 dB. Figure 10.5 shows the regions in which all coding gains must lie.

Simple analytic expressions for the coding gain are not known. The coding gain must be evaluated numerically for each code of interest. In contrast, the asymptotic coding gain given in the next definition has a simple definition but does not have as clear-cut a meaning. It is an approximation to the coding gain in the limit as E_b/N_0 goes to infinity and for t small compared to n. The asymptotic coding gain sharpens our understanding but should be used with great care because of the approximate analysis, which is sometimes meaningless, and because we are usually interested in moderate values of E_b/N_0.

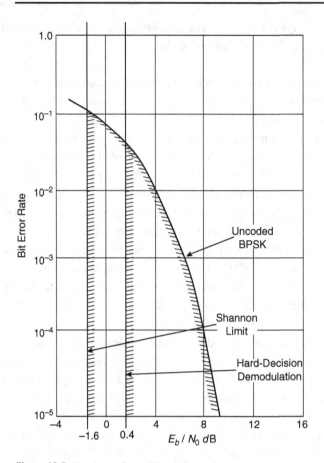

Figure 10.5. Regions of possible coding gain.

Definition 10.4.2 *A t-error-correcting block code of rate R has an asymptotic coding gain of*

$$G_a = R(t + 1)$$

when used with a hard-decision decoder, and

$$G = R(2t + 1)$$

when used with a soft-decision decoder.

The motivation for the definitions of asymptotic coding gain is a pair of approximate analyses that are valid asymptotically for large signal-to-noise ratios and if t is small compared to n. Let p_e denote the probability of bit error at the output of the hard-decision BPSK demodulator. Let p_{em} denote the probability of not decoding the block correctly, either because of block decoding error or because of block decoding default.

Then, starting with Theorem 10.4.1,

$$
P_{em} = \sum_{\ell=t+1}^{n} \binom{n}{\ell} p_e^{\ell} (1 - p_e)^{n-\ell}
$$

$$
\approx \binom{n}{t+1} p_e^{t+1}
$$

$$
\approx \binom{n}{t+1} \left[Q\left(\sqrt{\frac{2RE_b}{N_0}} \right) \right]^{t+1}
$$

using the expression for the probability of error of BPSK and recalling that the energy in each code bit is RE_b. For large x, $Q(x)$ behaves like $e^{-x^2/2}$, so for large x, $[Q(x)]^s \approx Q(x\sqrt{s})$. Therefore,

$$
P_{em} \approx \binom{n}{t+1} Q\left(\sqrt{R(t+1)\frac{2E_b}{N_0}} \right).
$$

Asymptotically as E_b/N_0 goes to infinity, the argument of Q is as if E_b were amplified by $R(t+1)$ – hence the term "asymptotic coding gain". However, the approximate analysis is meaningful only if the binomial coefficient appearing as a multiplier is of second-order importance. This means that t must be small compared to n. In this case, we may also write the asymptotic approximation

$$
P_{em} \sim e^{-Rd_{min}E_b/2N_0}
$$

to express the approximate asymptotic behavior of hard-decision decoding as a function of E_b/N_0.

For a soft-decision decoder, the motivation for defining asymptotic coding gain is an argument based on minimum euclidean distance between codewords. Suppose that there are two codewords at Hamming distance $2t + 1$. Then in euclidean distance they are separated by $(2t+1)RE_b$. By the methods of Chapter 3, the probability of decoding error, conditional on the premise that one of these two words was transmitted, is

$$
P_{em} \approx Q\left(\sqrt{R(2t+1)\frac{2E_b}{N_0}} \right).
$$

If the linear code has $N_{d_{min}}$ nearest codeword neighbors at distance $2t + 1$ from each codeword, then the union bound dictates that the probability of error can be at most $N_{d_{min}}$ times larger:

$$
P_{em} \approx N_{d_{min}} Q\left(\sqrt{R(2t+1)\frac{2E_b}{N_0}} \right).
$$

Again, for large E_b/N_0, the argument of the function $Q(x)$ is as if E_b were amplified by $R(2t+1)$, and again the term "asymptotic coding gain" describes this. The significance of this term, however, requires that the number of nearest neighbors $N_{d_{\min}}$ is not so exponentially large that it offsets the exponentially small behavior of $Q(x)$. In this case, we may again write an asymptotic approximation

$$p_{em} \sim e^{-Rd_{\min}E_b/N_0}$$

to express the asymptotic behavior in E_b/N_0 of soft-decision decoding. However, in any particular case, it is difficult to know whether $N_{d_{\min}}$ plays a critical role in determining the order-of-magnitude of p_{em}.

10.5   Convolutional codes for data transmission

In contrast to a block code for data transmission, which encodes a block of k data symbols into a block of n codeword symbols, is a *tree code* for data transmission, which encodes a stream of data symbols into a stream of codeword symbols. The stream of data symbols has no predetermined length. A data sequence is shifted into the encoder beginning at time zero and continuing indefinitely into the future, and a sequence of code symbols is shifted out of the encoder. We are interested in those tree codes for which the encoder is a finite-state machine whose state depends only on the past mk data symbols for some constants m and k. Such a tree code is called a *trellis code* because the codewords of such a code can be described by a labeled trellis. The trellis is a compact way to display the codewords of a trellis code. We saw the Ungerboeck codes as examples of trellis codes in Section 9.4. In this section we shall study the convolutional codes as another example of trellis codes.

The stream of data symbols entering the canonical encoder for a trellis code is broken into segments of k symbols each, called *dataframes*. The encoder can store the past m dataframes of the datastream. During each frame time, a new dataframe enters the encoder and the oldest dataframe is discarded by the encoder. At the end of any frame time the encoder has stored the most recent m dataframes; a total of mk data symbols. We can regard these data symbols as stored in an mk-stage shift register with memory cells corresponding to the symbol alphabet. At the beginning of a frame, the encoder knows only the new incoming dataframe and the m previously stored dataframes. From these $k(m+1)$ data symbols, the encoder computes a single codeword frame, n symbols in length by means of a fixed encoding function. This codeword frame is shifted out of the encoder as the next dataframe is shifted in. Hence the channel must transmit n codeword symbols for each k data symbols. The *rate R* of the trellis code is defined as $R = k/n$. The *constraint length v* of the convolutional code is defined as the number of memory cells needed by a minimal encoder – that is, an encoder with the fewest memory cells. The constraint length satisfies $v \leq mk$, the inequality occurring in codes for which it is not necessary to retain all mk input symbols. The *codewords* of the trellis code are the infinitely long sequences that can be produced by the encoder. These are the sequences that can be read along the paths through the trellis describing the code.

A special kind of trellis code is a *convolutional code*, which is the only kind of code we shall study in this section. A convolutional code is a trellis code over a Galois field with a linearity property on the codewords. If c and c' are two codewords of a convolutional code – that is, two infinitely long sequences that can be generated by the encoder – and α and β are any elements of the field of the code, then $\alpha c + \beta c'$ is also a codeword. The linear combination $\alpha c + \beta c'$ is to be understood as the infinite sequence $\alpha c_i + \beta c_i'$ for $i = 0, \ldots$. A convolutional code also has a shift-invariance property as do all trellis codes because the encoding rule depends only on the most recent mk data symbols. Convolutional codes are in common use for data transmission because of their simple structure and because – as a direct consequence of the linearity property – the codewords can be generated by the convolutions of finite-impulse-response filters in a Galois field, usually $GF(2)$.

A convolutional code is a set of codewords. The code itself should not be confused with the encoder for the convolutional code. Accordingly, we shall develop the structure of a convolutional code by starting with the trellis definition of the code, then finding encoders to fit the trellis and its labels. A simple example of a labeled trellis for a binary convolutional code with $k = 1$, $n = 2$, and $m = 2$ is shown in Figure 10.6. There are four states (denoted S_0, S_1, S_2, S_3) corresponding to the four possible two-bit patterns in the encoder memory. Every node of the trellis must have two branches leaving it so that one branch can be assigned to data bit zero and one branch can be assigned to data bit one. Each branch leads to the new state caused by that input bit. Each branch is labeled with the pair of codeword bits that will be generated if that branch is transversed. The convolutional code is the set of all semi-infinite binary words that may be read off by following any path through the trellis and reading the code bits as they are passed. The labeled trellis defines a linear code – and hence a convolutional code – because the $GF(2)$ sum of the symbol sequences along any pair of paths is another symbol sequence on another path.

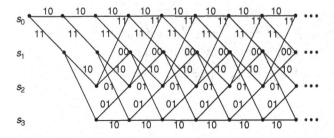

Figure 10.6. Trellis diagram for a convolutional code with constraint length 2.

To use the code requires an encoder. This means that at each node we must specify which branch is assigned to represent a data bit zero and which branch is assigned to represent a data bit one. In principle, this assignment is completely arbitrary, but in practice we want a rule that is easy to implement. We shall describe two encoders for this code, one that uses a two-bit feedforward shift register and is a nonsystematic encoder, and one that uses a two-bit feedback shift register and is a systematic encoder. In either case, the state corresponds to the two bits in the shift register and the encoder produces the same pair of code bits that label the trellis in Figure 10.6. The assignment of data bits to branches is what makes the encoders different.

To design an encoder that uses a feedforward shift register, we augment the trellis diagram as shown in Figure 10.7. The states are labeled with the contents of the shift register which, by assumption, are the two most recent data bits. The states are assigned to the trellis by placing the four two-bit patterns at the left of the trellis, the most recent bit to the left. Once this assignment of states is made, a correspondence is established between each specific value of the incoming data bit and one of the branches. This correspondence is explicitly recorded by attaching that data bit value as another label to each branch in Figure 10.7. The data bit on a branch is the input to the encoder, and the pair of code bits on the branch is the output of the encoder. The code bits can be regarded as a function of the two state bits and the incoming data bit. A circuit that implements the required binary logic is shown at the left in Figure 10.8. With this encoder, an arbitrary binary datastream is shifted into the encoder and the appropriate codestream is shifted out – two output code bits for each input data bit.

If there are no errors, the datastream is recovered by the circuit on the right side of Figure 10.8. The shift register circuit mimics the $GF(2)$ polynomial arithmetic given by

$$x(x^2 + x + 1) + (x + 1)(x^2 + 1) = 1.$$

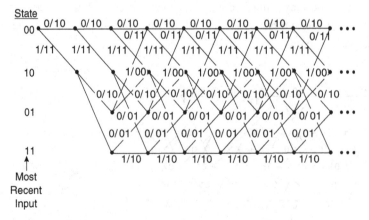

Figure 10.7. Trellis diagram annotated for a nonsystematic encoder.

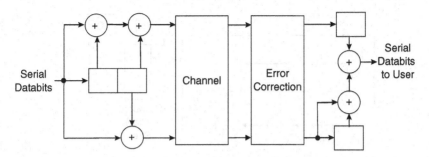

Figure 10.8. Application of a nonsystematic convolutional encoder.

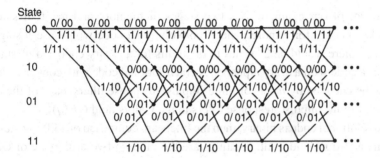

Figure 10.9. Trellis diagram annotated for a systematic encoder.

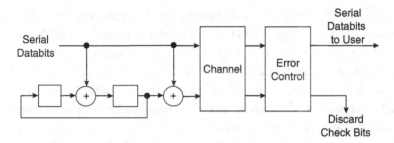

Figure 10.10. Application of a systematic convolutional encoder.

If there are errors, they are first corrected provided the error-correcting ability of the code has not been exceeded.

To design an alternative encoder that is systematic, we inspect the trellis of Figure 10.6 from a different point of view. On the pair of branches out of each node, the first bit of the label takes on each possible value, so we assign branches so that the first code bit is equal to the data bit. This leads to the trellis of Figure 10.9 and the encoder of Figure 10.10. The sets of codewords produced by the encoders of Figure 10.8 and Figure 10.10 are identical, but the way in which codewords represent data bits is different.

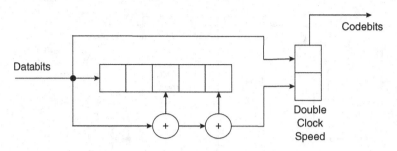

Figure 10.11. An encoder for another systematic convolutional code.

Each encoder of our example uses two stages of memory and has four states in the trellis. An encoder for a more general convolutional code would use v stages of memory. Then there would be 2^v states in the trellis for a binary code and q^v states in the trellis for a code with a q-ary alphabet. A convolutional code with constraint length $v = mk$ can be encoded by n sets of finite-impulse-response filters, each of the n sets consisting of k finite-impulse-response filters in the Galois field $GF(q)$.

If the convolutional code is binary, then the filters are composed of shift registers one bit wide; the additions and multiplications are the modulo-two arithmetic of $GF(2)$. The input to the encoder is a stream of databits at a rate of k bits per unit time and the output of the encoder is a stream of codebits to the channel at a rate of n bits per unit time.

A second example of an encoder for a binary convolutional code, one with constraint length five, is shown in Figure 10.11. This encoder is based on a feedforward shift register and is a systematic encoder because the data bits appear unaltered. In general, it is not possible to construct an encoder that is both systematic and uses a feedforward shift register. It is, however, always possible to construct a systematic encoder that uses a feedback shift register. We shall develop this statement by studying the structure of convolutional codes in a more formal way.

A mathematical description of a convolutional code can be formulated in the language of polynomial arithmetic. A polynomial of degree $n - 1$ over the field $GF(q)$ is a mathematical expression

$$f(x) = f_{n-1}x^{n-1} + f_{n-2}x^{n-2} + \cdots + f_1x + f_0$$

where the symbol x is an indeterminate, the coefficients $f_{n-1}, \ldots, f_0$ are elements of $GF(q)$, and the indices and exponents are integers.

A finite-impulse-response filter of length m can be represented by a polynomial of degree at most m in the field of the filter. The coefficients of the polynomial are the taps of the filter. These polynomials are called the *generator polynomials* of the convolutional code. The encoder shown in Figure 10.8 has the two generator polynomials

$$g_0(x) = x^2 + x + 1$$
$$g_1(x) = x^2 + 1.$$

The generator polynomials can be arranged into a matrix of polynomials

$$G(x) = [x^2 + x + 1 \quad x^2 + 1].$$

The encoder shown in Figure 10.11 has the two generator polynomials

$$g_0(x) = 1$$
$$g_1(x) = x^5 + x^3 + 1.$$

These generator polynomials also can be arranged into a matrix of polynomials

$$G(x) = [1 \quad x^5 + x^3 + 1].$$

In general, a convolutional encoder requires a total of kn generator polynomials to describe it, the longest of which has degree m. Let $g_{ij}(x)$ for $i = 1, \ldots, k$ and $j = 1, \ldots, n$ be the set of generator polynomials. These can be put together in a k by n matrix of polynomials, called the *generator matrix*, and given by

$$G(x) = [g_{ij}(x)].$$

A convolutional code with an n by k generator matrix is called an (n, k) convolutional code. When k is greater than one, some of the generator polynomials may be the zero polynomial. Figure 10.12 shows an example of a (3,2) binary convolutional code that has generator matrix

$$G(x) = \begin{bmatrix} 1 & x^2 & 1 \\ 0 & 1 & x + x^2 \end{bmatrix}.$$

Consider the input frame as k databits in parallel, and consider the sequence of input frames as k sequences of databits in parallel. These may be represented by k data polynomials $a_i(x)$ for $i = 1, \ldots, k$; or as a row vector of such polynomials

$$a(x) = [a_1(x), a_2(x), \ldots, a_k(x)].$$

If we break out the vector coefficients of the vector polynomials $a(x)$, we can write this as

$$a(x) = [a_{01}, a_{02}, \ldots, a_{0k}] + [a_{11}, a_{12}, \ldots, a_{1k}]x + [a_{21}, a_{22}, \ldots, a_{2k}]x^2 + \cdots.$$

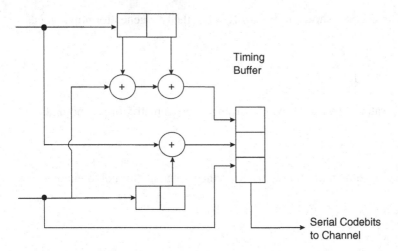

Figure 10.12. Encoder for a rate 2/3 convolutional code.

Each bracketed term displays one frame of input data symbols. Similarly the output codeword can be represented by n codeword polynomials $c_j(x)$ for $j = 1, \ldots, n$; or as a vector of such polynomials

$$\boldsymbol{c}(x) = [c_1(x), c_2(x), \ldots, c_n(x)].$$

The ℓth frame consists of the ℓth coefficients from each of these n polynomials. That is,

$$\boldsymbol{c}(x) = [c_{01}, c_{02}, \ldots, c_{0n}] + [c_{11}, c_{12}, \ldots, c_{1n}]x + [c_{12}, c_{22}, \ldots, c_{2n}]x^2 + \cdots$$

To form a serial stream of code symbols to pass through the channel, the coefficients of the n codeword polynomials are interleaved.

The encoding operation can now be described compactly as a vector-matrix product

$$\boldsymbol{c}(x) = \boldsymbol{a}(x)\boldsymbol{G}(x)$$

with jth component given by the polynomial expression

$$c_j(x) = \sum_{i=1}^{k} a_i(x) g_{ij}(x).$$

The definition of constraint length can be expressed succinctly in terms of the matrix of generator polynomials $\boldsymbol{G}(x)$. Given the matrix of generator polynomials $\boldsymbol{G}(x)$, the *constraint length* of the convolutional code is

$$\nu = \sum_{i=1}^{k} \max_{j}[\deg g_{ij}(x)].$$

Other ways of measuring the memory duration in the convolutional code may be defined, but are less useful.

A *systematic* generator matrix for a convolutional code is a matrix of polynomials of the form

$$G(x) = [I \vdots P(x)]$$

where I is a k by k identity matrix, and $P(x)$ is a k by $(n - k)$ matrix of polynomials.

A generator matrix for a block code always can be put into a systematic form by permutations of columns and elementary row operations. This is not possible in general for a matrix of polynomials because the division of a polynomial by a polynomial is not a polynomial in general. However, if we allow the matrix to include rational functions of the form $a(x)/b(x)$ as elements, then we can put the generator matrix into systematic form. For example, the generator matrix

$$G(x) = [x^2 + 1 \qquad x^2 + x + 1]$$

is equivalent to the systematic generator matrix

$$G'(x) = \begin{bmatrix} 1 & \dfrac{x^2 + x + 1}{x^2 + 1} \end{bmatrix}$$

$$= \begin{bmatrix} 1 & 1 + \dfrac{x}{x^2 + 1} \end{bmatrix}.$$

What this means is that a systematic encoder exists, but it includes feedback in the encoder due to the denominator $x^2 + 1$. Inspection of the systematic generator matrix shows how the encoder of Figure 10.10 was designed by laying out the feedback shift register corresponding to the matrix entry

$$g_{12}(x) = 1 + \frac{x}{x^2 + 1}.$$

Consequently, we have found a mathematical construction for passing between the two encoders shown in Figure 10.8 and Figure 10.10.

A *check matrix* $H(x)$ for a convolutional code is an $(n-k)$ by n matrix of polynomials that satisfies

$$G(x)H(x)^T = 0.$$

If $G(x)$ is a systematic generator matrix, a polynomial check matrix can be written down immediately as

$$H(x) = [-P(x)^T \vdots I]$$

where I here is a $(n-k)$ by $(n-k)$ identity matrix. It is straightforward to verify that

$$G(x)H(x)^T = 0.$$

We can always find a check matrix corresponding to any generator matrix by manipulating $G(x)$ into a systematic form (with rational functions) by permutations and elementary row operations, writing down the corresponding systematic check matrix, then multiplying out the denominators and inverting the permutations.

Just as for block codes, it is not correct to speak of systematic convolutional codes because the code exists independently of the method of encoding. It is only correct to speak of systematic *encoders* for convolutional codes. Systematic encoders for convolutional codes feel more satisfying because the data is visible in the encoded sequence and can be read directly if no errors are made. However, unlike block codes, not every convolutional code has a systematic convolutional encoder using only feedforward filters in the encoder. By using the feedback of polynomial division circuits, however, one can build a systematic encoder for any convolutional code.

Codewords that have not been systematically encoded do not display the data directly and must be designed so that the data can be recovered after the errors are corrected. Moreover they must be designed so that an uncorrectable error pattern will only destroy a finite amount of data. Let $k = 1$, so that the generator polynomials have only a single index. Then

$$G(x) = [g_1(x) \quad g_2(x) \quad \cdots \quad g_n(x)]$$

and

$$c_j(x) = a(x)g_j(x) \quad j = 1, \ldots, n.$$

At least one of the generator polynomials is not divisible by x because otherwise it corresponds to a pointless delay in each filter.

Definition 10.5.1 *An $(n, 1)$ convolutional code whose generator polynomials $g_1(x), \ldots, g_n(x)$ have greatest common divisor satisfying*

$$\mathrm{GCD}[g_1(x), \ldots, g_n(x)] = 1$$

is called a noncatastrophic convolutional code. Otherwise it is called a catastrophic convolutional code. The greatest common divisor is the polynomial of largest degree that divides $g_1(x), g_2(x), \ldots,$ and $g_n(x)$.

The reason for the term "catastrophic code" and the reason for dismissing such codes from further use lies in the fact that for such a code there are error patterns containing a finite number of channel errors that must produce an infinite number of errors in the decoder output. Let

$$A(x) = \text{GCD}[g_1(x), \ldots, g_n(x)]$$

and consider the data polynomial $a(x) = A(x)^{-1}$ by which we mean the polynomial obtained by formally dividing one by $A(x)$. If $A(x)$ is not equal to one, $a(x)$ will have infinite weight. But for each j, $c_j(x) = a(x)g_j(x)$ has finite weight because $A(x)$, a factor of $g_j(x)$, cancels $a(x)$. If $a(x)$ is encoded and the channel makes a finite number of errors, one in each place that one of the $c_j(x)$ is nonzero, the decoder will see all zeros out of the channel and erroneously conclude that $a(x) = 0$, thereby making an infinite number of errors.

For example, if $g_1(x) = x + 1$ and $g_2(x) = x^2 + 1$, then $\text{GCD}[g_1(x), g_2(x)] = x + 1$ over $GF(2)$. If $a(x) = 1 + x + x^2 + x^3 + \cdots$, then $c_1(x) = 1$, and $c_2(x) = 1 + x$ so the codeword has weight 3. Only three channel errors can change the codeword into the codeword for the all-zero dataword.

In the absence of errors, the datastream can be recovered from the codeword of any noncatastrophic convolutional code by using a corollary to the euclidean algorithm for polynomials. This corollary states that for any set of polynomials, if $\text{GCD}[g_1(x), \ldots, g_n(x)] = 1$, then there exist polynomials $b_1(x), \ldots, b_n(x)$ satisfying

$$b_1(x)g_1(x) + \cdots + b_n(x)g_n(x) = 1.$$

If the data polynomial $a(x)$ is encoded by

$$c_j(x) = a(x)g_j(x) \quad j = 1, \ldots, n$$

we can recover $a(x)$ by

$$a(x) = b_1(x)c_1(x) + \cdots + b_n(x)c_n(x)$$

as is readily checked by simple substitution.

For example, in $GF(2)$,

$$\text{GCD}[x^5 + x^2 + x + 1, x^6 + x^2 + x] = 1,$$

so there must exist polynomials $b_1(x)$ and $b_2(x)$ such that, in $GF(2)$

$$b_1(x)(x^5 + x^2 + x + 1) + b_2(x)(x^6 + x^2 + x) = 1.$$

These can be found to be

$$b_1(x) = x^4 + x^3 + 1$$
$$b_2(x) = x^3 + x^2 + 1.$$

A codeword composed of

$$c_1(x) = (x^5 + x^2 + x + 1)a(x)$$
$$c_2(x) = (x^6 + x^2 + x)a(x)$$

with generator polynomials $g_1(x) = x^5 + x^2 + x + 1$ and $g_2(x) = x^6 + x^2 + x$ can be inverted with the aid of $b_1(x)$ and $b_2(x)$

$$c_1(x)b_1(x) + c_2(x)b_2(x) = (x^5 + x^2 + x + 1)(x^4 + x^3 + 1)a(x)$$
$$+ (x^6 + x^2 + x)(x^3 + x^2 + 1)a(x)$$
$$= a(x).$$

A convolutional code corresponds to a set of coprime generator polynomials. It is not hard to find arbitrary sets of coprime polynomials. What is hard is to find sets that have good error-correcting ability.

10.6   Decoding of convolutional codes

When a convolutional codeword is passed through a channel, errors are made from time to time in the codeword symbols. The decoder must correct these errors by processing the senseword. However, the convolutional codeword is so long that the decoder can observe only a part of it at one time. Although the codeword is effectively infinite in length, all decoding decisions must be made on senseword segments of finite length. No matter how one chops out a part of the senseword for the decoder to work with, there may be useful information within other parts of the senseword that the decoder does not yet see and, in a well designed system, usually does not need.

To develop a decoding procedure for a convolutional code, consider the task of correcting errors in the first frame. If the first frame of the code can be corrected and decoded, then the first frame of the datastream will be known. Because the code is linear, the effect of the first dataframe on subsequent codeword frames can be computed and subtracted from subsequent frames of the senseword. Then the problem of decoding the second codeword frame is the same as was the problem of decoding the first codeword frame. Continuing in this way, if the first j frames can be successfully corrected, then the problem of decoding the $(j + 1)$th frame is the same as the problem of decoding the first frame. In this way, the decoder step by step regenerates frames of the codeword from which the datastream can be computed.

A *minimum Hamming distance* decoder is a decoder that finds the codeword that differs from a demodulated senseword in the fewest places. The following procedure, which is somewhat analogous to a decision-feedback demodulator, gives in principle a good minimum-distance decoder for a convolutional code. Choose an integer b that is sufficiently large. The integer b can be thought of as the width of a window through which the initial segment of the senseword is viewed. Usually the decoder window width is validated by computer simulation of the decoder. The decoder works with only the first b symbols of the senseword. Generate the initial segment of length b of every codeword and compare the first b symbols of the senseword to each of the codeword segments. Select the codeword that is closest to the senseword in Hamming distance in this segment. The first frame of the data sequence that produces the selected codeword is chosen as the regenerated first frame of the decoder output. This data frame is then re-encoded and subtracted from the senseword. The first n symbols of the senseword are now discarded and n new symbols are shifted into the decoder. The process is then repeated to find the next data frame.

The minimum-distance decoder would be quite complex if it were implemented in the naive way described in the preceding paragraph. Just as was the case in demodulating data in the presence of intersymbol interference, there is a great deal of structure in the computation that can be exploited to obtain an efficient method of implementing the minimum-distance decoder. The Viterbi algorithm is such an efficient method and can be used to search the trellis. For implementing a minimum-Hamming-distance decoder, the Viterbi algorithm uses Hamming distance in its computations.

At frame time b, the Viterbi algorithm determines the minimum-distance path to each node in that frame. The decoder then examines all surviving paths to see that they agree in the first frame. This frame defines a decoded data frame which is passed out of the decoder. Next, the decoder drops the first frame and takes in a new frame of the senseword for the next iteration. If again all surviving paths pass through the same node of the oldest surviving frame, then this data frame is decoded. The process continues in this way decoding frames indefinitely.

If b is chosen large enough, then a well-defined decision will almost always be made at each frame time. If the chosen code is properly matched to the channel this decision will very probably be the correct one. However, several things can go wrong. No matter how large b is chosen, occasionally the surviving paths might not all go through a common node in the current initial frame. This is a *decoding failure* or a *decoding default*. The decoder can be designed to put out an indication of decoding default to the user when this happens. Alternatively, the decoder can be designed to simply guess a bit.

Sometimes, the decoder will reach a well-defined decision, but a wrong one. This is a *decoding error*. It is usually rarer than a decoding default. When a decoding error occurs, the decoder will necessarily follow this with additional decoding errors.

A dependent sequence of decoding errors is called an *error event*. Not only does one want error events to be infrequent, but one wants their duration to be short when they do occur. A decoder will recover from an error event if the code is a noncatastrophic code, but a catastrophic code could have infinitely long error events.

As an example, consider the rate one-half convolutional code with generator polynomials $g_1(x) = x^2 + x + 1$ and $g_2(x) = x^2 + 1$. We choose a decoder with a decoding window width b equal to 15. Suppose that the binary senseword is

$$v = 0\ 1\ 1\ 1\ 1\ 0\ 1\ 0\ 0\ 1\ 0\ 1\ 0\ 0\ 0\ 0\ 0\ 0\ 0\ 0\ \ldots$$

The development of the candidate paths through the trellis is shown in Figure 10.13. At the third iteration, the decoder has already identified the shortest path to each node

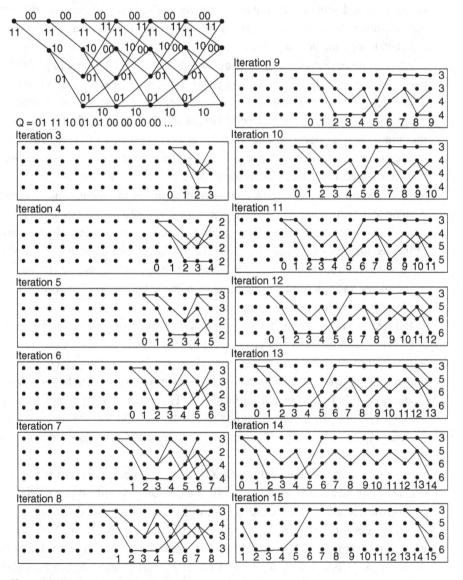

Figure 10.13. Sample of Viterbi algorithm.

of the third frame. Then, at iteration r, the decoder finds the shortest path to each node of the rth frame by extending the paths to each node of the $(r-1)$th frame and keeping the shortest path to each node. Whenever a tie occurs, as illustrated by the two ties at iteration 7, the decoder either may break the tie by guessing or may keep both paths in the tie. In the example, ties are retained until either they are eliminated or they reach the end of the decoder. As the decoder penetrates into deeper frames, the earlier frames reach the end of the decoder memory. If a path exists to only one node of the oldest frame, the decoding is complete. If several paths exist, then an uncorrectable error pattern has been detected. Either it can be flagged as a decoding default, or a guess can be made.

In the example, the data sequence has been decoded as

$$a = 1\,1\,1\,1\,0\,0\,0\,0 \ldots$$

because these are the data bits labeling the branches of the demodulated path through the trellis.

The decoder shown symbolically in Figure 10.13 might look much different in its actual implementation. For example, the active paths through the trellis could be represented by a table of four fifteen-bit numbers. At each iteration each fifteen-bit number is used to form two sixteen-bit numbers by appending a zero and a one. This forms eight sixteen-bit numbers forming four pairs of numbers and to each is attached the discrepancy of the path. Within each pair, only the minimum-distance number is saved, the other is discarded. Now the sixteen-bit numbers are shifted left by one bit to form a new table of four fifteen-bit numbers and the iteration is complete.

As the decoder progresses through any frames, the accumulating discrepancies continue to increase. To avoid overflow problems, they must be reduced occasionally. A simple procedure is periodically to subtract the smallest discrepancy from all of them. This does not affect the choice of the maximum discrepancy.

The Viterbi algorithm contains a fixed decoding window width b, which is the separation between the frame entering the decoder and the frame leaving the decoder. Technically, the optimum choice of b is unbounded because an optimum decision cannot be made until the surviving paths to all states share a common initial subpath, and this may take an arbitrarily long time. On occasion, however, little degradation occurs when the algorithm chooses a fixed decoding window width b that is sufficiently large. Moreover, there will always be imprecision in the channel model that makes overly precise distinction meaningless.

10.7 Performance of convolutional codes

A convolutional code is used to correct errors at the channel output. However, a convolutional codeword is infinitely long in principle and a demodulator with a nonzero probability of symbol error, no matter how small, will make an infinite number of errors in an infinitely long sequence. Thus, the convolutional decoder must correct an infinite number of errors in an infinitely long codeword. We should think of these errors as randomly, and sparsely, distributed along the codeword. The decoder begins working at the beginning of the senseword and corrects errors as it comes to them. Occasionally, by the nature of randomly distributed errors, there will be a cluster of errors that the code is not powerful enough to correct. A segment of the decoder output will either be incorrect, which we call an *error event* or undecodable, which we call a *default* or *erasure event*. An error event consists of an interval in which the decoded output bits appear to be normal but in fact are incorrect. A default event consists of an interval in which the decoder realizes that the senseword contains too many errors and cannot be corrected. The output bits in that interval are annotated to indicate that the interval contains too many bit errors to be corrected. We can think of each bit during such an event replaced by an erasure.

A decoder can be designed to reduce the number of error events by increasing the number of erasure events. It does this by detecting that during some interval the apparent number of errors to be corrected is unreasonable. For such a decoder we are interested in computing both the probability of decoding error (bit or event) and the probability of decoding erasure (bit or event); the sum of these two is the probability of decoding failure.

After a decoding error or decoding default occurs, a well-designed system will eventually return to a state of correct decoding. Otherwise, if the onset of a decoding failure can lead to a permanent state of failure, we say that the decoder is subject to infinite error propagation. Infinite error propagation may be due to the choice of a catastrophic set of generator polynomials, in which case it is called catastrophic error propagation. Infinite error propagation might also be caused by deficiencies in the choice of decoding algorithm, in which case it is called ordinary error propagation. A properly designed system will avoid both of these defects.

Loosely, to specify the performance of a convolutional code, we specify the bit error rate at the decoder output for some channel model, say an additive gaussian noise channel at a specified value of E_b/N_0. However, the task is deeper and more subtle than this. First of all, we cannot completely separate the performance of the code from the design of the decoder. Second, we need to distinguish between bit errors and bit erasures whenever the decoder includes an erasure output.

We shall study the performance of a convolutional code by first studying the distance structure of the code, then studying the relationship between the distance structure and the probability of error. The simplest decoding rule to study is the minimum-distance decoder with ties broken arbitrarily (which is the maximum-likelihood decoder is gaussian noise), because then a bit is either correct or in error; erasures do not occur. Two error probabilities are of interest. The *probability of bit error p_e* (or the bit error rate) is the limit, as ℓ goes to infinity, of the probability that the ℓth data bit at the output of the decoder is wrong. The probability of error event p_{ev} is the limit, as ℓ goes to infinity, of the probability that the decoder codeword lies along the wrong branch of the trellis for the branch corresponding to the ℓth data bit (this is the ℓth branch if $k = 1$). During an error event, some of the bits may be correct so $p_e < p_{ev}$. In fact for a typical binary code we may presume that $p_e \approx \frac{1}{2}p_{ev}$ under the argument that about half of the bits are correct during an error event.

Although the definitions of the probability of bit error and the probability of error event are precise, they may be difficult to compute. We will be satisfied with close approximations. Further, it would be a rare user that would care about the precise difference between a bit error and an event error. A chance correct bit embedded in an error burst has little value.

The many notions of decoder performance can be discussed in terms of the distance structure of the code, specifically in terms of the minimum distance of the code. A convolutional code has many minimum distances determined by the length of initial codeword segment over which minimum distance is measured. The distance measure given next is defined such that if two codewords both decode into the same first data frame, then they can be considered equivalent.

Definition 10.7.1 *The ℓth minimum distance d_ℓ of a convolutional code is equal to the smallest Hamming distance between any two initial codeword segments ℓ frames long that disagree in the initial frame. The nondecreasing sequence $d_1, d_2, d_3, \ldots,$ is called the distance profile of the convolutional code.*

A convolutional code is linear, so in searching over pairs of codewords for the minimum distance, one of the two codewords might just as well be the all-zero codeword. The ℓth minimum distance then is equal to the weight of the smallest weight codeword segment ℓ frames long that is nonzero in the first frame. This can be read off a labeled trellis.

For example, for the convolutional code whose trellis was shown in Figure 10.6, we see from inspection of the figure that $d_1 = 2$, $d_2 = 3$, $d_3 = 5$, and $d_i = 5$ for all i greater than 5.

Suppose that a convolutional code has ℓth minimum distance d_ℓ. If at most t errors satisfying

$$2t + 1 \leq d_\ell$$

occur in the first ℓ frames, then those that occur in the first codeword frame can be corrected.

The sequence d_ℓ is an increasing sequence of integers which will reach a largest value and not increase further. This largest value, which is the minimum sequence distance between any pair of infinitely long codewords, is conventionally called the free distance of the convolutional code. In this context, the terms "free distance" and "minimum distance" are synonymous.

Definition 10.7.2 *The free Hamming distance d_{min} of a convolutional code C is defined as the smallest Hamming distance between any two distinct codewords.*

To find the free distance of a noncatastrophic convolutional code, it is enough to check only pairs of paths that diverge from a common node and then rejoin at a common node after a finite time. Indeed, because the code is linear, it is enough that one of the two paths be the all-zero path. The free distance can be computed from the distance profile by

$$d_{min} = \max_{\ell} d_{\ell}.$$

In the example of Figure 10.6, the free Hamming distance equals 5.

The code symbols of a convolutional code may be modulated onto a BPSK waveform given by

$$c(t) = \sum_{\ell=0}^{\infty} c_{\ell} s(t - \ell T)$$

where now $c_{\ell} = \pm 1$ according to whether the ℓth bit of the convolutional codeword is a zero or a one. In such a case we may be interested in the euclidean distance between these BPSK waveforms in place of the Hamming distance between convolutional codewords. The euclidean distance is

$$d_E(c(t), c'(t)) = \left[\sum_{\ell=0}^{\infty} |c_{\ell} - c'_{\ell}|^2 \right]^{1/2}.$$

Clearly

$$d_E(c(t), c'(t)) = 2 d_H(\mathbf{c}, \mathbf{c}').$$

The euclidean distance between sequences is a generalization of the euclidean distance between the points of a real or complex signal constellation. The minimum distance between points of the signal constellation is replaced by the minimum distance between sequences.

Definition 10.7.3 *The free euclidean distance d_{min} of a set of sequences of real (or complex) numbers is the smallest euclidean distance between any two of the sequences.*

If the BPSK sequences arise by mapping the codebits of a convolutional codeword into ± 1, as in BPSK modulation, then it is evident that

$$(d_{min})_{euclidean} = 2(d_{min})_{Hamming}.$$

Just as the minimum distance between points of a signal constellation plays a major role in determining the probability of demodulation error in gaussian noise of a multi-level signaling waveform, so the minimum distance between sequences plays a major role in determining the probability of error in sequence demodulation. By reference to the asymptotic coding gain of block codes, we have the following definition of asymptotic coding gain for a convolutional code used with a soft-decision decoder.

Definition 10.7.4 *The asymptotic coding gain of a binary convolutional code C of rate R and free Hamming distance $d_{\min}$ modulated into the real (or complex) number system is*

$$G = Rd_{\min}.$$

By reference to the situation for block codes we may expect that an approximate formula for probability of decoding error is

$$p_e \approx N_{d_{\min}} Q\left(\sqrt{G\frac{2E_b}{N_0}}\right).$$

This formula can be verified by simulation. Alternatively we could show using the methods of Section 4.3 that this approximation follows roughly from a union bound. Thus, the asymptotic coding gain $G = Rd_{\min}$ is an approximate measure of the amount that E_b can be reduced because of the use of the code.

10.8   Turbo codes

The maximum-posterior principle can be used to form a decoder for a convolutional code. This is similar to the maximum-posterior demodulator for a signaling waveform with intersymbol interference, as was discussed in Section 7.7. The trellis structure of the convolutional code again makes it possible to implement a maximum-posterior decoder using a fast algorithm. However, the trellis structure of a convolutional code also results in the property that each data bit is directly coupled only to those nearby check bits that are computed from that data bit while it is still in the encoder, as is determined by the constraint length of the code. Other codeword bits affect the decoding of that data bit indirectly because those bits provide information to other near or far neighbors, which information then couples through the connecting structure. Bits that are far apart are only weakly and indirectly coupled. This means that only nearby bits provide significant helpful information for decoding a given bit; distant bits do not provide significant help. As a result, although maximum-posterior decoding of a convolutional code is superior to maximum-likelihood decoding, the difference in performance is too meager to justify the additional cost and complexity. In order to realize the performance potential of maximum-posterior decoding, the convolutional code, in its pure form, must be set aside.

On the other hand, because it enables the use of the two-way algorithm, without which the computations would be prohibitive, the trellis structure of a convolutional code is too important to discard. Although the one-dimensional nature of the trellis structure of the convolutional code is compatible with the fast computations of the two-way algorithm, the error performance of the convolutional code is limited by the one-dimensional nature of the trellis structure. Thus, there is good reason to retain the basic structure of the convolutional code, but it should be strengthened by additional structure.

A certain elaboration of a convolutional code, known as a *turbo code*, or a *Berrou code*, achieves this goal. It does this in a way that retains the linear structure of the convolutional code, but appends a second encoding of each databit into a second convolutional codeword. The new double-codeword structure greatly enhances the performance of the maximum-posterior decoder while retaining the feature of an affordable decoding algorithm.

To encode the data twice, a turbo encoder uses the same convolutional code twice in the way shown in Figure 10.14. It encodes the data both in its natural order, and in a scrambled order as provided by the permutation block. In this way each data bit is encoded twice, thereby providing the three streams of symbols that comprise the codeword; the first stream is the original datastream $c_0 = a$, the second stream is the first stream of check symbols c_1, and the third stream is the second stream of check symbols c_2. The complete turbo codeword is the interleave of these three streams $c = (c_0, c_1, c_2)$. The codeword will also be viewed as the two convolutional codewords $c' = (c_0, c_1)$ and $c'' = (\tilde{c}_0, c_2)$, where $\tilde{c}_0 = \pi(c_0)$ is a permutation of c. The structure combining two convolutional codes into one turbo code is referred to as *parallel concatenation*. If desired, because the code redundancy is too high, the two convolutional codes can be punctured by discarding some of the check bits.

In the first convolutional codeword, each data bit finds itself connected to the check bits according to the linearly-ordered structure of a trellis, and is strongly influenced in

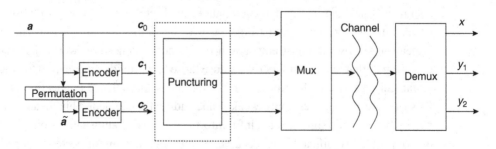

Figure 10.14. The encoder structure of a turbo code.

the decoder by the soft values of its neighboring check bits and data bits, but only weakly by the soft value of distant bits. In the second convolutional codeword, a given bit again finds itself connected to neighbors, but now the neighbors are different. For each data bit, each of its neighbors will have other neighbors in both codewords, resulting in a complex tangle of connections. The soft sensed value of each of the other bits can provide strong information helping to demodulate any given bit through the structure of the code.

10.9   Turbo decoding

The input to the turbo decoder is the soft senseword v observed as the filtered and sampled noisy turbo codeword at the output of the channel. The components of the soft senseword can be regarded as confidence information on each received symbol of the codeword. The senseword v will be written as $v = (x, y_1, y_2)$, where x corresponds to soft noisy data, y_1 corresponds to the soft noisy check bits of the first codeword, and y_2 corresponds to the soft noisy check bits of the second codeword. For the topic of turbo decoding the soft senseword symbols are the codeword symbols contaminated by additive white (memoryless) noise, usually gaussian. For the more comprehensive topic of *turbo equalization*, which we do not study, the received symbols of the convolutional code also have intersymbol interference.

The block conditional probability vector on the senseword v is

$$p(v|a) = p(x, y_1, y_2|a)$$

and, by Bayes' formula, the posterior is

$$p(a|x, y_1, y_2) = \frac{p(x, y_1, y_2|a)p(a)}{\sum_x \sum_{y_1} \sum_{y_2} p(x, y_1, y_2|a)p(a)}.$$

Although this statement of the problem is straightforward in principle, for a turbo code the right side is actually extremely complicated and unwieldy. Both y_1 and y_2 depend on a in a complicated way, and the blocklength n may be on the order of thousands. On the other hand, either of the two underlying convolutional codes has a posterior

$$p(a|x, y_i) = \frac{p(x, y_i|a)p(a)}{\sum_x \sum_{y_i} p(x, y_i|a)p(a)}$$

which is formulated just by ignoring the check symbols of the other convolutional code.

Either convolutional code can be decoded by any posterior decoder, such as the two-way algorithm. Moreover, if the decoder for the first convolutional code has a soft output on each symbol, as would be the case for a symbol by symbol posterior

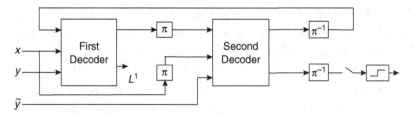

Figure 10.15. The structure of the turbo decoder.

computation, that soft output can be used as a prior by the decoder for the second convolutional codeword. But then the first convolutional code can be decoded again using the soft output of the second decoder as a prior. This leads to the notion of iterative decoding. The two convolutional codes are decoded alternately, with each vector of componentwise posteriors computed by one decoder feeding as a vector of componentwise priors to the other decoder for the next iteration.

As the decoders alternate in this way, the computed posteriors slowly change, each apparently moving closer either to zero or to one. After a moderate number of iterations, each bit can be demodulated. In this way, one obtains an approximation to the maximum-posterior decoder for the turbo code. Extensive experience shows that this iterative process works well. It has been found empirically for large binary Berrou codes, that ten to twenty iterations are usually sufficient for convergence. It has also been found empirically that by using a turbo code, bit error rates on the order of 10^{-6} can be obtained at very small values of E_b/N_0.

The structure of an iterative decoder is shown in Figure 10.15.

Problems for Chapter 10

10.1.   a. A Hamming $(7, 4)$ code is used to correct errors on a binary symmetric channel with error probability ϵ. What is the probability of decoding error? (A binary symmetric channel is a binary channel that has the same probability of error when a zero is transmitted as it does when a one is transmitted.)

b. If the code is used only to detect errors but not correct them, what is the probability of undetected error?

c. A Reed–Solomon $(7, 5)$ code is used to correct errors on an octal symmetric channel with error probability ϵ. What is the probability of decoding error?

10.2.   Let $n = 2^r - 1$, and consider a binary word of blocklength n. How many binary words are there that differ from the given word in not more than one place? Given 2^k distinct binary words of blocklength n with the property that any one of the words with one bit changed differs from every other word in at least two bit positions, how big can k be? Prove that no binary single-error-correcting code can have fewer check symbols than the Hamming code.

10.3. A sixteen-ary orthogonal signaling alphabet is used on an additive gaussian noise channel with an E_b/N_0 of 4 dB.

    a. Write down a probability transition matrix for this channel viewed as a discrete channel.

    b. A $(15,13)$ Reed–Solomon code is used to correct single errors. What is the probability of error of the overall system as a function of E_b/N_0?

10.4. A Reed–Solomon code can be defined alternatively as the set of words of blocklength n, that have zeros in the spectrum at C_j for $j = j_0, \ldots, j_0 + 2t - 1$. Using the modulation/translation properties of the Fourier transform, write out the proof that the Reed–Solomon code so defined will correct t errors. (It is common practice to use Reed–Solomon codes with $j_0 = 1$.)

10.5. An $(8, 6)$ Reed–Solomon code over the complex number system (with $C_1 = C_2 = 0$) is given. How many errors can be corrected? The senseword is $(1, 1, 1, 1, 0, 1, 1, 1)$. What is the error pattern?

10.6. The polynomial $p(x) = x^3 + x + 1$ is irreducible over $GF(2)$.

    a. Write out addition and multiplication tables for $GF(8)$.

    b. How many codewords are there in the $(7, 3)$ double-error-correcting Reed–Solomon code over $GF(8)$?

    c. Using this code, and with α denoting the field element x, the senseword in polynomial notation is

$$v(y) = \alpha^4 y^6 + \alpha^2 y^5 + \alpha^3 y^3 + \alpha^3 y^2 + \alpha^6 y + \alpha^4.$$

    Find the transmitted codeword.

    d. If the encoder is systematic, what are the data symbols?

10.7. A communication system uses a $(7,5)$ Reed–Solomon code in $GF(8)$, an eight-ary orthogonal family of waveforms, and a noncoherent demodulator.

    a. The waveforms are based on Hadamard sequences and cosine chips on the in-phase axis and are equal to zero on the quadrature axis. Find the relationship between the information rate and the width of the cosine chip.

    b. Is the choice of Reed–Solomon code well-matched to the choice of signaling waveform? Why?

    c. Set up and describe the sequence of equations that one would need to calculate p_e versus E_b/N_0 where p_e is the probability of symbol error. How does the solution change if p_e is the probability of bit error?

10.8. An *incomplete decoder* for a QAM signal constellation demodulates the received sample v into signal point c_m if $d(v, c_m) < d_{min}/2$, where d_{min} is the minimum distance between any two points in the signal constellation. Otherwise, the demodulator output is a special symbol called *data erasure*.

a. Prove that the probability that the decoder is incorrect is

$$1 - p_e = e^{-d_{\min}/\sigma^2}.$$

Express this in terms of E_b/N_0 for a sixteen-ary QASK signal constellation.

b. The probability of decoding failure can be subdivided into the probability of decoding error p_e, and the probability of decoding erasure p_r. Find approximate expressions for p_e and p_r as a function of E_b/N_0 for a sixteen-ary signal constellation.

c. Set up the expressions needed to compute probability of incorrect decoding as a function of E_b/N_0 for a Reed–Solomon (15,13) code used with this signal constellation and an errors-and-erasures decoder.

10.9.   A systematic generator matrix for a (2,1) convolutional code is given by

$$\left[1 \quad \frac{x^2 + 1}{x^2 + x + 1} \right].$$

a. What is the free Hamming distance of this convolutional code?
b. Sketch a systematic encoder based on this generator matrix.
c. Sketch a trellis labeled with data bits and codeword bits.

10.10.   A (7,4) Hamming code is used on a white gaussian noise channel with no intersymbol interference. Based on the vector of matched-filter output samples $v_0, v_1, \ldots, v_6$, a soft-decision demodulator finds the codeword c_r that is closest in euclidean distance to the senseword. That is, choose r such that

$$d(c_r, v) = \sum_{\ell=0}^{6} (c_{r\ell} - v_\ell)^2$$

is minimum. What is the probability of block decoding error as a function of E_b/N_0? What is the asymptotic coding gain with respect to BPSK? What is the relationship between the two answers?

10.11.   A convolutional code with rate $\frac{1}{3}$ and constraint length 2 has generator polynomials $g_1(x) = x^2 + x + 1$, $g_2(x) = x^2 + x + 1$, $g_3(x) = x^2 + 1$.
a. Construct the trellis for this code and find the free distance.
b. Give a circuit for recovering $a(x)$ from $c_1(x)$, $c_2(x)$, and $c_3(x)$ in the absence of errors.

10.12.   a. Draw a trellis for the convolutional code with generator polynomials

$$[x^2 + x + 1 \quad x^2 + 1].$$

b. Use the Viterbi algorithm to decode the senseword

$$v = 10001000001000000000 \ldots$$

10.13. A rate $\frac{1}{2}$ convolutional code over $GF(4)$ has generator polynomials $g_1(x) =$ $2x^3 + 2x^2 + 1$ and $g_2(x) = x^3 + x + 1$.

   a. Show that the minimum distance is 5.

   b. How many double-error patterns can occur within an initial senseword segment length of 8?

   c. Design a syndrome decoder for correcting all double errors.

10.14. a. A pulse (possibly complex) of energy E_p in white noise is passed through a matched filter. Show that the output signal-to-noise ratio is $2E_p/N_0$.

   b. An M-ary orthogonal set of waveforms with $M = 32$ is used to transmit data. What E_b/N_0 is needed to obtain a symbol error probability of 10^{-5}? What is the matched-filter signal-to-noise ratio?

   c. The same orthogonal signaling scheme is now used with a (31, 29) Reed–Solomon code in $GF(32)$. What is the required E_b/N_0 to obtain a probability of block decoding error of 10^{-5}? What is the required E_b/N_0 to obtain a probability of symbol error of 10^{-5}?

10.15. When the bandwidth of a channel is too small compared to the input symbol rate, "intersymbol interference" occurs. The case known as the "1+D channel" consists of a discrete-time channel in which each output voltage consists of the sum of the amplitudes of the last two input pulses.

   a. Describe this channel with binary inputs using a trellis, and describe the demodulator as a procedure for searching a trellis.

   b. For this channel construct your own example of a binary input sequence and a corresponding output sequence. Show how the Viterbi algorithm can be used for recovering the input sequence from a noisy version of the output sequence.

10.16. Prove that if the generator polynomials of a binary convolutional code each have an odd number of nonzero coefficients, then the code is transparent to a 180° phase ambiguity in the sense that if the codewords are complemented in the channel, the decoder will recover the complement of the datastream but is otherwise unaffected.

Notes for Chapter 10

We owe the idea of channel capacity and the recognition of the importance of data transmission codes to Shannon (1948, 1949). Shannon showed that the channel noise only sets a limit on the data rate, not on bit error rate. The bit error rate could be made as small as is desired by the use of a sufficiently strong data-transmission code. The first-known and simplest data transmission code is the Hamming code (1950). The best data-transmission codes in nonbinary finite fields were discovered by Reed and Solomon (1960) and independently by Arimoto (1961). The idea of the Fourier transform is implicit in the Reed–Solomon codes, and was used by many authors, but the first attempt to develop the Reed–Solomon code entirely from the Fourier transform point of view can be found in Blahut (1979).

The notion of a convolutional code was introduced by Elias (1954) and developed by Wozencraft (1957). Studies of the formal algebraic structure of convolutional codes were carried out by Massey and Sain (1968) and by Forney (1970). The Viterbi (1967) algorithm was first introduced as a pedagogical device whose practicality for convolutional codes of small constraint length was noticed by Heller (1968). The asymptotic tightness of the approximate formula for the probability of error of demodulating convolutional codes was confirmed by simulation by Heller and Jacobs (1971).

The use of iterative decoding was studied in a variety of situations by Hagenauer and his coauthors (1989, 1996). The notion of parallel concatenation in order to compensate for the limitation of the linear nearest neighbor structure or convolutional codes as applied to bitwise maximum posterior decoding is due to Berrou, Glaveiux, and Thitimajshima (1993). They also recognized the role of iterative decoding to combine the two partial decoders. The Berrou codes are usually called turbo codes because they are conceived so as to be suitable for turbo decoding.

11 Performance of Practical Demodulators

We have studied in great detail the effect of additive gaussian noise in a linear system because of its fundamental importance. Usually the ultimate limit on the performance of a digital communication system is set by its performance in gaussian noise. For this and other reasons, the demodulators studied in Chapter 3 presume that the received waveform has been contaminated only by additive gaussian noise. However, there are other important disturbances that should be understood. The demodulators studied in Chapter 4 extend the methods to include intersymbol interference in the received waveform. While additive gaussian noise and intersymbol interference are the most important channel impairments, the demodulator designer must be wary of other impairments that may affect the received signal. The demodulator must not be so rigid in its structure that unexpected impairments cause an undue loss of performance. This chapter describes a variety of channel impairments and methods to make the demodulator robust so that the performance will not collapse if the channel model is imperfect.

Most of the impairments in a system arise for reasons that are not practical to control, and so the waveform must be designed to be tolerant of them. Such impairments include both interference and nonlinearities. Sometimes nonlinearities may be introduced intentionally into the front end of the receiver because of a known, desired outcome. Then we must understand the effect of the nonlinearity in all its ramifications in order to anticipate undesirable side effects.

Of course, the first requirement is that there is adequate energy in the waveform to sustain the bit error rate in the presence of the expected impairments. It is customary to first budget sufficient energy to communicate on an additive gaussian noise channel. This budget provides a reference which can be adjusted if necessary to accommodate consideration of other impairments and uncertainties. We begin the chapter with a discussion of an energy budget.

11.1   Energy budgets

This book deals with modulation and coding for the digital communication of information. It does not deal with the electrophysics of waveform propagation and antenna theory, nor with detailed questions of the design of circuits or amplifiers. Rather, we study the relationship between the noise power and the required power in the signal that reaches the receiver to ensure reliable demodulation.

In this section, however, we would like to touch on the idea of an energy budget. Suppose that a digital communication system transmits R bits per second, and the power in the signal that reaches the receiver is S watts. Then in a long time, T, RT bits are received and the received energy is ST joules. The *energy per bit E_b* is defined as

$$E_b = \frac{ST}{RT}$$
$$= \frac{S}{R}.$$

The energy per bit E_b can also be stated for a message of finite duration. Given a message $m(t)$ of finite message duration T_m containing K information bits, the energy per bit is given by

$$E_b = \frac{E_m}{K}$$

where E_m is the message energy

$$E_m = \int_0^{T_m} m^2(t)dt.$$

The bit energy E_b is not an energy that can be measured directly by a meter. It must be calculated from the message energy and the number of information bits at the input of the encoder/modulator. At a casual glance, one may find a message structure at the input to the channel in which one may perceive a larger number of bits as channel symbols. The extra symbols may be check symbols for error control, or symbols for frame synchronization or some other channel protocol. These other symbols do not represent transmitted information, and their energy must be amortized over information bits. Only information bits are used in calculating E_b. One reason for defining E_b is so that two waveforms that are quite different in construction can be compared.

In addition to the message, the receiver is degraded by noise. Additive white gaussian noise arises in most applications and is also a standard noise against which digital communication systems are judged. White noise has a power density spectrum that is constant with frequency. The two-sided power density spectrum is

$$N(f) = \frac{N_0}{2} \qquad \text{(watts/hertz)},$$

and N_0 is called the (*one-sided*) *noise power density spectrum*, this terminology originating with the notion of power density at frequency $-f$ combined with that at frequency

$+f$. The unit of watts/hertz is actually a measure of energy because it is dimensionally equivalent to the energy unit of joules.

Because white noise is constant over all frequencies, it has infinite power. Consequently, this model of white noise can lead to mathematical and physical absurdities if it is not handled with care. The white noise model, however, is usually a good fit over the range of frequencies of interest and is remarkably useful because of its simplicity. If a baseband channel has an ideal rectangular transfer function

$$H(f) = \begin{cases} 1 & |f| \leq W \\ 0 & |f| > W, \end{cases}$$

and $W = B$, or if a passband channel has an ideal rectangular transfer function[1]

$$H(f) = \begin{cases} 1 & |f + f_0| \text{ or } |f - f_0| \leq W \\ 0 & |f + f_0| \text{ and } |f - f_0| > W \end{cases}$$

and $W = B/2$, then the received noise power N is given by

$$N = (2B)\frac{N_0}{2} = BN_0.$$

In both cases, N is computed by integrating $H(f)N(f)$ from negative infinity to positive infinity.

The noise power N in a band B can be measured. However, N_0 is the more fundamental quantity because it does not presume a definition for the transfer function $H(f)$. We shall usually prefer to express fundamental results in terms of N_0.

On a linear channel, the reception of the signal cannot be affected if both the signal and the noise are doubled. It is only the ratio E_b/N_0 that affects the bit error rate. Usually E_b/N_0 is expressed in decibels, defined as

$$\left(\frac{E_b}{N_0}\right)_{decibels} = 10 \log_{10} \frac{E_b}{N_0}.$$

Two different signaling schemes are compared by a comparison of their respective graphs of bit error rate versus required E_b/N_0, usually on a decibel scale.

We shall frequently quantify the performance of a digital communication waveform by stating a ratio E_b/N_0 at which that waveform can be satisfactorily demodulated. Simple waveforms will require large values of E_b/N_0 to operate satisfactorily, perhaps $E_b/N_0 = 12$ dB or more. On the other hand, sophisticated waveforms may be designed

---

[1] Notice that B is defined as the *full width* of the ideal passband channel centered about f_0, and the *half width* of the ideal baseband channel centered about $f = 0$. This definition yields the formula $N = BN_0$ in both cases, but it opens the possibility of confusion when using the complex baseband representation of the passband channel. On the other hand, W is always a half width of the ideal channel and does not change when a passband channel is converted to a complex baseband representation.

to operate satisfactorily with E_b/N_0 less than 5 dB, and ultimately, even as low as -1.6 dB. This, in fact, may be heralded as one of the great successes of the modern theory of digital communication – the savings of more than a factor of ten in energy by the use of a sophisticated waveform.

For free space propagation of electromagnetic waves, we can calculate the energy per bit at the receiver $E_b = E_{bR}$ in terms of the energy per bit at the transmitter E_{bT} by the range equation

$$E_{bR} = \left(\frac{\lambda^2}{4\pi} G_R \right) \left(\frac{1}{4\pi R_0^2} \right) G_T E_{bT}$$

where G_T and G_R are the gains of the transmitting and receiving antennas in the direction of propagation, R_0 is the range from transmitter to receiver, and λ is the wavelength of the electromagnetic wave. The equation may also be expressed in terms of power rather than of energy by replacing E_{bR} and E_{bT} by S_R and S_T, respectively.

The range equation has been introduced with the terms arranged to tell the story of the energy bookkeeping. Transmitted energy, E_{bT}, appears at the output of the transmitting antenna as the effective radiated energy $G_T E_{bT}$. The energy spreads in spherical waves. The term $4\pi R_0^2$ is the area of a sphere of radius R_0. By dividing $G_T E_{bT}$ by this term, we have the energy per unit area that passes through a surface at a distance of R_0 from the transmitter. The receiving antenna can be characterized simply by its effective area A_r, which depends on the direction from which the antenna is viewed. The antenna collects all of the radiation impinging on a region of area A_r and delivers that energy to the receiver. Characterizing an antenna by its effective area obviates any need here to study the antenna in detail. The relationship between antenna gain and effective area is

$$A_r = \frac{\lambda^2}{4\pi} G_r.$$

The product of these several terms yields the range equation.

Thermodynamics tells us that a receiver at absolute temperature T_R will always contaminate the received signal with additive white gaussian noise of power density spectrum $N_0 = kT_R$ where k is Boltzmann's constant ($k = 1.38 \times 10^{-23}$ joules per degree Kelvin). This noise is called *thermal noise* and, in electronic receivers, is due to unavoidable thermal fluctuations of electrons in the first stage of amplification. Practical receivers will have a somewhat larger value of noise power density spectrum expressed as $N_0 = FkT_R$, where F is a number known as the *noise figure* of the receiver. Thermodynamics requires that F is not less than 1. Ideally, $F = 1$, but in practice it is larger. A noise figure on the order of 3 or 4 dB is typical of a high-quality receiver.

We now can calculate the ratio E_b/N_0 from the data rate and from information provided by the designers of the antenna, the power amplifier in the transmitter, and the first-stage amplifier in the receiver. The ratio E_b/N_0 will prove to be the most meaningful measure of signal strength at the receiver. Using the principles of this book, we must

Table 11.1. *Sample energy budget*

Transmit power	350 kilowatt		
Bit rate	5×10^5		
E_b	7 millijoule (mj)	38.45 dB	(mj)
G_t	1000	30 dB	(isotropic)
Range loss $(4\pi R^2)^{-1}$	$R = 10^5$ meter	-110.99 dB	(m^2)
$\frac{\lambda^2}{4\pi} G_R$	$G_R = 1, \lambda = .1$m	-40.62 dB	(m^2)
		-93.16 dB	(mj)
kT	Room temperature	-114 dB	(milliwatts/per hertz)
Noise figure	4	6 dB	
		-108 dB	(milliwatts/per hertz)
	$(E_b/N_0)_{\text{act}}$	14.84 dB	
	$(E_b/N_0)_{\text{req}}$	8.00 dB	
	Margin	6.94 dB	

design a waveform that can be demodulated at the value of E_b/N_0 seen at the receiver. Denote by $(E_b/N_0)_{\text{act}}$ the value of E_b/N_0 (in decibels) predicted by the range equation, and by $(E_b/N_0)_{\text{req}}$ (in decibels) the value of E_b/N_0 required to demodulate the chosen waveform. The design margin is then given by

$$\text{design margin} = (E_b/N_0)_{\text{act}} - (E_b/N_0)_{\text{req}}.$$

The design margin must be nonnegative. A positive design margin provides insurance against unforeseen circumstances in the operation of the communication system.

One can construct either a power budget or an energy budget; the two are equivalent. The energy budget has the advantage that the figure of merit is E_b/N_0 rather than S/N. To compute received noise power N from N_0 requires that a definition of system bandwidth be entered into the power budget. For an energy budget, however, the system bandwidth, which is clumsy to define, does not enter the calculation and does not even need to be introduced.

A sample energy budget for a communication system is shown in Table 11.1. The calculations are made in terms of decibels so that addition and subtraction replace multiplication and division. Units are annotated to check that the result is dimensionless. For purposes of this example, $(E_b/N_0)_{\text{req}} = 8$ dB has been arbitrarily selected. A major task of this book is to compute $(E_b/N_0)_{\text{req}}$ for many communication waveforms of interest: indeed, to design waveforms so that $(E_b/N_0)_{\text{req}}$ is small.

11.2 Channel capacity

Is it possible to improve the performance of a digital communication system indefinitely by the use of ever more complicated modems? An answer to this question is given by the subject of information theory, and we outline that conclusion in this section.

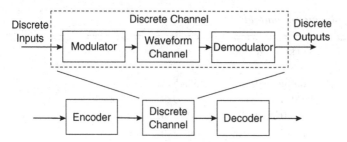

Figure 11.1. Making a discrete channel from a waveform channel.

Any communication system in which a sequence of symbols chosen from a fixed alphabet of symbols $\{s_0, \ldots, s_{M-1}\}$ can be transmitted from one point to another is called a *discrete channel*. A discrete channel is made from a waveform channel by the modem as indicated in Figure 11.1. We saw examples in the previous chapter of modems whose inputs and outputs are streams of bits, and also modems whose inputs and outputs are streams of symbols in some larger alphabet. Thus we may have a discrete binary channel or a discrete M-ary channel.

A *noiseless discrete channel* is one for which the output symbol is completely determined by the input symbol. A *noisy discrete channel*, which is studied in this chapter, is one for which the output symbol is not completely determined by the input symbol; only some probability distribution on the set of output symbols is determined by the input symbol. The noisy binary channel is an important example of a noisy channel.

If the probability distribution on the output symbol is independent of previous inputs or outputs of the channel, the channel is called a *memoryless channel*. The probability of error is independent from symbol to symbol for a memoryless channel. In the context of this book, this means that there is no intersymbol interference, and the noise is independent from sample to sample.

Information can be sent reliably through a discrete noisy channel by the use of an elaborate cross-checking technique known as a data transmission code or an error-control code. A binary data transmission code encodes k data bits into n code bits so as to combat channel errors, as was discussed in Section 10.1. A general data transmission code encodes k data symbols into n code symbols. The ratio k/n is called the *rate* of the code. It is easy to fall into the misconception that a data transmission code is "wasting" some of the channel bits (or symbols) because the channel must convey more code bits than data bits. Actually the data rate is possible only because the error-control code is used. The more enlightened view is that the use of the code makes it possible to achieve performance that would not be possible without a code. The code may allow more bits per second to be run through the channel but then imposes a tax. For a rate one-half code, for example, the tax is equal to half of the channel bits. We would be willing to pay this tax whenever the data transmission code allows the channel bit rate to be increased by more than a factor of two.

Consider an arbitrary noisy waveform channel used with the most general encoder/modulator. To use the channel to transmit a message, a waveform of duration T representing that message is transmitted. The waveform comes from the set of all possible waveforms that can be created by the encoder/modulator. The demodulator/decoder then tries to determine which waveform was sent so that the message can be recovered. Because the channel is noisy, there is a probability of error in recovering the correct message. Let p_{em} denote the average probability of message decoding error averaged over all messages. For this purpose, all messages are assumed to be equally likely. The probability of bit error p_e (or the probability of symbol error) at the output of the demodulator will be less than the probability of message error because not every bit will be wrong in a wrong message. At the output of the demodulator, bit errors may not be independent but may tend to be clustered by the structure of the waveform, so we may also speak of error events. The performance of a code for data transmission is judged by the bit error rate or the message error rate at the decoder output, or by the probability of an error event.

Figure 11.2 shows an arbitrary and very general family of N waveforms of duration T. Any modulation scheme for data transmission can be described in principle by describing, for each T, all of the N waveforms of duration T that it can produce. We would like to design the waveforms so that N is as large as possible so that $\log_2 N$ bits can be transmitted in time T. Clearly, with a largest permissible probability of message error p_{em} specified, N cannot be increased indefinitely. Suppose that $N(T, p_{em})$ is the largest number of waveforms of duration T in any such set that can be distinguished with average probability of error p_{em}. In precise terms, this means that it is possible to construct $N(T, p_{em})$ waveforms of duration T such that the waveforms can be distinguished by the demodulator/decoder with average probability of error p_{em}, and that it is not possible to construct more than $N(T, p_{em})$ waveforms satisfying these requirements. In practice, we have little hope of ever computing $N(T, p_{em})$, but such a function exists in principle. Given a message with k databits, one can assign each of the 2^k messages to one of the $N(T, p_{em})$ channel waveforms provided $2^k \leq N(T, p_{em})$. This means that RT data bits can be transmitted in time T with probability of message error p_{em} provided

$$RT \leq \log_2 N(T, p_{em}).$$

The largest rate at which information can be transmitted reliably through a noisy channel – that is, with very small probability of message error, or of bit error – is called the capacity of that noisy channel. Specifically, we can define the capacity C (in units of bits per second) of a noisy channel operationally by

$$C = \lim_{p_{em} \to 0} \lim_{T \to \infty} \left[\frac{\log_2 N(T, p_{em})}{T} \right].$$

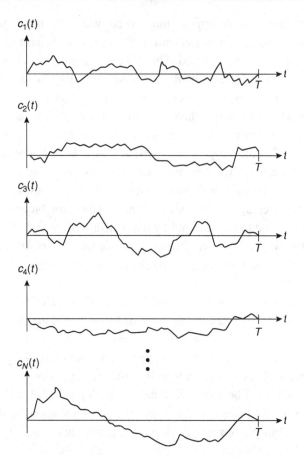

Figure 11.2. A family of waveforms.

It is not at all obvious that C is nonzero. This definition of channel capacity implies that if C is positive, then we can choose any probability of message error p_{em} no matter how small and there will be a T, perhaps very large, such that there are about 2^{CT} waveforms of duration T that will be confused only with probability smaller than p_{em}. Thus, we can send approximately C bits per second at any probability of error we choose, no matter how small. In principle, we can choose the bit error rate to be thousands of times smaller than the probability of failure of the channel or of the communication equipment.

It is not obvious that information can be transmitted reliably through a discrete noisy channel at any nonzero rate, and it is not obvious that the capacity C is larger than zero. A remarkable deduction of information theory tells us that for most channels, C is positive. As long as we are content to signal through the channel at a rate R smaller than the capacity C, the probability of message decoding error p_{em}, and hence the probability of decoded bit error p_e, can be specified arbitrarily small. Good codes exist with arbitrarily small decoded bit error rate and we can choose the code rate R and the decoded bit error rate p_e independently. The required blocklength will depend on these choices. Conversely, no codes of good performance exist if data is to be transmitted at a data rate larger than the channel capacity. If the data rate is larger than the channel capacity, no matter how we encode, the bit error rate will be large. In fact, if we try to transmit data at a high rate, the channel will be sure to make enough errors on average so that the actual rate of data transmission will be smaller than the channel capacity.

The promise of the channel capacity is not without a price; to get small p_{em} one must encode in large blocks of length n or duration T and the cost of such encoders might be prohibitive. Further, for most channels, we do not know practical procedures for designing optimum or nearly optimum codes and encoders of large blocklength, though we do know many procedures that are quite good. The definition of capacity applies just as well to discrete-time channels or continuous-time channels. In the latter case, we would speak of "input waveforms of duration T" and express capacity in units of bits per second, while in the former case, of "input sequences of duration T" (or of blocklength n) and express capacity in units of bits per symbol.

The definition of channel capacity that we have given is not in a form that is practical for computing C. Instead, it asks that we design an optimum communication system for every T and every p_{em}, calculate $N(T, P_{em})$, and then take the indicated limits. It is a major achievement of information theory (which we do not study in this book), that simple formulas for C are obtained and these formulas are obtained in a way that completely sidesteps the enormously difficult problem of optimum waveform design. We shall examine some examples of formulas for the capacity of some channels, but we forgo the general task of deriving these formulas. The binary symmetric channel is a binary channel whose probability of error is independent of whether the symbol zero or one is transmitted, and is memoryless if the probability of error is independent from bit to bit. The capacity (in units of bits per symbol) of a memoryless binary symmetric channel with single bit error probability p_e at the channel output is

$$C = 1 + p_e \log_2 p_e + (1 - p_e) \log_2(1 - p_e)$$

bits per symbol. The binary symmetric channel is the kind of discrete channel that is created by a binary phase-shift keying modulator and demodulator. As viewed from modulator input to demodulator output, BPSK forms a discrete memoryless channel with a probability of bit error p_e. An encoder and decoder surrounding the modulator and demodulator can reduce the probability of bit error as much as desired, provided the rate of the code is not larger than C.

An M-ary orthogonal waveform alphabet creates another kind of discrete channel, known as an M-ary symmetric channel. An M-ary symmetric channel makes independent symbol errors with probability p_e. Thus, a correct symbol is received with probability $(1 - p_e)$ and each of the $M - 1$ wrong symbols is received with probability $p_e/(M - 1)$. The capacity of a memoryless M-ary symmetric channel with single symbol error probability of p_e is

$$C = \log_2 M + p_e \log_2 \frac{p_e}{M - 1} + (1 - p_e) \log_2(1 - p_e)$$

bits per channel symbol. Again, in principle, an encoder/decoder can reduce the bit error rate as much as desired provided the code rate is not larger than C.

11.3   Capacity of gaussian channels

The binary symmetric channel and the M-ary symmetric channel have an input alphabet that is finite. Now, in this section, we shall study channels whose input alphabet is continuous. First we study the important discrete-time, continuous channel with additive and memoryless gaussian noise.[2] Then we study waveform channels, both baseband and passband, with additive gaussian noise.

Let S be the average input power to the additive gaussian-noise channel, and let N be the average noise power. The capacity (in units of bits per sample) of this channel is given by the Shannon capacity formula

$$C = \frac{1}{2} \log_2 \left(1 + \frac{S}{N}\right)$$

in units of bits per channel symbol.

A baseband waveform channel with the ideal rectangular transfer function

$$H(f) = \begin{cases} 1 & |f| \leq W \\ 0 & \text{otherwise} \end{cases}$$

as shown in Figure 11.3, can be converted into a discrete-time channel by using the Nyquist sampling theorem, but with a shift in the point of view. Instead of using

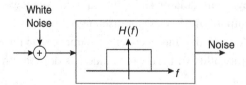

Figure 11.3. Ideal rectangular baseband channel.

---

[2] In discrete time, white noise is also called memoryless noise.

the sampling theorem in its original setting to go from a continuous-time signal to a discrete-time signal, we use the interpolation formula to go from a discrete-time signal to a continuous-time signal. This provides the modulator. The channel adds white gaussian noise that is confined to the ideal passband characteristic by filtering in the receiver. The receiver recreates a discrete-time signal by sampling. There are $2W$ Nyquist samples per second consisting of samples of the input signal contaminated by noise. Because the noise is white gaussian noise, the noise samples are independent. Because the $2W$ noise samples per second are independent, the discrete-time channel is a memoryless noise channel. Hence the Shannon capacity formula can be applied to give the capacity expressed in units of bits per sample. In units of bits per second, it becomes

$$C = W \log_2 \left(1 + \frac{S}{N}\right)$$

because there are $2W$ independent samples per second.

A complex baseband channel also has a capacity. The capacity of the complex baseband channel with an ideal rectangular passband of bandwidth W is

$$C = 2W \log_2 \left(1 + \frac{S}{N}\right)$$

the factor of two occurring because the in-phase and quadrature channels are independent. The ratio S/N refers to the ratio of the signal power per component to the noise power per component. Equivalently, S/N is the ratio of the total signal power to the total noise power.

A similar discussion applies to the passband channel. A passband waveform channel with an ideal rectangular transfer function

$$H(f) = \begin{cases} 1 & |f \pm f_0| \leq W \\ 0 & \text{otherwise} \end{cases}$$

as is shown in Figure 11.4, also can be converted into a discrete-time channel by sampling. Now there are $4W$ Nyquist samples per second – $2W$ in-phase samples per second and $2W$ quadrature samples per second. Taken together, these can be written

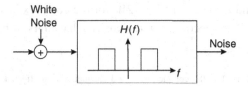

Figure 11.4. Ideal rectangular passband channel.

as $2W$ complex Nyquist samples per second. For the passband channel, the Shannon capacity formula becomes

$$C = 2W \log_2\left(1 + \frac{S}{N}\right)$$

just as for the complex baseband channel.

The Shannon capacity formula for the real baseband channel in additive gaussian noise

$$C = W \log_2\left(1 + \frac{S}{N}\right)$$

leads to precise statements about the values of E_b/N_0 for which good waveforms exist. For an infinite-length message of rate R information bits per second, it is meaningful to define E_b provided the average message power S is constant when averaged over large time intervals. Then

$$E_b = \frac{S}{R}.$$

This expression can be obtained by breaking the message into large blocks, each block of duration T having energy ST and containing RT bits.

Let the signal power be written $S = E_b R$ and let the noise power N be written $N_0 W$. The rate R must be less than the channel capacity C, so the Shannon capacity formula yields the inequality

$$\frac{R}{W} < \frac{C}{W} = \log_2\left(1 + \frac{R E_b}{W N_0}\right).$$

For the corresponding complex baseband channel or for the corresponding passband channel, the inequality is

$$\frac{R}{2W} < \frac{C}{2W} = \log_2\left(1 + \frac{R E_b}{2 W N_0}\right).$$

Define the spectral bit rate density r by

$$r = \frac{R}{B}$$

where $B = W$ for the real baseband channel and $B = 2W$ for the complex baseband channel or the passband channel.[3] The spectral bit rate density r has units of bits per second per hertz.

---

[3] The intuition here is that the passband channel or the complex baseband channel has, in effect, twice the bandwidth of the real baseband channel. When normalized by bandwidth, the capacity of all three cases is the same.

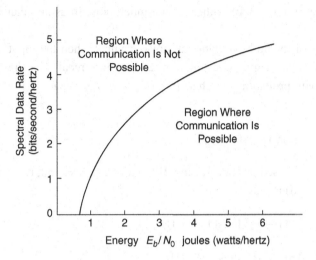

Figure 11.5. Capacity of baseband gaussian noise channel.

The spectral bit rate density r and E_b/N_0 are the two most important figures of merit of a digital communication system. In terms of these two parameters, the inequality above becomes

$$\frac{E_b}{N_0} > \frac{2^r - 1}{r}.$$

By designing a sufficiently sophisticated digital communication system, the ratio E_b/N_0 can be made arbitrarily close to the bound. The inequality, shown in Figure 11.5, tells us that increasing the bit rate per unit bandwidth increases the required energy per bit. This is the basis of the energy/bandwidth trade of digital communication theory, where increasing bandwidth at a fixed information rate can reduce power requirements.

Every communication system for a gaussian noise channel can be described by a point lying below the curve of Figure 11.5. For any point below the curve one can design a communication system that has as small a bit error rate as one desires. The history of the digital communication industry can be described in part as a series of attempts to move ever closer to this limiting curve with systems that have very low bit error rate. The successful systems are those that employ judicious combinations of modulation techniques and data transmission codes.

Waveforms that achieve most of the capacity of the additive gaussian noise channel will use a wide range of input and output amplitude values. If the inputs and outputs are restricted in their values, as in BPSK, then the capacity curves of the additive gaussian noise channel without such a restriction no longer apply. Now the model of the channel must be changed and the capacity recomputed. It will, of course, be smaller because a constraint cannot make the situation better. For an ideal rectangular gaussian channel (used with no intersymbol interference) whose output samples are demodulated to ± 1

(or to $\pm A$), called a hardlimited output, there is not much point in using other than a binary input.

To find the capacity of the additive gaussian noise channel when the input values are constrained to the binary signal constellation ± 1, we first recall that the binary memoryless channel with probability of channel bit error p_e has capacity (in units of bits per symbol)

$$C = 1 + p_e \log_2 p_e + (1 - p_e) \log_2(1 - p_e).$$

In units of bits per second based on transmitting $2W$ symbols per second, the capacity (in units of bits per second) is

$$C = 2W(1 + p_e \log_2 p_e + (1 - p_e) \log_2(1 - p_e))$$

and the spectral bit rate density satisfies $r \leq C/W$.

Because we want this inequality for binary signaling expressed as a function of E_b/N_0, we must express p_e in terms of E_b/N_0. The energy E_b is the average energy per user databit. It is not necessarily equal to the energy per BPSK channel bit because the definition of capacity makes no statement that the data bits are modulated into the BPSK sequence with one data bit becoming one channel pulse. The energy per user bit E_b is related to the energy per BPSK channel pulse E_p by

$$E_p = \frac{R}{2W} E_b = \frac{r}{2} E_b$$

with R the rate in bits per second and $2W$ the number of pulses per second. (If it were possible for every user bit to be a channel bit, and $R/2W$ were equal to 1, then E_p and E_b would be equal. In the presence of noise, however, reliable transmission will require that $r < 2$ and $E_b > E_p$.) The probability of bit error for BPSK is

$$p_e = Q\left(\sqrt{\frac{2E_p}{N_0}}\right)$$

$$= Q\left(\sqrt{r \frac{E_b}{N_0}}\right).$$

Combining this with the inequality

$$r \leq 2(1 + p_e \log_2 p_e + (1 - p_e) \log_2(1 - p_e))$$

for the spectral bit rate density r gives an inequality relationship between r and E_b/N_0.

We can gain another important insight for the additive gaussian noise channel by replotting Figure 11.5 with E_b/N_0 expressed in decibels as given in Figure 11.6. Now

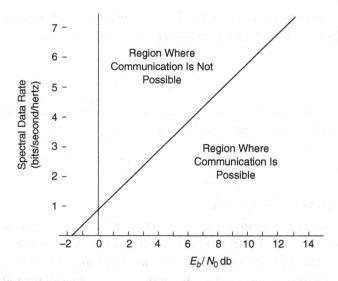

Figure 11.6. Capacity of baseband gaussian noise channel.

the capacity curve appears as nearly a straight line. The reciprocal slope at $r = 6$ is about 2.3 dB/bit per second per hertz. This says that with optimal signaling in the region of six bits per second per hertz, if we increase the spectral bit rate density by one bit per second per hertz, then *all* bits transmitted must have their energy increased by about 2.3 dB. Asymptotically, the slope of this curve approaches one bit per second per hertz per 3 dB, which can be seen by differentiating

$$\left(\frac{E_b}{N_0}\right)_{dB} = \log_{10}\frac{2^r - 1}{r}$$

to get

$$\frac{d}{dr}\left(\frac{E_b}{N_0}\right)_{dB} = \left[\frac{2^r}{2^r - 1} - \frac{1}{r\log_e 2}\right]\log_{10} 2$$

for large r. This means that, when r is large, increasing the spectral bit rate density by one bit per second per hertz requires increasing the energy of every transmitted bit by about 3 dB. Increasing the data rate by increasing the spectral bit rate density is exponentially expensive in power.

If bandwidth W is a plentiful resource but energy is scarce, then one can let W go to infinity. Then the spectral bit rate density r goes to zero. Figure 11.5 shows graphically, however, that E_b/N_0 remains finite, as is reaffirmed by the following theorem.

Theorem 11.3.1 *For an additive white gaussian noise channel,*

$$\frac{E_b}{N_0} \geq \log_e 2 = 0.69.$$

Moreover, for an ideal rectangular additive white gaussian noise channel used without intersymbol interference whose Nyquist samples are demodulated to ± 1,

$$\frac{E_b}{N_0} \geq \frac{\pi}{2} \log_e 2 = 1.18.$$

Proof   Over nonnegative values of r, the function $(2^r - 1)/r$ has its minimum when r is equal to zero. By L'Hopital's rule, it follows that

$$\frac{E_b}{N_0} > \frac{2^r - 1}{r} \geq \log_2 e$$

which proves the first line of the theorem.

To prove the second line of the theorem, we first observe that if the channel consists of an additive gaussian noise channel followed by a hardlimiter, there is evidently no reason to use other than a binary input. Thus, we can start with the formula for the capacity of the binary memoryless channel. Because $R \leq C$, we recall the following inequality:

$$r \leq 2(1 + p_e \log_2 p_e + (1 - p_e) \log_2(1 - p_e))$$

developed earlier. By series expansion of the right side around the point $p_e = \frac{1}{2}$,

$$r \leq 4 \left(p_e - \frac{1}{2} \right)^2 \log_2 e + o\left(\left(p_e - \frac{1}{2} \right)^4 \right)$$

and

$$p_e = Q \left(\sqrt{r \frac{E_b}{N_0}} \right).$$

The $o\left(\left(p_e - \frac{1}{2} \right)^4 \right)$ term will not play a role and we will not need to study it closely. The reason is as follows. The required E_b/N_0 is certainly least when r is very small, because imposing a constraint on the bandwidth cannot possibly decrease the required E_b/N_0. (Otherwise, when bandwidth is large, the optimum waveform would use only part of the bandwidth.) This means that the required E_b/N_0 is least when the probability of error of a channel bit p_e is very close to one-half.

Let $z = rE_b/N_0$, and notice that $Q(z)$ satisfies

$$p_e = \frac{1}{2} - \int_0^z \frac{1}{\sqrt{2\pi}} e^{-x^2/2} dx$$

$$\geq \frac{1}{2} - \frac{z}{\sqrt{2\pi}}.$$

Consequently,

$$r \leq 4 \left(p_e - \frac{1}{2} \right)^2 \log_2 e + o\left(\left(p_e - \frac{1}{2} \right)^4 \right)$$

$$\leq 4 \frac{z^2}{2\pi} \log_2 e + o(z^4)$$

$$= \frac{2}{\pi} r \frac{E_b}{N_0} \log_2 e + o\left(\left(r \frac{E_b}{N_0} \right)^2 \right).$$

As r is made small, the second term on the right can be neglected. In any case it can only lead to a larger lower bound on E_n/N_0.

Therefore,

$$1 \leq \frac{2}{\pi} \frac{E_b}{N_0} \log_2 e$$

and

$$\frac{E_b}{N_0} \geq \frac{\pi/2}{\log_2 e} = \frac{\pi}{2} \log_e 2$$

as was to be proved.   ∎

The second inequality of the theorem, when referenced to the first inequality, says that the energy penalty for using a hard decision in the demodulator is $\pi/2$. The first inequality of the theorem is a fundamental limit. Expressed in terms of decibels, this inequality states that the ratio E_b/N_0 is not less than -1.6 dB for any digital communication system in gaussian noise. On the other hand, for any E_b/N_0 larger than -1.6 dB, one can communicate with as small a bit error rate as desired, but the communication system might be outrageously expensive if one demands an unreasonably small bit error rate. The second inequality of the theorem, which is closely related to the hardlimiter to be studied in Section 11.8, says that E_b/N_0 cannot be less than 0.4 dB if a hard-decision demodulator is used. It is important to recognize here that the 2 dB penalty of the hard decision applies to the case of a discrete-time channel, which may have been created from a continuous-time channel at the Nyquist rate. A continuous-time channel allows zero crossings at any time and, for this reason, our analysis does not apply. A continuous-time channel might not suffer the same 2 dB loss in performance due to a hard decision.

A hardlimiter may be viewed as a quantizer at the output of a channel. We may instead consider quantization at the input to the channel with a continuous alphabet at the output of the channel. A discrete-time channel, real or complex, can be converted into a discrete-alphabet input channel by choosing a finite set of input amplitudes. Such sets of input amplitudes are the signal constellations that were studied in Chapter 2. Figure 2.6

showed some signal constellations for real signaling, and Figure 5.17 showed some signal constellations for complex signaling. The resulting discrete-time waveforms are easily turned into continuous-time passband waveforms for passage through the channel, as by using a Nyquist pulse and the modulation theorem.

When the input amplitude at each sample time is restricted to values in a signal constellation with 2^k points, then we expect that not more than k bits can be transmitted per input symbol; this occurs reliably only when the signal-to-noise ratio is sufficiently high. This means that when the channel input is restricted to take values only in a specific signal constellation, the channel capacity is reduced as compared to the capacity of the additive gaussian noise channel without such a constraint. This is because many possible inputs have been lost, so there are fewer possible waveforms that can be transmitted. The loss in channel capacity is the price paid for the simplification of a signal constellation. We would like to choose the signal constellation so that reduction in channel capacity is inconsequential. The methods of information theory can be used to compute the channel capacity when the input is specified to lie in a given signal constellation. The channel capacity as a function of signal-to-noise ratio, computed numerically, is shown in Figures 11.7 and 11.8 for a variety of signal constellations whose symbols are used equiprobably, as well as the capacity of the unconstrained gaussian channel. For low

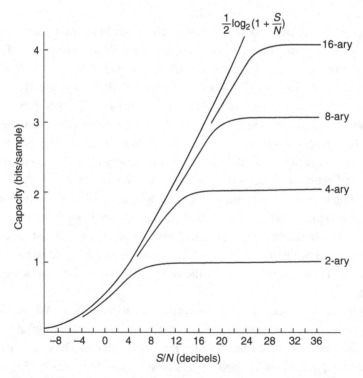

Figure 11.7. Channel capacity for some one-dimensional constellations.

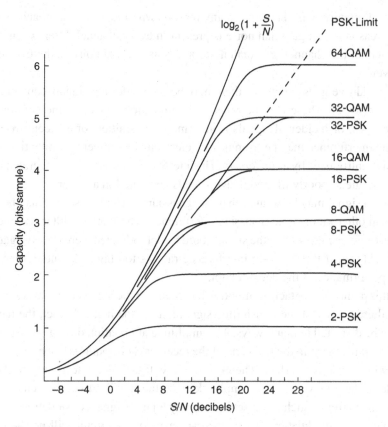

Figure 11.8. Channel capacity for some two-dimensional constellations.

enough signal-to-noise ratio, each signal constellation has a capacity nearly that of the unconstrained channel but falling a little short. The constraint that the symbols are used equiprobably is useful in practice, but the capacity curves would improve slightly if that constraint were removed. Figures 11.7 and 11.8 can be used to determine which signal constellation closely enough achieves the available capacity. For example, by examining Figure 11.8, we see that the capacity of a gaussian passband channel with a 10 dB signal-to-noise ratio is only slightly degraded by choosing the sixteen-QAM signal constellation, but the eight-PSK signal constellation would give up more than 0.5 bits/sample.

11.4 Signal impairments and noise

In common practice, the architecture of a modem is developed initially for an additive gaussian noise channel, and then the design is embellished to protect against other kinds of noise and impairments. There are many reasons why the task is approached in this way. One reason is that gaussian noise is present in every channel. Even when it does not arise within the channel, gaussian noise arises as thermal noise in the front end of the receiver.

Besides additive gaussian noise, there may be other additive signals contaminating the received signal. These may be other communication signals in the environment, perhaps echoes, or incidental signals from remote transmitters or adjacent frequency bands. The interference may be from nearby electronic equipment, or intentional jamming signals introduced by an adversary. The interference may even be generated within the receiver, due to poorly filtered image signals generated in a mixer.

A received signal may be weaker than the gaussian thermal noise and may be much weaker than other kinds of nongaussian interference. The matched filter will increase the signal-to-noise ratio so that the signal could be demodulated were only the gaussian noise present but, if the signal-to-interference ratio is too high, the interference may limit the performance of the demodulator.

Up to this point another tacit assumption has been that the received signal is a faithful, perhaps filtered, copy of the transmitted signal but for the noise. In fact, the received signal can be distorted in many ways. The amplitude attenuation during propagation is offset by amplifier gain in the front end of the receiver. This requires some form of gain control, which is never perfect. Therefore there will be some uncertainty in the true amplitude of the received signal. Amplitude uncertainty will affect the demodulation error in some systems, such as those that employ a large signal constellation.

If a coherent demodulator is used, the carrier recovery circuit will not be perfect because it will recover the carrier only with some residual phase error. There will also be a residual time synchronization error due to the time recovery circuit, with the result that the output of the matched filter will not be sampled at the correct instant. All of these impairments will affect the demodulator's performance. We should recalculate p_e versus E_b/N_0 for the signaling waveform of interest in the presence of whatever significant impairments might be encountered.

Nonlinearities may be present within the receiver as well – some intentional and some not. In this chapter, we shall study the effects of nonlinearities, especially the hardlimiter. Many kinds of interference can be suppressed by placing a hardlimiter or other nonlinearity prior to the matched filter. We shall want to know what effect the nonlinearity has on the signal-to-interference ratio, and also what effect a nonlinearity has on the signal-to-noise ratio in case the interference is not present.

11.5   Amplitude uncertainty

A signal constellation has a scale factor and to demodulate a received symbol to the closest point of the signal constellation requires that the amplitude of the signal be

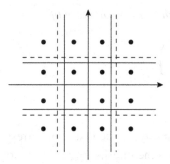

Figure 11.9. The effect of amplitude offset on decision regions.

known so that the received signal can be expressed in the same units as the signal constellation. But the amplitude of the signal entering a receiver is generally not known. It must be estimated from the signal itself. If the amplitude is known, then the amplitude or the scale factor of the decision regions can be adjusted and the demodulator remains optimal with respect to the current value of the scale factor.

There is also the possibility that, while the amplitude varies, the variation is not precisely known to the receiver. Then the decision regions will be misplaced, and the true performance will not be described either by the true E_b/N_0 or by the apparent E_b/N_0. Figure 11.9 shows a severe case for sixteen-ary quadrature-amplitude modulation. The dotted lines show the correct decision regions. However, the construction of actual decision regions is based on the minimum distance to a sixteen-ary QAM constellation with the wrong scale factor. Because the decision regions are wrong, the probability of error will be larger than it should be.

If, instead the example were the sixteen-ary PSK signal constellation, there would be no sensitivity to amplitude uncertainty because in that case the decision regions depend only on phase. Although the demodulator does not depend on the amplitude parameter, the probability of error does.

To illustrate how the probability of bit error is affected by uncertainty in the received amplitude, we shall study the OOK signal constellation. Except for the BPSK signal constellation, which would not adequately illustrate the point, the simplest signal constellation is the binary OOK signal constellation. We shall find the probability of demodulation error for OOK in additive gaussian noise for a pulse whose actual amplitude at the output of the matched filter is the random variable $A = a\bar{A}$, with mean $\bar{A}$ and variance $\bar{A}^2\sigma_a^2$. Intuitively, changing the amplitude A of a pulse by δA should have the same effect as adding noise of the same magnitude δA. This is the substance of the following theorem.

Theorem 11.5.1 *With a binary OOK waveform, whose amplitude is gaussian distributed with mean $\bar{A}$ and variance $\bar{A}^2\sigma_a^2$, by setting the decision threshold at $\bar{A}/2$, it is* *possible to achieve a bit error rate in additive white gaussian noise of*

$$p_e = \frac{1}{2}Q\left(\sqrt{\frac{E_b}{N_0}}\right) + \frac{1}{2}Q\left(\sqrt{\frac{E_b/N_0}{(1 + (2E_b/N_0)\sigma_a^2)}}\right)$$

where $2E_b/N_0 = (\bar{A}/\sigma)^2$.

Proof A data bit zero is transmitted with probability $\frac{1}{2}$. The first term corresponds to the case of a transmitted zero. It depends only on the threshold $\Theta = \bar{A}/2$, not on the actual value taken by A.

A data bit one is transmitted with probability $\frac{1}{2}$. The second term corresponds to the case of a transmitted one. One can calculate this term directly. It is easier, however, to bypass the algebraic manipulations by noticing that increasing A by δA changes the term $A + n$ by the same amount as if the noise instead were increased by δA. In particular, the probability of error should depend only on the sum $\sigma^2 + \bar{A}^2\sigma_a^2$, rather than on either term individually. Consequently, we reach the expression stated in the theorem.   ∎

The performance of binary OOK in the presence of random unknown amplitude fluctuations is shown in Figure 11.10. When E_b/N_0 is large, the expression in

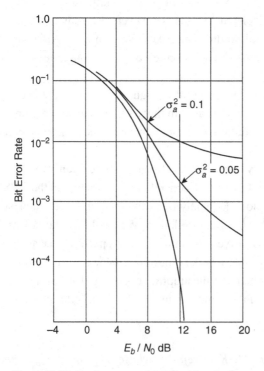

Figure 11.10. Performance of OOK with unknown amplitude fluctuations.

Theorem 11.5.1 is approximated by

$$p_e \sim \frac{1}{2} Q\left(\sqrt{\frac{1}{2}\sigma_a^{-2}}\right),$$

so the performance is completely dominated by the random amplitude fluctuations. There is an important lesson here. When p_e is very small, one should be wary of the elementary performance curves that relate p_e to E_b/N_0 in additive gaussian noise because phenomena that were neglected in deriving the performance curves may be significant.

To set the threshold, the amplitude A must be estimated. If A changes very slowly in comparison to the bit rate, we may estimate A from the received signal. Let

$$v(t) = \sum_{\ell=0}^{n-1} a_\ell s(t - \ell T) + n(t)$$

where $n(t)$ is white gaussian noise, $a_\ell = 0$ or A, and $s(t)$ is such that there is no intersymbol interference at the output of the matched filter. An elementary estimate for A is based on the assumption that half of the databits are ones. This leads to a simple average of the matched-filter outputs

$$\widehat{A} = 2\left[\frac{1}{n}\sum_{\ell=0}^{n-1} u_\ell\right].$$

A better estimator is the maximum-likelihood estimator. The generalized likelihood function for a single bit that is averaged over the two data values is

$$\Lambda_\ell(A) = \frac{1}{2}e^{-(x_\ell - A)^2/2\sigma^2} + \frac{1}{2}e^{-x_\ell^2/2\sigma^2},$$

and the averaged log-likelihood function for the full record is

$$\Lambda(A) = \sum_{\ell=0}^{n-1} \log\left[e^{-(x_\ell - A)^2/2\sigma^2} + e^{-x_\ell^2/2\sigma^2}\right].$$

By factoring out and dropping additive terms that do not depend on A, we can redefine the likelihood statistic as

$$\Lambda(A) = \sum_{\ell=0}^{n-1} \log\left[e^{2x_\ell A/2\sigma^2 - A^2/2\sigma^2} + 1\right].$$

The maximum-likelihood estimate is that value of A for which the derivative of $\Lambda(A)$ is zero. Thus, the estimate $\widehat{A}$ satisfies

$$\sum_{\ell=0}^{n-1}(x_\ell - \widehat{A})\left[\frac{e^{(2x_\ell\widehat{A}-\widehat{A}^2)/2\sigma^2}}{e^{(2x_\ell\widehat{A}-\widehat{A}^2)/2\sigma^2} + 1}\right] = 0.$$

To understand what this equation tells us, observe that the bracketed term is close to one if x_ℓ is close to $\widehat{A}$, and it is close to zero if x_ℓ is close to zero. Thus we can approximate the bracketed term by one or by zero, and so approximate the estimator by

$$\widehat{A} = \frac{1}{n'}\sum_{\widehat{x}_\ell=1}x_\ell$$

where the sum is over only those terms with $x_\ell \geq \widehat{A}/2$, and n' is the number of terms in the sum. This approximation is still an implicit expression for an estimator – which can be solved by a search procedure – but the sense of it is fairly transparent.

For a larger signal constellation, the maximum-likelihood estimator of amplitude can be prohibitively complicated to use. However, one may examine the form of the maximum-likelihood estimator as a vehicle for prompting the design of ad hoc estimators that are practical to implement.

11.6   Phase uncertainty

Any unintentional phase modulation that is introduced into the received signal either by the transmitter, by the propagation medium, or by the receiver is a source of phase uncertainty. This may be due to the motion of the transmit antenna or the receive antenna because of vibration; to phase errors arising in the local oscillators; to residual phase errors arising in the carrier recovery circuitry; or to phase errors due to inhomogeneities in the propagation medium. That portion of the phase error that fluctuates at a rate comparable to or faster than the symbol rate is called *phase noise*.

Phase synchronization was discussed in detail in Chapter 8, as was the variance of the residual phase error. All that remains to discuss is how the residual phase error folds into the probability of demodulation error.

Let θ be a random variable denoting the residual phase error. Usually we will model θ as a gaussian random variable with a zero mean and variance σ_θ^2. When represented at complex baseband the received signal with a phase error is

$$v(t) = e^{j\theta(t)}c(t) + n(t).$$

We can usually assume that the phase error changes slowly enough with time that it can be treated as a constant over each symbol. This means that we can use the approximation

$$v(t) = \sum_{\ell=-\infty}^{\infty} a_\ell e^{j\theta(\ell T)} s(t - \ell T) + n(t).$$

The expected real part of the ℓth output sample of the matched filter is $a_\ell \cos\theta$. The coherent demodulator determines the data symbol as if θ were zero and the probability of error depends on θ. Let $p_{e|\theta}$ denote the probability of symbol error conditional on θ. The expected probability of error is

$$p_e = E[p_{e|\theta}]$$

which will be a function of E_b/N_0 and σ_θ^2.

As an example, we will develop an expression for the probability of error of BPSK in the presence of gaussian-distributed phase noise. Given a phase error of θ, the amplitude of the real part of the matched-filter output is $\pm A \cos\theta$, and E_b is degraded by $\cos^2\theta$. Thus

$$p_{e|\theta} = Q\left(\sqrt{\frac{2E_b}{N_0} \cos^2\theta}\right)$$

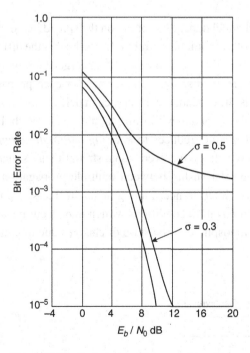

Figure 11.11. Performance loss caused by phase noise.

and

$$p_e = \int_{-\infty}^{\infty} \frac{1}{\sqrt{2\pi\sigma_\theta^2}} e^{-\theta^2/2\sigma_\theta^2} Q\left(\sqrt{\frac{2E_b}{N_0}} \cos^2\theta\right) d\theta.$$

This integral can be evaluated by numerical integration to produce curves such as those shown in Figure 11.11. These curves show that phase noise with variance σ_θ^2 equal to $(0.3)^2$ radians-squared requires about 2 dB larger E_b/N_0 to maintain a bit error rate of 10^{-5}. A phase noise with variance σ_θ^2 equal to $(0.5)^2$ radians-squared cannot maintain a bit error ratio of 10^{-5} with any reasonable increase in E_b/N_0.

11.7 Multipath channels and fading

A *multipath channel* is one in which there are multiple propagation paths, as shown in Figure 11.12. Each path is of a different length, and so each copy of the received signal has a different propagation delay. Multipath arises in free-space propagation because of reflections from objects in the environment, or for some carrier frequencies because of multiple reflection layers in the ionosphere. Multipath is common in two-way communication systems, such as mobile telephone circuits in urban environments, because of multiple echoes caused by varied and unplanned reflections between transmitter and receiver.

The character of the multipath channel depends strongly on the spread in the various delays as compared to the duration of a modulation symbol. It may be that the difference in propagation delay on the various paths is comparable to, or larger than, a symbol duration. Then a single symbol is received several times, once for each propagation path, but overlapping other symbols. If, instead, the difference in delay is much shorter than a symbol duration and the delay changes with time, then a single symbol interferes with itself and the multipath channel reduces to a *fading multipath channel*. A fading multipath channel is one in which the received signal strength varies markedly with time because of the changing relationship between multiple propagation paths causing constructive and destructive interference. Fading arises in free-space propagation because of changes in the propagation conditions, which may occur because of changes in reflection layers in the ionosphere, or because of changes due to motion of the transmitter or receiver.

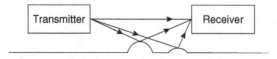

Figure 11.12. A simplified multipath geometry.

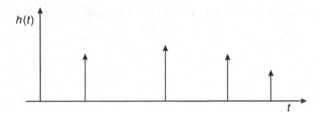

Figure 11.13. Impulse response of a multipath channel.

The impulse response of a multipath channel is of the form

$$h(t) = \sum_{i=1}^{I} h_i(t - \tau_i)$$

where $h_i(t)$ is the impulse response and τ_i is the delay of the ith propagation path. In some cases, the propagation paths may have slowly varying attenuations and slowly varying delays. Then we should write $\tau_i(t)$ in place of τ_i.

Figure 11.13 shows the impulse response of a multipath channel for which each propagation path has an impulse response whose duration is small compared to the relative delay between propagation paths, and each path is modeled as a pure delay.

If the duration of a channel symbol is small compared to the separation between impulses as shown in Figure 11.13, then a multipath channel can be regarded as a kind of unintentional diversity communication channel. To achieve superior performance, it is important to collect and combine the energy of all significant multipath signals. If the multipath response can be learned by the receiver by observation of the received signal, then it may be possible to coherently or noncoherently combine the diversity "fingers" if they can be separated. A receiver that has a response at each multipath component is called a *rake receiver*.

A slowly fading channel is one for which the received signal strength changes slowly in comparison to the symbol rate. The received signal is

$$v(t) = a(t)e^{j\theta(t)}c(t) + n(t)$$

where $a(t)$ and $\theta(t)$ vary slowly in comparison to $c(t)$. We shall consider both coherent and noncoherent demodulators that estimate $a(t)$, and possibly $\theta(t)$, from the received signal. If the phase angle $\theta(t)$ is slowly varying, it can be recovered in the demodulator and stripped from the carrier. Of course, in practice, neither $\theta(t)$ nor $a(t)$ will be estimated precisely, and this will cause further degradation as was studied in Sections 11.5 and 11.6.

Before studying the demodulators, we shall first motivate the study of slowly fading channels by considering a multipath model and the typical circumstances that can lead to a fading channel. A passband multipath channel may be given in the complex baseband representation

$$h(t) = \sum_{i=1}^{I} h_i(t - \tau_i)e^{j\theta_i}$$

where I is very large, and the phase term (or part of the phase term) is $\theta_i = 2\pi f_0 \tau_i$ which arises due to a delay of τ_i on the carrier. If the impulse response and the delays τ_i of each multipath channel is short compared to the duration of the transmitted signal, then we can write this as

$$h(t) = \delta(t) \sum_{i=1}^{I} h_i e^{j\theta_i}$$

where the h_i are random and independent. By reference to the central limit theorem, when I is large we can approximate this by a complex gaussian random variable. Changing this representation of the complex gaussian random variable to an amplitude and phase representation, we have

$$h(t) = Ae^{j\theta}\delta(t)$$

where A is a rayleigh random variable and θ is uniform as was shown in Section 6.4. In the general case the multipath situation is changing with time, and we write the channel output as

$$v(t) = A(t)e^{j\theta(t)}\delta(t)$$

with the assumption that $A(t)$ and $\theta(t)$ are changing slowly in comparison with $c(t)$, but are rayleigh and uniform random variables, respectively, at each t. This means that over one or several channel symbols, the channel can be considered as fixed. This channel is known as a slowly fading rayleigh channel. We may choose to use a phase-recovery technique to estimate and remove the slowly varying $\theta(t)$. We then have a received signal of the form

$$v(t) = A(t)c(t) + n(t),$$

which can be coherently demodulated. Because $A(t)$ is slowly varying, it can be estimated from the received waveform, and each symbol can be demodulated as if $A(t)$ were constant and known. The probability of symbol error then depends on the actual signal-to-noise ratio for that symbol, and on the accuracy of the estimate of $A(t)$.

For example, in the case that $A = a\bar{A}$, where a is a fixed and known attenuation, BPSK has an error probability given by

$$p_{e|a} = Q\left(\sqrt{a^2\frac{2E_b}{N_0}}\right).$$

If, instead, a is a random variable with a probability density function $p(a)$, the average probability of bit error is given by the expectation

$$p_e = E[p_{e|a}]$$

$$= \int_0^\infty p(a)p_{e|a}da.$$

For rayleigh fading, $p(a)$ is a rayleigh probability density function. Then

$$p_e = \int_0^\infty \frac{a}{b^2}e^{-a^2/2b^2}p_{e|a}da$$

where the parameter b should be chosen such that

$$E[a^2E_b] = E_b.$$

This requires that $2b^2 = 1$. Then

$$p(a) = 2ae^{-a^2}.$$

Now we are ready for the following theorem.

Theorem 11.7.1 *The bit error rate of BPSK on a phase-coherent slowly rayleigh fading channel is*

$$p_e = \frac{1}{2}\left[1 - \sqrt{\frac{E_b}{N_0 + E_b}}\right].$$

Proof   Conditional on a, the probability of error is

$$p_{e|a} = \int_0^\infty \frac{1}{\sqrt{2\pi}\sigma}e^{-(x-a\bar{A})^2/2\sigma^2}dx$$

$$= \int_{a\bar{A}}^\infty \frac{1}{\sqrt{2\pi}\sigma}e^{-x^2/2\sigma^2}dx$$

so that, by reference to the discussion prior to the theorem statement,

$$p_e = \int_0^\infty 2ae^{-a^2}\int_{a\bar{A}}^\infty \frac{1}{\sqrt{2\pi}\sigma}e^{-x^2/2\sigma^2}dx\,da.$$

Replace x/σ by y and $a\sqrt{2}$ by z to make the exponents compatible. Then

$$p_e = \int_0^\infty z e^{-z^2/2} \int_{\bar{A}z/\sqrt{2}\sigma}^\infty \frac{1}{\sqrt{2\pi}} e^{-y^2/2} dy \, dz.$$

The integration is over a wedge-shaped region in the y, z plane, which is well-suited to working in polar coordinates. Let

$$y = r \cos \phi$$

$$z = r \sin \phi$$

$$dy \, dz = r \, dr \, d\phi.$$

The lower limit on the y integration is

$$y = \frac{\bar{A}}{\sqrt{2}\sigma} z,$$

which translates into

$$\phi = \tan^{-1} \left(\frac{\sqrt{2}\sigma}{\bar{A}} \right).$$

Therefore

$$p_e = \int_0^\infty \int_0^{\tan^{-1}(\sqrt{2}\sigma/\bar{A})} \frac{1}{\sqrt{2\pi}} r^2 \sin \phi \, e^{-r^2/2} dr \, d\phi,$$

which separates into two elementary integrals that can be evaluated to give

$$p_e = \frac{1}{\sqrt{2\pi}} \left(\frac{\sqrt{2\pi}}{2} \right) \left[1 - \cos \tan^{-1} \left(\frac{\sqrt{2}\sigma}{\bar{A}} \right) \right]$$

$$= \frac{1}{2} \left[1 - \frac{\bar{A}}{\sqrt{2\sigma^2 + \bar{A}^2}} \right]$$

$$= \frac{1}{2} \left[1 - \sqrt{\frac{E_b/N_0}{1 + E_b/N_0}} \right],$$

as was to be proved. ∎

We could have proved Theorem 11.7.1 in a slightly different way by first deriving an unconditional probability density function on x as

$$p(x) = \int_0^\infty p(a)p(x|a)da$$

$$= \int_0^\infty 2ae^{-a^2} \frac{1}{\sqrt{2\pi}\sigma} e^{-(x-a\bar{A})^2/2\sigma^2} da$$

and then integrating $p(x)$ from 0 to ∞.

Theorem 11.7.2 *The bit error rate of a coherently demodulated binary FSK on a slowly rayleigh fading channel is*

$$p_e = \frac{1}{2}\left[1 - \sqrt{\frac{E_b}{2N_0 + E_b}}\right].$$

Proof The proof is nearly the same as the proof of Theorem 11.7.1. ∎

Theorem 11.7.3 *The bit error rate of DPSK on a slowly rayleigh fading channel is*

$$p_e = \frac{1}{2 + 2E_b/N_0},$$

and the bit error rate of noncoherently demodulated binary FSK on a slowly rayleigh fading channel is

$$p_e = \frac{1}{2 + E_b/N_0}.$$

Proof The starting point is the expression for the probability of error of DPSK

$$p_e = \tfrac{1}{2}e^{-E_b/N_0}$$

$$= \tfrac{1}{2}e^{-A^2/2\sigma^2}.$$

Therefore for the proof at hand,

$$p_{e|a} = \tfrac{1}{2}e^{-a^2\bar{A}^2/2\sigma^2}$$

and

$$p_e = \int_0^\infty 2ae^{-a^2}\tfrac{1}{2}e^{-a^2\bar{A}^2/2\sigma^2} da$$

$$= \int_0^\infty ae^{-a^2(1+\bar{A}^2/2\sigma^2)} da$$

$$= \frac{1}{2 + 2E_b/N_0},$$

as was to be proved. The second half of the theorem can be proved by noting that noncoherent FSK requires 3 dB more energy than DPSK.   ∎

Whereas the probability of error on an additive gaussian-noise channel decreases exponentially with E_b/N_0, the probability of error on a slowly rayleigh fading channel goes as $(E_b/N_0)^{-1}$. This is a considerable degradation in performance. To recover, a diversity technique is commonly used. The most important technique is the noncoherent combining of an M-ary FSK diversity waveform. We shall derive the maximum-likelihood diversity combining rule for this case, where we shall need to deal with the probability density function of the magnitude of the matched-filter output. Surprisingly, even though the ricean probability density function cannot be expressed in terms of elementary functions, when averaged over a rayleigh amplitude distribution, the probability density function takes on a simple form, as follows.

Theorem 11.7.4   *For each a, let $p(x|a)$ be the ricean density function*

$$p(x|a) = xe^{-(x^2+a^2)/2}I_0(ax),$$

and let a be a rayleigh random variable. Then $p(x)$ is rayleigh with the density function

$$p(x) = \frac{2x}{3}e^{-x^2/3}.$$

Proof   The unconditional probability density function is

$$p(x) = \int_0^\infty p(a)p(x|a)da$$

$$= \int_0^\infty 2ae^{-a^2}[xe^{-(x^2+a^2)/2}I_0(ax)]da.$$

Substitute the definition of $I_0(z)$ and interchange the order of integration

$$p(x) = \int_0^{2\pi}\int_0^\infty \frac{ax}{\pi}e^{-((x^2-2xa\cos\phi+a^2)/2)-a^2}da\,d\phi.$$

With the change of variables, $a\cos\phi = u$, $a\sin\phi = v$, and $a\,da\,d\phi = du\,dv$, this becomes

$$p(x) = \frac{x}{\pi}\left[\int_{-\infty}^\infty e^{-(3u^2-2ux+x^2)/2}du\right]\left[\int_{-\infty}^\infty e^{-3v^2/2}dv\right].$$

Completing the square and redefining the variable of integration in the first integral gives

$$p(x) = \frac{x}{\pi} e^{-x^2/3} \left[\int_{-\infty}^{\infty} e^{-3v^2/2} dv \right]^2,$$

$$= \frac{2x}{3} e^{-x^2/3}$$

which completes the proof of the theorem. ∎

Theorem 11.7.5   *The maximum-likelihood demodulator for a noncoherently received L-ary diversity, M-ary FSK waveform, with independently fading rayleigh amplitudes in gaussian noise, chooses that m for which*

$$\Lambda(m) = \sum_{\ell=0}^{L-1} u_{m\ell}^2$$

is largest, where $u_{m\ell}$ is the output of the mth matched filter at the ℓth diversity sample.

Proof   Let $p_S(x_{m\ell})$ and $p_N(x_{m\ell})$ be as in the paragraph leading to Theorem 11.7.1. The maximum-likelihood principle leads us to maximize the function

$$\Lambda''(m) = \sum_{\ell=0}^{L-1} \log \frac{p_S(x_{m\ell})}{p_N(x_{m\ell})}.$$

When there is only noise, each output magnitude is described by a rayleigh random variable.

$$p_N(x) = \frac{x}{\sigma^2} e^{-x^2/2\sigma^2}.$$

When there is a rayleigh distributed signal in gaussian noise, the magnitude again is described by a rayleigh random variable, as asserted by Theorem 11.7.4.

$$p_S(x) = \frac{2x}{1+2\sigma} e^{-x^2(1+2\sigma^2)}.$$

Therefore

$$\Lambda''(m) = \sum_{\ell=0}^{L-1} \left[\log \frac{2\sigma^2}{1+2\sigma^2} + \frac{x_{m\ell}^2}{2\sigma^2(1+2\sigma^2)} \right].$$

This is maximized by the same m that maximizes

$$\Lambda(m) = \sum_{\ell=0}^{L-1} x_{m\ell}^2,$$

and the proof is complete. ∎

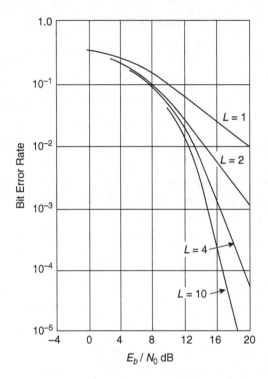

Figure 11.14. Performance of L-ary diversity binary FSK on a rayleigh fading channel.

To evaluate the performance of L-ary diversity signaling on a slowly rayleigh-fading channel, one must compute the probability of demodulation error. The probability of demodulation error is computed initially by finding a probability density function on the decision statistic in the presence of a signal plus noise and in the presence of noise only. This requires an L-fold convolution of probability densities. The probability of error is then written in the form of an equation similar to that appearing in Theorem 6.8.1. Numerical integration gives performance curves, such as those shown in Figure 11.14 for the case $M = 2$, which is binary FSK. The curve labeled $L = 1$ is described by Theorem 11.7.3.

11.8   The hardlimiter at baseband and passband

In applications in which the amplitude of the received signal prior to the matched filter is small compared to the additive gaussian noise, one may elect to pass the received signal through a hardlimiter. The reason for this is that strong impulsive interfering

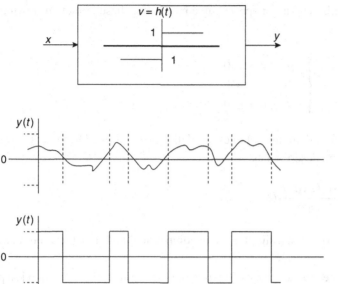

Figure 11.15. The hardlimiter.

signals are greatly attenuated, and as we shall see in some cases, the signal itself is only slightly attenuated.

The hardlimiter is defined by the function

$$y = h(x) = \begin{cases} 1 & x > 0 \\ 0 & x = 0 \\ -1 & x < 0. \end{cases}$$

The equality condition is of little importance, sometimes it is merged with either of the inequality conditions, as by redefining $h(x) = -1$ when $x \leq 0$. Then the hardlimiter can be thought of as a one-bit scalar quantizer.

The hardlimiter is represented symbolically in Figure 11.15. Also shown is an example of the output of the hardlimiter when the input is a baseband waveform. The only attribute of the input waveform that is retained in the output waveform is the location of the zero crossings.

We are interested in the hardlimiter primarily when it is used with passband waveforms as inputs. When restricted in this way, the hardlimiter is called a *passband hardlimiter*. A passband hardlimiter has been introduced intentionally into many kinds of communication receivers, particularly in satellite relays and in jam-resistant communication systems, to prevent a strong signal from masking a weak signal.

There is an indirect way of specifying a hardlimiter based on the following definite integral

$$\int_{-\infty}^{\infty} \frac{\sin(\xi a)}{\xi} d\xi = \begin{cases} \pi & \text{if } a > 0 \\ 0 & \text{if } a = 0 \\ -\pi & \text{if } a < 0, \end{cases}$$

which can be found in most tables of definite integrals. Using this integral, the hardlimiter can be described as

$$y(t) = \frac{1}{\pi} \int_{-\infty}^{\infty} \frac{\sin(\xi x(t))}{\xi} d\xi.$$

This description of the hardlimiter looks complicated and unnatural, but it will prove to be useful.

We are interested in the output of a passband hardlimiter when the input is a passband waveform in additive gaussian noise. Thus

$$x(t) = s(t) + n(t)$$

where both $s(t)$ and $n(t)$ are passband waveforms.

The proof of the next theorem requires reference to tables of definite integrals. Specifically, we shall use the well-known definite integrals

$$\int_0^{\infty} J_0(\xi x) x e^{-x^2/2} dx = e^{-\xi^2/2}$$

where $J_0(x)$ is a Bessel function of the first kind, and

$$\int_{-\infty}^{\infty} \sin(2xy) e^{-x^2} \frac{dx}{x} = \pi \operatorname{erf}(y)$$

where the error function $\operatorname{erf}(y)$ is defined as

$$\operatorname{erf}(y) = \frac{2}{\sqrt{\pi}} \int_0^y e^{-z^2} dz.$$

Theorem 11.8.1 (Reed–Jain Theorem)   *The expected output* $E[y(t)]$ *of a passband hardlimiter when the input is the passband signal* $s(t)$ *in additive stationary passband gaussian noise of variance* N *is given by*

$$E[y(t)] = \operatorname{erf}\left(\frac{s(t)}{\sqrt{2N}}\right)$$

Proof   The proof begins by expressing $y(t)$ as a function of $x(t)$ using the definite integral

$$y(t) = \frac{1}{\pi} \int_{-\infty}^{\infty} \sin[\xi x(t)] \frac{d\xi}{\xi}.$$

When $x(t) = s(t) + n(t)$ is inserted into this integral, the sine of a sum appears. It may be expanded into the sum of products

$$y(t) = \frac{1}{\pi} \int_{-\infty}^{\infty} \sin[\xi s(t)] \cos[\xi n(t)] \frac{d\xi}{\xi} + \frac{1}{\pi} \int_{-\infty}^{\infty} \cos[\xi s(t)] \sin[\xi n(t)] \frac{d\xi}{\xi}.$$

The passband gaussian noise at the hardlimiter input may be expressed as

$$n(t) = r(t) \cos(2\pi f_0 t + \phi(t))$$

where the envelope $r(t)$ and the phase $\phi(t)$ are slowly time-varying random variables having, at each instant, rayleigh and uniform probability densities, respectively.

Substitute the above statement of the noise as the noise term in the previous expression and use the general Fourier expansions

$$\cos(z \cos p) = J_0(z) + 2 \sum_{m=1}^{\infty} (-1)^m J_{2m}(z) \cos 2mp$$

$$\sin(z \cos p) = 2 \sum_{m=0}^{\infty} (-1)^m J_{2m+1}(z) \cos(2m+1)p,$$

in terms of the Bessel functions $J_\ell(z)$. Then the following expression for the output of the hardlimiter is obtained:

$$y(t) = \frac{1}{\pi} \int_{-\infty}^{\infty} \sin[\xi s(t)] \cdot J_0(\xi r(t)) \frac{d\xi}{\xi}$$

$$+ \frac{2}{\pi} \sum_{m=1}^{\infty} (-1)^m \int_{-\infty}^{\infty} \sin[\xi s(t)] J_{2m}(\xi r(t)) \frac{d\xi}{\xi} \cos 2m(2\pi f_0 t + \phi(t))$$

$$+ \frac{2}{\pi} \sum_{m=0}^{\infty} (-1)^m \int_{-\infty}^{\infty} \cos[\xi s(t)] J_{2m+1}(\xi r(t)) \frac{d\xi}{\xi} \cdot \cos(2m+1)(2\pi f_0 t + \phi(t)).$$

All terms in the above equation are random because the noise amplitude $r(t)$ and phase $\phi(t)$ are present. Because all terms in the second two integrals contain a random phase, those two integrals will not contribute to the average and can be dropped when taking the expectation of $y(t)$. Only the first integral will yield an average output. Therefore

$$E[y(t)] = \frac{1}{\pi} \int_{-\infty}^{\infty} \sin[\xi s(t)] E[J_0(\xi r)] \frac{d\xi}{\xi}.$$

The expected value of $J_0(\xi r)$ is given by

$$E[J_0(\xi r)] = \int_0^\infty J_0(\xi r)p(r)dr$$

where $p(r)$, the probability density function of the noise amplitude, is a rayleigh density function

$$p(r) = \frac{r}{\sigma^2}e^{-r^2/2\sigma^2}.$$

Therefore

$$E[J_0(\xi r)] = \int_0^\infty J_0(\xi r)\frac{r}{\sigma^2}e^{-r^2/2\sigma^2}dr$$

$$= e^{-\frac{1}{2}\xi^2\sigma^2}$$

by the first definite integral given prior to the statement of the theorem. Substitution into the expression for $E[y(t)]$ gives

$$E[y(t)] = \frac{1}{\pi}\int_{-\infty}^\infty \sin[\xi s(t)]e^{-\frac{1}{2}\xi^2\sigma^2}\frac{d\xi}{\xi}$$

$$= \mathrm{erf}\left(\frac{s(t)}{\sqrt{2N}}\right)$$

by the second definite integral given at the start of the proof. This completes the proof of the theorem. ∎

When the signal is weak compared to the noise, the output of the hardlimiter may be described more simply. We shall establish a model in which the signal $s(t)$ passes through the hardlimiter attenuated but otherwise unchanged. The noise, on the other hand, is dispersed in several ways, as shown in Figure 11.16. The noise power is clustered at the harmonics of f_0. At each harmonic, part of the noise power appears in the form of a scaled replica of the original noise signal, and part of it is smeared out in a more complicated way.

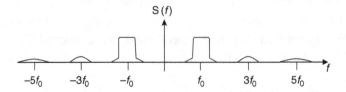

Figure 11.16. Spectrum of hardlimited passband white gaussian noise.

Corollary 11.8.2 *A weak passband signal $s(t)$ in additive passband gaussian noise has its signal-to-noise ratio reduced by $2/\pi$ when passed through a passband hardlimiter. That is,*

$$\left(\frac{S}{N}\right)_{out} \cong \frac{2}{\pi}\left(\frac{S}{N}\right)_{in}.$$

Moreover, if the hardlimiter output is passband-filtered to its original bandwidth, then

$$\left(\frac{S}{N}\right)_{out} > \frac{\pi}{4}\left(\frac{S}{N}\right)_{in}.$$

Proof The error function is defined as

$$\mathrm{erf}(x) = \frac{2}{\sqrt{\pi}} \int_0^y e^{-x^2} dx.$$

When x is small, the integrand is approximately equal to one, which leads to the approximation

$$\mathrm{erf}(x) \approx \frac{2}{\sqrt{\pi}} x.$$

Therefore, the expected value of the output of the hardlimiter, by Theorem 11.8.1 is

$$\mathrm{E}[z(t)] = \sqrt{\frac{2}{\pi}} \frac{s(t)}{\sigma},$$

so the output power is $2/(\pi\sigma^2)$ times the input power. Because the noise power at the hardlimiter output is essentially the same as the total output power – which equals one – the first part of the corollary is proved.

The second statement is a consequence of how the noise power is distributed on the frequency axis. With $s(t)$ equal to zero, the input of the passband hardlimiter is passband gaussian noise, which can be put in the form

$$n(t) = r(t) \cos[2\pi f_0 t + \phi(t)].$$

The hardlimiter discards $r(t)$ and turns the phase-modulated cosine waveform into a phase-modulated square waveform. Let

$$\xi = f_0 t + \frac{1}{2\pi}\phi(t).$$

The hardlimiter output $n_0(t)$ is a square wave in the variable ξ. Using the Fourier series expansion of a square wave in ξ gives

$$n_0(t) = \frac{4}{\pi} \sum_{k=0}^{\infty} \frac{(-1)^k \cos(2k+1)2\pi\xi(t)}{2k+1}$$

$$= \frac{4}{\pi} \sum_{k=0}^{\infty} \frac{(-1)^k \cos(2k+1)(2\pi f_0 t + \phi(t))}{2k+1}.$$

If f_0 is large, then the spectrum of the first term does not overlap with the spectra of the other terms. All terms but the first can be rejected by a passband filter that is wide enough to pass only the first harmonic. Then the expression for $n_0(t)$ is given by

$$n_0(t) = \frac{4}{\pi} \cos[2\pi f_0 t + \phi(t)].$$

Because the average of $\cos^2 x$ equals $\frac{1}{2}$, the average output noise power in the first harmonic is

$$N_0 = \frac{1}{2}\left(\frac{16}{\pi^2}\right).$$

Because of the phase modulation $\phi(t)$, part of the noise power in the first harmonic will actually lie outside the original passband. This part of the noise also will be rejected, so the output noise satisfies the strict inequality

$$N_{out} < \frac{8}{\pi^2}.$$

Combining this inequality with the first part of the corollary proves the second part of the corollary. ∎

Next, we look deeper into the effect of the passband hardlimiter on the spectrum of the gaussian noise. For this purpose, we take the input to the hardlimiter to be noise only. We can anticipate something of the behavior. By rewriting the noise $n(t)$ in two parts,

$$n(t) = n'(t) + n''(t),$$

where $n''(t)$ is the portion of the noise lying in a narrow-frequency interval about an arbitrary frequency f, and $n'(t)$ is the remainder of the noise. If the frequency interval is small, then $n''(t)$ is small compared to $n'(t)$. We can regard $n''(t)$ to be the signal and apply Theorem 11.8.1 and Corollary 11.8.2. They say that $n''(t)$ passes through the filter attenuated. Because this is true for every frequency component of $n(t)$, we conclude that, in some sense, the output of the hardlimiter preserves some sort of attenuated replica of $n(t)$, though contaminated by other terms.

The next several theorems will provide a more precise description of the relationship between the output and the input of a hardlimiter excited by gaussian noise.

Theorem 11.8.3 (Van Vleck–Middleton Theorem) *If the input to a hardlimiter is a zero-mean stationary gaussian process $x(t)$ with autocorrelation function $R_{xx}(t)$, then the output has the autocorrelation function*

$$R_{yy}(\tau) = \frac{2}{\pi} \sin^{-1}\left[\frac{R_{xx}(\tau)}{R_{xx}(0)}\right].$$

Proof We begin with the representation

$$y(t) = \frac{1}{\pi} \int_{-\infty}^{\infty} \sin[\xi x(t)] \frac{d\xi}{\xi}.$$

Then

$$R_{yy}(\tau) = E[y(t)y(t+\tau)]$$

$$= \frac{1}{\pi^2} E \int_{-\infty}^{\infty} \int_{-\infty}^{\infty} \sin[\xi x(t)] \sin[\eta x(t+\tau)] \frac{d\xi \, d\eta}{\xi \, \eta}.$$

Use Cauchy's formula to expand each sinusoid into two terms

$$R_{yy}(\tau) = -\frac{1}{4\pi^2} E \int_{-\infty}^{\infty} \int_{-\infty}^{\infty} [e^{j\xi x(t)} - e^{-j\xi x(t)}][e^{j\eta x(t+\tau)} - e^{-j\eta x(t+\tau)}] \frac{d\xi \, d\eta}{\xi \, \eta}.$$

Expanding the product in the integrand gives four terms, and the expectation can be distributed across these four terms. We shall work through the expectation of one of these terms. Let

$$M(\xi, \eta) = E[e^{j\xi x(t)+j\eta x(t+\tau)}].$$

This can be recognized as a bivariate characteristic function. Because $x(t)$ and $x(t+\tau)$ are jointly gaussian, the bivariate characteristic function can be readily evaluated. It has the form of a two-dimensional Fourier transform, and can be evaluated as such. It is

$$M(\xi, \eta) = e^{-(\xi^2 + 2\rho\xi\eta + \eta^2)/2}$$

where ρ is the correlation coefficient of $x(t)$ and $x(t+\tau)$.

There will be four such terms, but only two different terms. They are equal in pairs. The expression for $R_{yy}(\tau)$ becomes

$$R_{yy}(\tau) = -\frac{1}{2\pi^2} \int_{-\infty}^{\infty} \int_{-\infty}^{\infty} [e^{-(\xi^2+2\rho\xi\eta+\eta^2)/2} - e^{-(\xi^2-2\rho\xi\eta+\eta^2)/2}] \frac{d\xi}{\xi} \frac{d\eta}{\eta}$$

$$= \frac{1}{2\pi^2} \int_{-\infty}^{\infty} \int_{-\infty}^{\infty} e^{-(\xi^2+\eta^2)/2} [e^{\rho\xi\eta} - e^{-\rho\xi\eta}] \frac{d\xi}{\xi} \frac{d\eta}{\eta}.$$

Change the integral of a difference into the difference of two integrals. Then in the second integral, change the dummy variable ξ to $-\xi$. The second integral changes sign and is now seen to be the same as the first. Then the two combine to give

$$R_{yy}(\tau) = \frac{1}{\pi^2} \int_{-\infty}^{\infty} \int_{-\infty}^{\infty} e^{-(\xi^2+2\rho\xi\eta+\eta^2)/2} \frac{d\xi}{\xi} \frac{d\eta}{\eta}.$$

To integrate this, we use the method of differentiating under the integral sign. Let $I(\rho)$ denote the integral as a function of ρ, and notice that $I(0)$ equals zero. The derivative of $I(\rho)$ is

$$\frac{dI(\rho)}{d\rho} = \frac{1}{\pi^2} \int_{-\infty}^{\infty} \int_{-\infty}^{\infty} e^{-(\xi^2+2\rho\xi\eta+\eta^2)/2} d\xi \, d\eta.$$

This is in the form of the integral of a two-dimensional gaussian density function and would be equal to one if normalized properly. Hence by inspection of the required normalizations, we find that

$$\frac{dI(\rho)}{d\rho} = \frac{2}{\pi} \frac{1}{\sqrt{1-\rho^2}}.$$

Consequently,

$$I(\rho) = \frac{2}{\pi} \int_0^\rho \frac{dx}{\sqrt{1-x^2}}$$

$$= \frac{2}{\pi} \sin^{-1} \rho,$$

which, because $\rho = R_{xx}(\tau)/R_{xx}(0)$ completes the proof of the theorem. ∎

This theorem and the next theorem are companions: Theorem 11.8.3 gave the autocorrelation at the output of the hardlimiter, while Theorem 11.8.4 gives the cross-correlation between the input and the output.

Theorem 11.8.4 *If the input to a hardlimiter is a stationary, unbiased gaussian process with autocorrelation function $R_{xx}(\tau)$, then the cross-correlation function between input and output is given by*

$$R_{xy}(\tau) = \sqrt{\frac{2}{\pi}} \frac{R_{xx}(\tau)}{\sqrt{R_{xx}(0)}}.$$

Proof We begin with the representation

$$y(t) = \frac{1}{\pi} \int_{-\infty}^{\infty} \sin[\xi x(t)] \frac{d\xi}{\xi}.$$

Then

$$R_{xy}(\tau) = E[x(t+\tau)y(t)]$$

$$= \frac{1}{\pi} E \int_{-\infty}^{\infty} [x(t+\tau)] \sin[\xi x(t)] \frac{d\xi}{\xi}.$$

Use Cauchy's formula to expand the sine into two terms

$$R_{xy}(\tau) = \frac{1}{2j\pi} E \int_{-\infty}^{\infty} x(t+\tau)[e^{j\xi x(t)} - e^{-j\xi x(t)}] \frac{d\xi}{\xi}.$$

Expanding the product gives two terms: we will work through the expectation of one of these terms by recalling the bivariate characteristic function

$$M(\xi, \eta) = E[e^{j\xi x(t) + j\eta x(t+\tau)}],$$

we can write

$$\left.\frac{\partial M(\xi, \eta)}{\partial \eta}\right|_{\eta=0} = E[jx(t+\tau)e^{j\xi x(t)}],$$

which has the form of the integrand above. But we know that

$$M(\xi, \eta) = e^{-(\xi^2 + 2\rho(\tau)\xi\eta + \eta^2)/2}$$

where $\rho = \rho(\tau)$ is the correlation coefficient of $x(t)$ with $x(t+\tau)$. Therefore

$$jE[x(t+\tau)e^{j\xi x(t)}] = -\rho(\tau)\xi e^{-\xi^2/2}.$$

Because there are two such terms, the expression for $R_{xy}(\tau)$ becomes

$$R_{xy}(\tau) = \frac{1}{\pi} \int_{-\infty}^{\infty} \rho(\tau)e^{-\xi^2/2} d\xi$$

$$= \sqrt{\frac{2}{\pi}}\rho(\tau) = \sqrt{\frac{2}{\pi}} \frac{R_{xx}(\tau)}{R_{xx}(0)}.$$

This completes the proof of the theorem. ∎

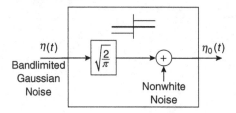

Figure 11.17. Second-order model of hardlimited gaussian noise.

The picture that emerges from Theorems 11.8.3 and 11.8.4 is shown in Figure 11.17. Up to second-order moments, we can model the effect of the hardlimiter on gaussian noise $x(t)$ in terms of the equation

$$y(t) = \sqrt{\frac{2}{\pi}} \frac{x(t)}{\sqrt{R_{xx}(0)}} + e(t)$$

where $e(t)$ is a noise term uncorrelated with $x(t)$. We call it "self-noise" because it is noise created by the signal $x(t)$. To validate this equation, correlate both sides with $x(t)$, then with $y(t)$. This gives the correlation functions

$$R_{xy}(\tau) = \sqrt{\frac{2}{\pi}} \frac{R_{xx}(\tau)}{\sqrt{R_{xx}(0)}} + E[e(t)x(t+\tau)]$$

and

$$R_{yy}(\tau) = \frac{2}{\pi} \frac{R_{xx}(\tau)}{R_{xx}(0)} + \sqrt{\frac{2}{\pi R_{xx}(0)}} E[e(t)x(t+\tau) + x(t)e(t+\tau)] + E[e(t)e(t+\tau)].$$

The first equation and Theorem 11.8.4 lead to the conclusion that

$$E[e(t)x(t+\tau)] = 0$$

and the expression for $R_{yy}(\tau)$ becomes

$$R_{yy}(\tau) = \frac{2}{\pi} \frac{R_{xx}(\tau)}{R_{xx}(0)} + E[e(t)e(t+\tau)].$$

Comparing this equation to Theorem 11.8.3 gives the autocorrelation function of the self-noise

$$E[e(t)e(t+\tau)] = \frac{2}{\pi} \left[\sin^{-1} \frac{R_{xx}(\tau)}{R_{xx}(0)} - \frac{R_{xx}(\tau)}{R_{xx}(0)} \right].$$

This model is meaningful up to second moments. Even though $e(t)$ is uncorrelated with $x(t)$, it is not independent of $x(t)$. In fact, $e(t)$ is completely determined by $x(t)$.

The power density spectrum of the self-noise is obtained by taking the Fourier transform of the autocorrelation function. Using the Taylor series

$$\sin^{-1} x - x = \frac{1}{6}x^3 + \frac{3}{40}x^5 + \frac{5}{112}x^7 + \cdots,$$

we see that the autocorrelation function of the self-noise is

$$R_{ee}(\tau) = \frac{2}{\pi}\left[\frac{1}{6}R_{xx}(\tau)^3 + \frac{3}{40}R_{xx}(\tau)^5 + \frac{5}{112}R_{xx}(\tau)^7 + \cdots\right]$$

where the function has been normalized so that $R_{xx}(0) = 1$. Then the power density spectrum is

$$S_{ee}(f) = \frac{2}{\pi}\left[\frac{1}{6}S(f)^{(*3)} + \frac{3}{40}S(f)^{(*5)} + \frac{5}{112}S(f)^{(*7)} + \cdots\right],$$

with the notation defined by

$$S(f)^{(*n)} = S(f) * S(f) * \cdots * S(f)$$

where there are n copies of $S(f)$ on the right and $S(f)$ is the Fourier transform of $R_{xx}(f)$. This is the background that is behind Figure 11.16. The rearrangement of the passband noise spectrum at the input $S(f)$ to form the noise spectrum at the output, comprised of $S_{yy}(f)$ and $S_{ee}(f)$, is illustrated in Figure 11.18. Other terms at harmonics of f_0 are not shown here, but were shown in Figure 11.16. The total power density spectrum near f_0 is shown in Figure 11.19. If the output is filtered back to its original passband, the small tails of $S_{ee}(f)$ will be suppressed.

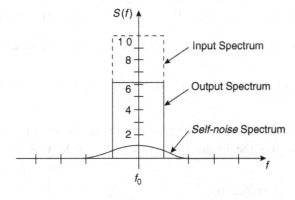

Figure 11.18. Rearrangement of spectrum at the output of a hardlimiter.

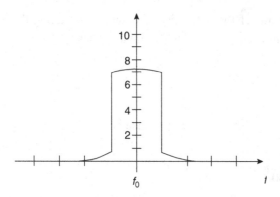

Figure 11.19. Passband spectrum at the output of a hardlimiter.

11.9   Smooth nonlinearities

A *smooth nonlinearity* $h(x)$ is a continuous function that operates on the instantaneous value of the input waveform $x(t)$ to produce the output $y(t) = h(x(t))$. Among the smooth nonlinearities are monotonic nonlinearities, such as softlimiters and companders, and nonlinearities that are not monotonic, such as square law devices. We shall examine the effect that a nonlinearity can have on a pulse $s(t)$ or on the pulse spectrum $S(f)$. Generally, the nonlinearity will change the way the energy is distributed on the frequency axis, placing energy in regions of the frequency axis where there was none prior to the nonlinearity. This is sometimes loosely referred to as energy "splattered" out of the original band by the nonlinearity. The first principle we will develop shows, however, a sense in which this newly created signal spectrum is uninformative.

Theorem 11.9.1   *Let $h(x)$ be a strictly increasing function of x such that $h[s(t)]$ has finite energy whenever $s(t)$ does. Suppose that $s_1(t)$ and $s_2(t)$ are finite energy baseband pulses whose spectra are zero only at frequencies satisfying $|f| \le W$. If $h(s_1(t))$ and $h(s_2(t))$ have spectra that are equal for $|f| \le W$, then $s_1(t)$ equals $s_2(t)$ for all t.*

Proof   Let $r_1(t) = h(s_1(t))$ and $r_2(t) = h(s_2(t))$. By Parseval's theorem, we can write

$$\int_{-\infty}^{\infty} [R_1(f) - R_2(f)][S_1^*(f) - S_2^*(f)]df = \int_{-\infty}^{\infty} [r_1(t) - r_2(t)][s_1^*(t) - s_2^*(t)]dt.$$

The left side is zero because the first factor of the integrand is zero when $|f| \le W$, and each term in the second factor of the integrand is zero when $|f| > W$. Consequently,

$$\int_{-\infty}^{\infty} [r_1(t) - r_2(t)][s_1^*(t) - s_2^*(t)]dt = 0.$$

But $h(x)$ is a strictly increasing function, so the first factor of the integrand has the same sign as the second factor, thus the integrand is never negative. Because the integral equals zero, the integrand must be zero. But both factors are nonzero if $s_1(t)$ does not equal $s_2(t)$. Therefore $s_1(t)$ equals $s_2(t)$.   ∎

The point of the theorem is this: obviously because h is strictly increasing, when $s_1(t) \neq s_2(t)$, then $h(s_1(t)) \neq h(s_2(t))$. What is not so obvious, but is asserted by the theorem, is that $h(s_1(t))$ and $h(s_2(t))$ can be passband filtered to the original bandwidth and will still be different. The difference in the outputs of the filtered pulses, however, need not be as large as the difference in the original pulses. Although a monotone nonlinearity, followed by a passband filter, may reduce the euclidean distance between two distinct waveforms in that passband, it can never reduce that distance to zero. In additive gaussian noise, the effect of the nonlinearity can be described as an effective loss in E_b/N_0. In gaussian noise, it must be a loss, in general, rather than a gain because the maximum-likelihood demodulators do not call for any nonlinearity. Determining the magnitude of this loss is rarely analytically tractable, if ever; it usually requires a computer simulation to determine.

The effect of a smooth nonlinearity on a passband signal is also of interest. This effect often can be described separately for the modulation and the carrier. The outgoing signal of any smooth passband nonlinearity, such as a softlimiter, can be described in principle by a method slightly similar to the Van Vleck–Middleton theorem. The smooth nonlinearity affects both the modulation and the carrier, the latter by creating harmonics. Let the smooth nonlinearity $h(x)$ have input $x(t) = A\cos(2\pi f_0 t + \theta)$. Because the input is periodic, the output will be periodic. Therefore, the output can be expanded in a Fourier series, which must have the form

$$h(x(t)) = \sum_{\ell=0}^{\infty} h_\ell(A)\cos(2\pi \ell f_0 t + \ell\theta)$$

where $h_\ell(A)$ is the ℓth Fourier coefficient, which is a function of A. More generally when the amplitude and phase are varying slowly in comparison with $\cos 2\pi f_0 t$,

$$x(t) = A(t)\cos(2\pi f_0 t + \theta(t)),$$

and we have the quasi-stationary approximation

$$h(x(t)) = \sum_{\ell=0}^{\infty} h_\ell(A(t))\cos(2\pi \ell f_0 t + \ell\theta(t)).$$

In general, it would be tedious to compute the $h_\ell(A)$. However, for integer power-law nonlinearities, standard trigonometric identities allow a closed-form expression. For example, if $h(x) = x^3$, we can state explicitly

$$h(x(t)) = A^3(t) \cos^3(2\pi f_0 t + \theta(t))$$

$$= \frac{3}{4}A^3(t) \cos(2\pi f_0 t + \theta(t)) + \frac{1}{4}A^3(t) \cos 3(2\pi f_0 t + \theta(t)).$$

The passband signal at f_0 can be filtered from the other harmonics if f_0 is large. The effect of the cubic nonlinearity on the passband signal is almost as if the amplitude only were passed through the cubic nonlinearity, and then modulated. Other odd-power nonlinearities of passband waveforms behave similarly. Even-power nonlinearities, however, have no output modulation component at frequency f_0.

The quasi-stationary approximation regards the passband waveform as narrow in frequency with respect to f_0, and the nonlinearity affects the carrier and the modulation separately. The cubic nonlinearity does expand bandwidth because $A(t)^2$ has a wider bandwidth than $A(t)$, not because the dominant carrier broadens.

This approximation breaks down if the passband waveform is not narrow. This is demonstrated by considering the case of two sinusoids. When the input to a smooth linearity is the sum of two passband signals, the output of the nonlinearity may be predominantly the sum of two passband signals. Consider the passband signal

$$x(t) = x_1(t) + x_2(t)$$

$$= a_0(t) \cos 2\pi f_0 t + a_1(t) \cos 2\pi f_1 t.$$

The cubic nonlinearity will form cubes of the terms $x_1(t)$ and $x_2(t)$ individually, which can be treated as before, and cross terms such as

$$3x_1(t)^2 x_2(t) = (a_0(t)^2 \cos^2 2\pi f_0 t)(a_1(t) \cos 2\pi f_1 t)$$

$$= \frac{3a_0(t)^2 a_1(t)}{2}\left[\cos 2\pi f_1 t + \frac{1}{2}\cos 2\pi(2f_0 + f_1)t + \frac{1}{2}\cos 2\pi(2f_0 - f_1)t\right].$$

If f_0 and f_1 are not too different, the new carrier at $2f_0 - f_1$ will lie in the modulation band.

Problems for Chapter 11

11.1. a. Is the performance of an optimal coherent demodulator for M-ary orthogonal signaling affected by amplitude variations in the received signal (slow variations compared to the symbol duration)? Does the answer depend on whether the variations are known to the demodulator? Explain.

b. Repeat for a noncoherent demodulator.

11.2. a. Suppose that a coherent demodulator for M-ary orthogonal signaling contains a noisy reference. The phase error θ_e is modeled as a zero-mean gaussian random variable of variance σ_θ^2. Assuming that $\sigma_\theta^2 \ll 1$, set up an integral for p_e as a function of E_b, N_0, and σ_θ^2.

   b. For $M = 2$, express p_e using the error integral $Q(x)$.

11.3. Give an ad hoc estimator for the amplitude of a coherent binary FSK waveform in additive gaussian noise when it is known that the amplitude does not change during a block of n bits. Find the maximum-likelihood estimator for estimating amplitude and demodulating data simultaneously. Find the maximum-likelihood estimator for estimating amplitude only with data treated as a random nuisance parameter and averaged out. How do these three estimators compare?

11.4. a. Prove Theorem 11.5.1 by a direct argument as follows. Begin with the equations

$$p_{e|a} = \frac{1}{2}\int_{A/2}^{\infty} \frac{1}{\sqrt{2\pi}\sigma}e^{-x^2/2\sigma^2}\,dx + \frac{1}{2}\int_{-\infty}^{A/2}\frac{1}{\sqrt{2\pi}\sigma}e^{-(x-a)^2/2\sigma^2}\,dx$$

$$p_e = \int_{-\infty}^{\infty}\frac{1}{2\pi A\sigma_a}e^{-(aA-\bar{A})^2/2\bar{A}2\sigma a^2}\,da$$

   and evaluate p_e by interchanging integrations, combining exponents, and completing the square in the exponent.

   b. Is it possible to redefine the threshold to reduce the average bit error rate?

11.5. Give a tight approximation for the probability of bit error for a sixteen-ary square signal constellation if the signal amplitude is multiplied by a gaussian random variable a of mean one and variance σ_a^2, with σ_a much smaller than one.

11.6. Let x be the magnitude (normalized by $(2N_0)^{\frac{1}{2}}$) of the matched-filter output of a pulse output of a rayleigh fading channel. The probability density function of x is

$$p(x) = \frac{x}{1 + 2b^2(E_p/N_0)}e^{-x^2/(2+4b^2E_p/N_0)}.$$

   a. Let $y = x^2$. Show that the probability density function of y is

$$p(y) = \frac{1}{1 + 2b^2(E_p/N_0)}e^{-y/(1+2b^2E_p/N_0)}$$

   when the signal is present, and

$$p(y) = e^{-y}$$

   when the signal is absent.

b. Given L independent diversity samples with statistic $z = \sum_{\ell=1}^{L} x_\ell^2$, show that

$$p(z) = \frac{z^{L-1}}{(L-1)!(1+2b^2(E_p/N_0))} e^{-z/(1+2b^2 E_p/N_0)}$$

when the signal is present, and

$$p(z) = \frac{z^{L-1}}{(L-1)!} e^{-z}$$

when the signal is absent.

c. Show that the bit error rate of noncoherently combined L-ary diversity binary FSK on a slowly fading rayleigh channel is given by

$$p_e = \sum_{\ell=0}^{L-1} \frac{(-1)^\ell (2L-1)!}{\ell!(L-1)!(L-1-\ell)!(L+\ell)} \left(2 + 2b\frac{E_p}{N_0}\right)^{-L-\ell}.$$

Verify that this reduces to the correct equation when $L = 1$.

11.7. In contrast to the abrupt transition of a hardlimiter, a softlimiter makes a gradual transition between the limits of ± 1. One possible choice of amplitude characteristic for the softlimiter is the error function

$$y = \text{erf}\left(\frac{x}{\sqrt{2\alpha}}\right).$$

Prove that when a signal $s(t)$ in gaussian noise of variance σ^2 is applied to this softlimiter, the expected output is

$$E[y(t)] = \text{erf}\left[\frac{s(t)}{\sqrt{2(\sigma^2 + \alpha^2)}}\right].$$

11.8. Prove that whereas a weak signal in the presence of gaussian noise is attenuated with respect to the noise by 1 dB by a passband hardlimiter, a weak signal in the presence of an interfering pure sinusoid is attenuated with respect to the sinusoid by 6 dB by a passband hardlimiter. Specifically, let

$$v(t) = \cos 2\pi f_1 t + a \cos 2\pi f_0 t$$

where $\cos 2\pi f_1 t$ is the interfering sinusoid, $a \cos 2\pi f_0 t$ is the signal, and a is small compared to one. Show that the input to the hardlimiter is

$$v(t) = \sqrt{(1 + a\cos 2\pi(f_0 - f_1)t)^2 + (a\sin 2\pi(f_0 - f_1)t)^2} \cos(2\pi f_1 t + \phi(t)).$$

The output of the hardlimiter, when filtered to the passband, is approximately

$$v(t) = \cos(2\pi f_1 t + \phi(t))$$

$$= \frac{\cos 2\pi f_1 t + a \cos 2\pi f_0 t}{\sqrt{1 + a^2 + 2a \cos 2\pi (f_0 - f_1)t}}$$

$$\approx \cos 2\pi f_1 t + \frac{a}{2} \cos 2\pi f_0 t - \frac{a}{2} \cos 2\pi (2f_1 - f_0)t.$$

If a is replaced by a pulse, $as(t)$, describe what happens at the output of a filter matched to the passband pulse $\widetilde{s}(t)$.

11.9. A diversity communication system sends the same BPSK waveform through two additive gaussian-noise channels. A malicious source of interference, called a *jammer*, randomly chooses one of the two channels independently from bit to bit and inserts additional gaussian noise into that channel. Therefore the outputs of a pair of matched filters on the pair of channels both have a mean output of $\pm A$, depending on the value of the data bit, and have variances N_0, $N_0 + J_0$ or $N_0 + J_0$, N_0, depending on which channel the jammer chooses.

    a. What is the maximum-likelihood decision rule for the ℓth data bit in terms of the outputs of the two matched filters at time ℓT, denoted x_ℓ and y_ℓ?

    b. Write down a likelihood function for the ℓth data bit with the jammer state averaged out. What is the decision rule maximizing this function, given that $J_0 \gg N_0$?

    c. Now suppose that there are three diversity channels and the jammer randomly chooses one of the three to jam at each bit time. What is the maximum-likelihood decision rule?

    d. Suppose that there are two diversity channels but that the jammer state remains the same for a block of n data bits. What is the likelihood function for the jammer state averaged over data? What is the estimate for the jammer state that maximizes this function?

11.10. A diversity communication system uses two diversity channels and binary FSK signaling. The two channels have identical but unknown phase errors. Sketch an optimal demodulator and give an expression for the probability of bit error as a function of E_b/N_0, given that both channels are functioning and are of equal amplitude.

11.11. a. Given that random variable X has a probability density function $p(x)$, what is the probability density function for X^2?

    b. Given that random variables X_ℓ have probability density functions $p(x_\ell)$, what is the probability density function for $\sum_{\ell=0}^{L-1} X_\ell^2$?

    c. Write down the set of equations that must be numerically integrated to compute the probability of bit error as a function of E_b/N_0 for an eight-ary orthogonal signaling waveform with 4-ary noncoherent diversity and square-law combining.

11.12.   Find the maximum-likelihood demodulator for a noncoherently received L-ary diversity OOK waveform with independent rayleigh fading amplitudes in additive gaussian noise.

11.13.   Let $\epsilon > 0$ be given. Show that, for any d, one can find a nonlinearity $h(x)$ and two pulses $s_1(t)$ and $s_2(t)$ satisfying Theorem 11.9.1, such that

$$\int_{-\infty}^{\infty} (S_1(f) - S_2(f))^2 df = d$$

and

$$\int_{-W}^{W} (R_1(f) - R_2(f))^2 df < \epsilon$$

where $r_1(t)$ and $r_2(t)$ are the outputs of the nonlinearity.

11.14.   A mixer, used to change carrier frequency from f_0 to f_1, is constructed from a nonlinear element

$$y = a_0 + a_1 x + a_2 x^2 + a_3 x^3$$

by setting

$$x(t) = v(t) + \cos 2\pi (f_0 - f_1)t$$

and filtering out the component of $y(t)$ at frequency f_1. Suppose $f_0 = 100\,\text{MHz}$, $f_1 = 10\,\text{MHz}$ and $v(t)$ has bandwidth equal to 1 MHz.
   a. Sketch the spectrum of $y(t)$ and specify the filter.
   b. Suppose that $v(t)$ is contaminated by an interfering signal at frequency f_0'. At what values of f_0' will there be a problem of interference at the intermediate frequency?

11.15.   A satellite transponder may contain a nonlinearity that will require extensive numerical computation to study it completely. The flavor of such a problem may be appreciated, however, by looking at a simple discrete-time problem. A discrete-time BPSK waveform with intersymbol interference is

$$c_k = a_k + 0.1 a_{k-1}$$

(where $a_k = \pm 1$) and is contaminated by additive gaussian noise and passed through a hardlimiter

$$v_k = \text{sgn}\,[c_k + n_k]$$

where n_k is memoryless zero-mean gaussian noise with variance $\sigma^2 = 1$.

a. Is a decision-feedback demodulator a suitable demodulator? Is the Viterbi algorithm suitable?

b. What is the maximum-likelihood demodulator?

c. Construct an example of a data sequence and a received sequence and show how the data sequence can be recovered by a recursive procedure.

11.16. A coherent 32-ary diversity communication channel contains a hardlimiter prior to the matched filter in each diversity channel and the signal energy is divided evenly among the channels.

a. Suppose that all channels are equally noisy and that the signal is less than the noise at the hardlimiter input. What is the penalty in E_b/N_0 because of the presence of the hardlimiters?

b. Now suppose that the noise is twice as strong on sixteen of the diversity channels as on the others. By how much must E_b be increased to compensate? Compare this to the case in which there is no hardlimiter.

c. Finally, suppose that the noise is eight times as strong on four of the diversity channels as on the others. By how much must E_b be increased to compensate? Compare this to the case in which there is no hardlimiter.

Notes for Chapter 11

The analysis of the performance degradation due to phase and amplitude imbalance is a standard calculation and was surveyed by Franks (1980). The method known as a rake receiver for dealing with multipath was demonstrated by Price and Green (1958) based on decision-theoretic fundamentals. The optimal diversity combining technique for a rayleigh fading channel was derived by Pierce (1958). Price (1954, 1956) had also studied diversity combining. Cheun (1997) calculates the bit error rate of a rake receiver in various applications.

The effect of a hardlimiter on gaussian noise is important in communication systems, in radar systems, and in control systems. It has been well-studied, starting with the work of Davenport (1953) and of Van Vleck and Middleton (1966). The Van Vleck–Middleton formula can also be obtained as a consequence of Price's theorem (1958). The effect of a hardlimiter on a signal in gaussian noise was studied by Reed (1958) and Jain (1972). The cross-correlation between the input and output of a nonlinear device was studied by Bussgang (1952). The theorem that a monotonic nonlinearity cannot reduce the distance to zero between two waveforms within a passband was given by Beurling and reported by Landau (1960). Continuing work on the study of the effect of nonlinearities includes contributions by Manasse, Price, and Lerner (1958); Cahn (1961); Jones (1963); Blachman (1964, 1971); and Davisson and Milstein (1972). Robust methods for signal processing were studied by Kassam and Poor (1985).

The deep insight that every stationary channel is described by a constant, called its *capacity*, is one of the many great contributions of Shannon (1948). The formula for the capacity of the additive gaussian noise channel is a fundamental formula of information theory, and is derived in most textbooks of that subject.

Berger (1966) compared the performance of a linear equalizer to performance bounds of information theory. Price (1972) extended this comparison to decision-feedback demodulation, thereby establishing the intrinsic superiority of that technique.

12 Secure Communications

The communication problem can be given a new dimension of complexity by the introduction of an adversary. The adversary may have a variety of goals. The goal may be to interrupt communication, to detect the occurrence of communication, to determine the specific message transmitted, or to determine the location or the identity of the transmitter. The communication problem now takes on aspects of the theory of games. The transmitter and receiver comprise one team while the adversary comprises the other team.

An adversary may try to interrupt communication by falsifying the messages or by inserting noise into the channel. In the former case, the adversary is called a *spoofer* while in the latter case, the adversary is called a *jammer*. An adversary who intends to read the specific message transmitted is called a *cryptanalyst*. An adversary who intends to determine the location or the identity of the transmitter or to detect the occurrence of communication is called a *signal exploiter*.

Waveform techniques to counter a jammer or an exploiter are similar; both try to spread the waveform over a wide bandwidth. Such waveforms are called *antijam* waveforms or *antiexploitation* waveforms. Techniques to counter a spoofer or a cryptanalyst tend to be similar: these may use a secret permutation on the set of messages to represent the actual message by a surrogate message formed in an agreed, invertible way based on a secret key. Techniques to counter a spoofer are called *authentication* or *signature verification*.

12.1 The jammer channel

The jammer channel is a continuous-time waveform channel in which the received signal is given by

$$v(t) = c(t) + n(t) + e(t)$$

where $n(t)$ is additive noise and $e(t)$ is a waveform chosen maliciously by an adversary who intends to disrupt communication. The jammer channel occurs in military and political situations. The jammer selects a noiselike waveform from the set of waveforms at its disposal; usually, in theoretical analyses, the only limitation on the jammer's waveform is average transmitted power.

The jamming problem is not meaningful unless the resources available to each side are specified. If both sides have unlimited resources, then a meaningful solution to the problem cannot be found because the opponents will continually increase, without limit, the sophistication of their respective equipment and their transmitted power. If resource limitations are specified, then the problem has a meaningful solution and will usually be expressed in terms of a maximum data rate under these constraints that the optimal communicator can maintain in the presence of the optimal jammer. For the purpose of waveform design, the jammer's resources are usually defined as the average jammer power measured at the receiver; the transmitter's resources are defined as the average signal power measured at the receiver. The transmitter usually is also constrained in the total occupied bandwidth.

The jammer always has the option of transmitting gaussian noise. It can also transmit other kinds of noise waveforms, including random pulses and tones. We can view the problem in two ways. In the first case, which is the best case for the designer of the communication system, the jammer first selects the noise waveform. We then find the best communication waveform to penetrate that form of noise. In the second case, which is the worst case for the designer of the communication system, we first select the communication waveform. The jammer then selects a noiselike waveform that is the most disruptive to that communication waveform. The communicator's strategy is to turn the second case into the first case by designing the system to make the worst-case jammer tactic the least disruptive. This assumption, that the jammer always uses the optimum jamming waveform, is a premise that underlies all of the theory, although real jammers need not be so diabolical.

Information theory gives us the strong statement, without conditions, that white gaussian noise is the most difficult noise to communicate through, in the sense that the achievable bit rate at a given power level is least. Therefore, in the first case, to reduce capacity, the jammer will select broadband gaussian noise. In the second case, the jammer has the option of transmitting broadband gaussian noise. However, the optimal jammer may select a more damaging jamming tactic based on its knowledge of the communication waveform. The designer tries to construct a waveform such that, even though the waveform is known, the jammer can find no tactic more damaging than white gaussian noise.

The *jammer saddle point* is the name for the equilibrium condition given by that pair of strategies wherein the transmitter waveform is so clever that it forces the jammer to use white gaussian noise as its optimum strategy, and the jammer is wise enough to realize that this is the optimum strategy left open to it, and so it uses gaussian noise. The jammer saddle point provides a convenient point of reference to describe the quality of an antijam system. The vulnerability of an antijam waveform is judged by the ratio of the power that the optimum jammer needs to jam the system to the power that a white gaussian-noise jammer needs to jam the system. The waveform design is deficient to the extent that this ratio is less than one, as it will be for any practical waveform.

The method of spread-spectrum signaling, to be studied in Section 12.3, is central to most antijam communication systems. The motivation can be developed by recalling that, for any fixed signaling technique in white gaussian noise, the probability of symbol error depends on E_b/N_0. Consequently, when the jammer transmits white gaussian noise of two-sided power density spectrum $J_0/2$, the probability of symbol error depends on E_b/J_0. To improve performance, one can increase E_b, decrease J_0, or use a waveform allowing demodulation at smaller E_b/N_0. To increase E_b, one can increase the effective transmitted power or reduce the data rate. To decrease J_0, one can use spread spectrum. Spread spectrum is that class of waveform design techniques that takes a "low" bandwidth signal and converts it to a high bandwidth signal in an invertible way that is not known to the jammer. The jammer must then cover this wide bandwidth with its fixed jammer power J. Hence the jammer power density spectrum J_0 is reduced by the ratio of the bandwidths.

The most pessimistic assumption affecting the design of an antijam communication system is that the jammer will have full knowledge of the design of the transmitter and receiver. It will only lack knowledge of the secret key controlling the generation of certain parameters such as the value of a pseudorandom sequence. This jammer will attack the receiver at its most vulnerable point. Instead of a direct attack trying to mask the output of the matched filters, the jammer may try to disrupt the carrier synchronization or time synchronization functions, or it may try to exploit a weakness introduced by a nonlinearity in the receiver. Indeed, every special defensive mechanism or circuit introduced into a receiver to counter one jammer tactic needs to be examined to ensure that it has not introduced a new vulnerability to a different jammer tactic.

Nonlinearities are always present in a practical receiver. Phase and time synchronization circuits will always contain nonlinear elements. The demodulator may include nonlinear elements either because of unavoidable limitations of the dynamic range, or because the designer has some other specific reason for including them. To counter certain jammer tactics, such as partial-message jamming or pulse jamming, a nonlinearity may be introduced. Therefore the communication system designer must always be wary of any vulnerability to new jammer tactics created by the nonlinearity, and must ensure that the jammer cannot find a new strategy that attacks an unrecognized vulnerability.

12.2   Partial-message jamming

A *partial-message jammer* works on the principle that, in most applications, destroying only a small part of a message invalidates the entire message. The jammer may not have enough average power to jam the entire message, but if it concentrates the available power on part of the message, the jammer may be able to destroy that part. It is pointless to inquire about whether this is a sensible strategy for the jammer. The design specification for most jam-protected communication equipment will specify a maximum bit error rate. The communication system must be designed to meet that specification for any jammer tactic, limited only by jammer average power. In particular, the system must operate when the jammer uses partial-message jamming.

A simple tactic for the partial-message jammer is to transmit bandlimited white gaussian noise with power density spectrum J_0/ρ in the communications band during a fraction ρ of the time,[1] and to transmit nothing during the remaining time. On average, the power density spectrum of the jammer is still J_0. The receiver sees a total noise power density spectrum of $N_0 + J_0/\rho$ for a fraction ρ of the time, and N_0 for the remaining time. The parameter ρ is called the *duty factor* of the jammer.

We know that, for BPSK in additive white gaussian noise, the probability of bit error is given by

$$p_e = Q\left(\sqrt{\frac{2E_b}{N_0}}\right).$$

Similar expressions have been derived in Chapter 3 for other signaling waveforms, and the following discussion is easily modified to apply to these as well.

The average probability of error in the presence of a partial-message jammer is obtained by replacing N_0 by $N_0 + J_0/\rho$ during the time when the jammer is on. (We are ignoring as inconsequential the consideration that the jammer may turn on midway through a bit interval.) Therefore, for BPSK,

$$p_e = (1 - \rho)Q\left(\sqrt{\frac{2E_b}{N_0}}\right) + \rho Q\left(\sqrt{\frac{2E_b}{N_0 + J_0/\rho}}\right).$$

When N_0 is small in comparison to J_0, we can write

$$p_e \approx \rho Q\left(\sqrt{\frac{2E_b\rho}{J_0}}\right).$$

The intelligent jammer will maximize this expression by choice of ρ. By numerical methods, we can find that the function $zQ(z^{\frac{1}{2}})$ has its maximum value of 0.1657 at $z = 1.44$. Therefore, provided that it is not larger than one, the maximizing choice of

---

[1] Technically, the power density spectrum is not defined for a nonstationary waveform. To avoid imprecision, we could replace the jammer signal by $w(t)e(t)$ where $w(t)$ is zero or one, and $e(t)$ is stationary noise.

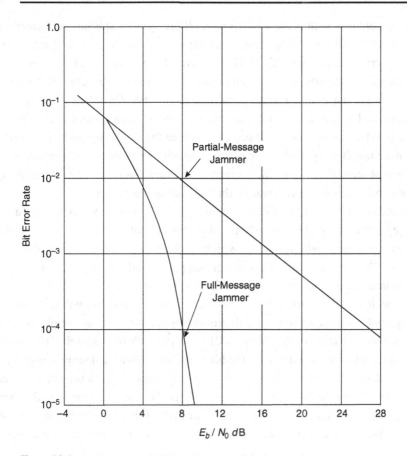

Figure 12.1. Performance of PSK against a partial-message jammer.

ρ is $1.44/(2E_b/J_0)$ and

$$p_e = \frac{0.1657}{2E_b/J_0}.$$

This expression goes as $(E_b/J_0)^{-1}$. The probability of error decreases much more slowly than it does in white gaussian noise, which decrease is asymptotically exponential in E_b/N_0. A comparison is shown graphically in Figure 12.1. At a bit error rate of 10^{-5}, the partial-message jammer needs to use only a thousandth of the average power of a full-message jammer.[2]

The tactic of partial-message jamming gives the jammer an enormous advantage against simple BPSK and, for similar reasons, against most other simple waveforms. The communication waveform designer must consider the strategy of partial-message jamming and must try to build a defensive mechanism into the waveform design. The

---

[2] However, there comes a point at which the average jammer power may not be a meaningful limitation on a jammer if the duty factor is so small that the peak jammer power is unreasonably large.

two basic defenses within the waveform itself are symbol splitting and error-control codes. Symbol splitting is a diversity technique, repeating the same symbol L times, as was described in Section 7.5. To be effective, the symbol energy must be split into distinctly separated parts of the waveform but recombined in the receiver in such a way that there is not more than a small penalty in required E_b/N_0 when the noise is white and gaussian. The goal of symbol splitting is to recover each symbol with a probability of error governed primarily by the copies that have the largest signal-to-jamming ratio, and to have nearly the performance of a system without symbol splitting whenever the jammer is absent. This forces the jammer to divide its noise power so that nearly all symbol copies receive approximately the same jamming power.

To evaluate the performance of a symbol-splitting diversity system against a partial-message jammer, we will idealize the problem. Specifically, we assume that the receiver can always recognize without failure which symbols are contaminated by the jamming signal. We then say that the demodulator has perfect side information. For a partial-message jammer, the jamming signal is either strong or absent, so the approximation is a good one. In practice, there is a small probability that the receiver will fail to recognize that a symbol is jammed, but this small probability is ignored in our simplified analysis.

The simplest diversity waveform is BPSK with each bit transmitted L times in a pseudorandomly scrambled way in the bit sequence so that the jammer cannot predict where the L copies of the same bit will appear. To demodulate a bit, the receiver will examine the L received copies of the bit, identify the jammed copies, and coherently sum the matched-filter output of all copies of that bit that are not jammed, as shown in Figure 12.2. The detection decision is made once on the sum of the matched-filter outputs.

Let the L' unjammed copies of the bit under discussion be at times $k_1 T, k_2 T, \ldots, k_{L'} T$. Then the L' pulses comprising these L' copies can be considered to form a single composite pulse, given by

$$(\pm a)s_c(t) = (\pm a) \sum_{\ell=1}^{L'} s(t - k_\ell T).$$

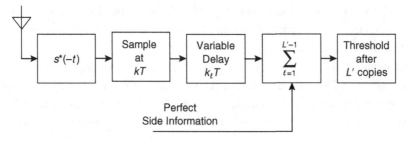

Figure 12.2. Elementary combining of diversity BPSK.

This can be regarded as a single pulse in white gaussian noise and demodulated as such. The demodulator passes the received copy through the filter matched to $s_c(t)$ and detects the sign of the output at $t = 0$. The appropriate sampled output of the filter $s_c^*(-t)$ is

$$u_c = \int_{-\infty}^{\infty} v(\xi)s_c^*(\xi)d\xi$$

$$= \int_{-\infty}^{\infty} v(\xi)\sum_{\ell=1}^{L'} s^*(\xi - k_\ell T)d\xi$$

$$= \sum_{\ell=1}^{L'} \int_{-\infty}^{\infty} v(\xi)s^*(\xi - k_\ell T)d\xi$$

$$= \sum_{\ell=1}^{L'} u(k_\ell T).$$

This tells us that the output of matched filter $s_c^*(-t)$ can be realized by summing the appropriate output samples of filter $s^*(-t)$. When there is no jammer, all copies of the pulse will be integrated, and the demodulation decision will use all of the energy dedicated to that bit. Each copy of that bit has energy E_b/L and there are L copies. Consequently, there will be no loss in E_b/N_0 when there is no jammer.

When some of the matched-filter outputs $u(k_\ell T)$ are known to be jammed, then it is appropriate to discard those outputs or to attenuate their contribution to the sum. Assuming for the moment that the hypothetical receiver has perfect side information describing the jammer tactic, we may use for the decision statistic the weighted sum

$$u_c(0) = \sum_{\ell=1}^{L'} \frac{N_0}{N_\ell} u(k_\ell T)$$

where N_ℓ is the total power due to the jammer plus noise at the ℓth copy of the bit. In the simplest case

$$N_\ell = \begin{cases} N_0 + J_0/\rho \\ N_0 \end{cases}$$

according to whether or not the ℓth copy of the bit is jammed. If J_0/ρ is large compared to N_0, we have the approximation

$$u_c(0) \approx \begin{cases} \sum_{\text{unjammed}} u(k_\ell T) \\ \sum_{\ell=1}^{L'} \frac{N_0}{N_0+J_0/\rho} u(k_\ell T), \end{cases}$$

according to whether some copies are unjammed or all are jammed. This approximation makes it easy to write down an approximation to the probability of symbol error.

Let ρ be the probability that a given copy of the bit is jammed. The probability of jamming ℓ copies is $\binom{L}{\ell} \rho^{\ell} (1 - \rho)^{L-1}$. When ℓ copies are jammed, the received bit energy is reduced to $E_b(L - \ell)/L$. Therefore

$$p_e = \sum_{\ell=0}^{L-1} \binom{L}{\ell} \rho^{\ell} (1 - \rho)^{L-\ell} Q\left(\sqrt{\frac{2E_b(L - \ell)/L}{N_0}}\right) + \rho^L Q\left(\sqrt{\frac{2E_b}{N_0 + J_0/\rho}}\right)$$

where the last term accounts for the case in which all L copies are jammed. If N_0 is negligibly small compared to J_0/ρ, the probability of error can be approximated as

$$p_e \approx \rho^L Q\left(\sqrt{\frac{2E_b\rho}{J_0}}\right).$$

For fixed L, E_b, and J_0, the partial-message jammer will choose ρ to maximize p_e. To find the maximum of p_e, we need to maximize the function $z^L Q(\sqrt{z})$. For each L, this function has a unique maximum that can be found numerically or graphically. The probability of error versus E_b/J_0 is shown in Figure 12.3. This figure tells us that, within

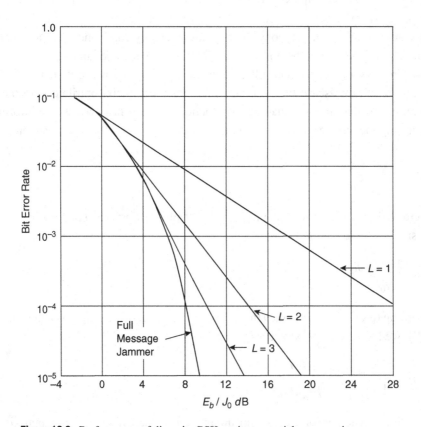

Figure 12.3. Performance of diversity PSK against a partial-message jammer.

the limits of the approximation, four-way or five-way symbol splitting will ensure that a partial-message jammer is not much worse than a full-message jammer.

The performance, shown in Figure 12.3, is an approximation because the thermal noise N_0 is neglected and because the side information is assumed to be perfect. Practical methods for combining diversity pulses must estimate the needed side information from the received signal itself. The estimate should be robust with respect to variations in the jammer signal. Otherwise, the jammer may attempt a back-door attack on the demodulator by trying to create misleading side information.

Diversity transmission can be thought of as a degenerate kind of error-control code. Such a code is known as a *repetition code*. If one adopts this view, it must be regarded as an $(L, 1)$ block code demodulated by using full (soft) side information in the form of likelihood statistics on the unjammed symbols and erasures on the jammed symbols.

One can also choose to use a general (n, k) block code, but usually in the role of an outer code. The individual codeword symbols are demodulated as such and errors or erasures may occur in the symbols that are severely jammed. Erasures correspond to symbols that are known to be jammed, errors correspond to symbols that are jammed; but are thought not to be.

12.3   Bandwidth expansion

Bandwidth expansion is at the heart of most antijam communication waveforms, and spread spectrum is the most common method of bandwidth expansion. Spread spectrum achieves its performance by forcing the jammer to cover a wider spectrum than necessary, thereby diluting the jammer's power density spectrum. Merely spreading the bandwidth of the communication waveform is not sufficient. The spreading must be done in such a way that the jammer cannot mimic the spreading in its waveform. Thus there must be a secret function associated with the spreading waveform; otherwise the spread-spectrum waveform will have no jamming advantage against an optimum jammer.

In this section, we shall explore the relationship between bandwidth expansion and channel capacity, and then describe some general methods of bandwidth expansion. In the next two sections, we shall study in some detail specific methods of spreading the spectrum.

The performance of any digital communication waveform in gaussian noise depends on E_b/N_0. The required value of E_b/N_0 depends on the chosen modulation technique, but once the modulation technique is fixed, the pulses can be redesigned in many ways without changing E_b. In particular, the pulse can be chosen to have a very large bandwidth in comparison to the data rate because the energy in a pulse and the bandwidth of the pulse can be chosen separately. The reason for choosing a pulse with a large

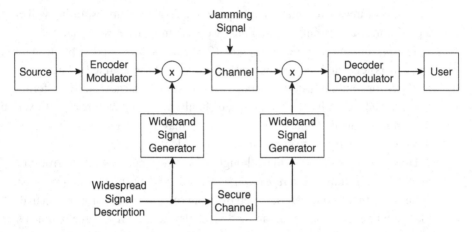

Figure 12.4. Spread-spectrum signaling.

bandwidth is to force the jammer to occupy a large bandwidth also. If J_0 is the power density spectrum of a white noise jammer that has a fixed total power J, then J_0 is made smaller by increasing the bandwidth W because $J_0 = J/W$. This is the essential idea of the spread spectrum strategy. It reduces J_0 by increasing W when J is fixed.

The term "spread spectrum" describes any of a class of techniques that take a signal of small bandwidth B and converts it into a signal of large bandwidth W where the ratio W/B is much larger than one. As shown in Figure 12.4, this is done by modulating the information waveform with a wideband noiselike waveform that is unknown to the jammer. The wideband signal is transmitted through the channel, received, compressed back into the original low bandwidth signal by demodulating the information waveform from the wideband waveform, and filtering. This scheme requires that the transmitter and the receiver share knowledge of the spreading waveform through a secure (but not necessarily contemporaneous) channel. If the jammer does not know the wideband waveform, its strategy is to use gaussian noise of bandwidth W as the jamming signal. In any case, after the signal spectrum is compressed, the jammer signal will still have bandwidth W, as shown in Figure 12.5. After passband filtering of the despread received signal, the jammer noise power will be reduced by the factor B/W.

The signal power S is equal to $E_b R$ where R is the data rate and the jammer power J is equal to $J_0 W$ where W is the jammer bandwidth. Then we can write

$$\frac{E_b}{J_0} = \frac{W/R}{J/S}.$$

This relationship is used in the form

$$\left(\frac{E_b}{J_0}\right)_{dB} = \left(\frac{W}{R}\right)_{dB} - \left(\frac{J}{S}\right)_{dB}.$$

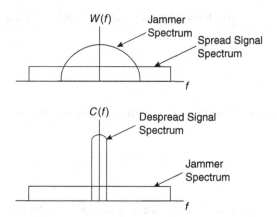

Figure 12.5. Heuristic notion of spread spectrum.

The ratio W/R is referred to as the *processing gain* of the spread-spectrum system. The reciprocal of the processing gain is just the spectral bit rate density $r = R/W$. A spread-spectrum waveform with a very large processing gain, such as 60 dB, has a very small spectral bit rate density, such as 10^{-6} bits per hertz.

The value of J/S for which E_b/J_0 equals its minimum acceptable value is referred to as the *jamming margin*. The jamming margin is the difference between the actual value of J/S and the required value of J/S to achieve the bit error rate. Thus

$$\text{jamming margin} = \left(\frac{J}{S}\right)_{\text{dB act}} - \left(\frac{J}{S}\right)_{\text{dB req}}$$

$$= \left(\frac{W}{R}\right)_{\text{dB act}} - \left(\frac{E_b}{N_0}\right)_{\text{dB req}}.$$

Spread-spectrum systems are a straightforward way to expand bandwidth, but are not an optimum use of the wide bandwidth. This can be seen by looking at the channel capacity

$$R \geq C = W \log_2 \left(1 + \frac{E_b}{J_0}\frac{R}{W}\right).$$

For the channel of bandwidth W, this equation can be written

$$\frac{E_b}{J_0} \geq \frac{2^{R/W} - 1}{R/W}.$$

The total jammer power is fixed at J. With N_0 replaced by J/W, one can compare three inequalities of interest.

(i) Bandwidth occupancy of W hertz by a direct choice of wideband modulation waveform

$$E_b \geq \frac{J}{W} \frac{2^{R/W} - 1}{R/W}.$$

(ii) Spread spectrum expansion to W hertz from a narrowband modulation waveform of bandwidth B hertz

$$E_b \geq \frac{J}{W} \frac{2^{R/B} - 1}{R/B}.$$

(iii) No bandwidth expansion from a narrowband waveform of bandwidth B hertz

$$E_b \geq \frac{J}{B} \frac{2^{R/B} - 1}{R/B}.$$

A comparison of the second and third options shows that spectrum spreading reduces the required energy by the value of the processing gain W/B. This ratio is a performance improvement, but is only linear in the bandwidth ratio. Additional improvement is possible by a more general bandwidth expansion as indicated by the first inequality. What this means in practice is that part of the available bandwidth should be used to support a stronger waveform such as an M-ary orthogonal signaling waveform, or to transmit the check symbols of a data transmission code. The coding gain provided by a good code is always larger than the increased bandwidth ratio that will be needed by the code. The rest of the available bandwidth can then be used for spectrum spreading.

We can make this point another way by considering the probability of error rather than the channel capacity. For an uncoded BPSK waveform, the probability of error is

$$p_e = Q\left(\sqrt{\frac{2E_b}{N_0}}\right).$$

For a coded system, with code rate R_c and minimum distance $2t + 1$, the probability of error is approximated by

$$p_e \approx Q\left(\sqrt{\frac{2E_b}{N_0} R_c(2t + 1)}\right).$$

If, further, the system is then spread to bandwidth W with processing gain W/R, the probability of error in the presence of a jammer can be approximated as

$$p_e \approx Q\left(\sqrt{2\frac{W/R}{J/S} R_c(2t + 1)}\right)$$

by recalling that $J_0 = J/W$ and $E_b = S/R$. While one may observe that the processing gain W/R could be increased by a factor of R_c if there were no code, this improvement would be more than offset by the loss of the term $R_c(2t + 1)$. For small amounts of excess bandwidth, coding gain is a better investment of bandwidth than processing gain. Eventually, coding gain can increase with bandwidth not much faster than could processing gain so when the excess bandwidth is large, part of it should be used for coding gain and the rest for processing gain. Of course, this conclusion holds only for wideband gaussian noise. For other jammer tactics, coding can have even more spectacular advantages and will be far superior to simple spectrum spreading.

The use of M-ary orthogonal signaling and error-control codes will both improve performance significantly faster than linear in the additional bandwidth required, at least when the bandwidth expansion is small – a factor of two or three. In addition, error-control codes are necessary in a spread-spectrum system to protect against partial-time jamming tactics. A well-designed spread-spectrum system will use an error-control code and a modulation waveform that operate at a small E_b/N_0 and then spread the spectrum of this waveform to fill out the available frequency band.

It is common practice to separate the function of spectrum spreading from the function of modulation. Rather than choosing a pulse shape $s(t)$ that is itself wideband, the modulator uses a more conventional pulse to create a waveform $c(t)$, which then is subjected to a second level of modulation to create the wideband signal.

The two spread-spectrum techniques commonly used in practice, known as *direct-sequence spread spectrum* and *frequency-hopping spread spectrum*, are discussed in detail in the next two sections. Both methods can be subsumed by a single general discussion that describes the spreading function by a phase-modulated wideband carrier

$$c_s(t) = \cos(2\pi f_0 t + \theta(t))$$

where the phase modulation $\theta(t)$ is selected so that $c_s(t)$ is wideband. The communication waveform $c(t)$ is then modulated onto this carrier to produce the wideband signal $w(t)$ by the multiplication

$$w(t) = c(t)c_s(t)$$
$$= c(t)\cos(2\pi f_0 t + \theta(t)).$$

If $c_s(t)$ is chosen so it has a much wider bandwidth than $c(t)$, then the signal $w(t)$ will also have a much wider bandwidth than $c(t)$.

Though the jammer may be aware of the general strategy, it is ignorant of the specific wideband signal $c_s(t)$. To ensure that this is true, many applications require $\theta(t)$ to change rapidly with time and in a way that is unpredictable to anyone without a secret key. To demodulate the signal, the receiver will need a copy of $c_s(t)$ or of the secret key used to generate $c_s(t)$. This means that there is a secret channel between the transmitter and the receiver, as shown in Figure 12.4, over which the secret key is passed. This channel need not operate at the same time as the main communication channel. Indeed, the key distribution may have taken place years before its use.

The receiver has a time-synchronized local replica $c'_s(t)$ of the phase-modulated carrier $c_s(t)$, given by

$$c'_s(t) = \cos(2\pi f_0 t + \theta(t) + \theta_0).$$

The phase offset θ_0 will appear in the local replica carrier $c'_s(t)$ in those applications where it is deemed to be impractical to lock the phases of the carriers and noncoherent demodulation is used. We will consider only coherent reception in which θ_0 is zero, and $c'_s(t) = c_s(t)$.

The jammer adds a jamming signal $e(t)$ to the transmitted signal before it reaches the receiver. The received wideband signal is

$$v_s(t) = c(t)\cos(2\pi f_0 t + \theta(t)) + e_R(t)\cos 2\pi f_0 t - e_I(t)\sin 2\pi f_0 t.$$

The receiver mixes the local-replica carrier with the wideband received signal $v_s(t)$ to obtain the narrowband received signal

$$v(t) = c_s(t)v_s(t).$$

Writing this out gives

$$v(t) = c(t)\cos^2(2\pi f_0 t + \theta(t))$$
$$+ [e_R(t)\cos 2\pi f_0 t - e_I(t)\sin 2\pi f_0 t]\cos(2\pi f_0 t + \theta(t)).$$

Expanding the trigonometric functions, and rejecting the terms at frequency $2f_0$, gives

$$v(t) = c(t) + e_R(t)\cos\theta(t) - e_I(t)\sin\theta(t).$$

The signal has now been collapsed back to the original narrowband signal $c(t)$. The jamming signal, however, now is wideband even if the original jamming signal were narrowband. This is because $\cos\theta(t)$ and $\sin\theta(t)$ are wideband signals. By passing the signal $v(t)$ through a narrowband filter matched to $c(t)$, most of the jamming power is rejected. Consequently, the jamming power is reduced by the ratio of the bandwidths.

Of course, if the jammer were omniscient, or has access to the key, it would then use

$$e_R(t) = n(t)\cos\theta(t)$$
$$e_I(t) = n(t)\sin\theta(t)$$

or something closely related, where $n(t)$ is narrowband noise. Then the jamming signal would also collapse in the receiver into the narrowband signal, and the spectrum spreading would be defeated. It is for this reason that $\theta(t)$ must be a secret function known in complete detail only to the transmitter and receiver.

12.4 Direct-sequence spread spectrum

Let $c(t)$ be the baseband modulation waveform that is to be spread. For a direct-sequence spread-spectrum system, a BPSK waveform consisting of a sequence of positive and negative pulses works well. The BPSK baseband modulation waveform can be expressed as

$$c(t) = \sum_{\ell=-\infty}^{\infty} a_\ell s(t - \ell T)$$

where $s(t)$ is the baseband pulse shape, and a_ℓ is either $+A$ or $-A$ depending on the data bit to be sent in the ℓth interval.

A direct-sequence spread-spectrum communication system employs a waveform of the form $w(t) = c_s(t)c(t)$ where $c(t)$ is the baseband modulation waveform, possibly complex, and $c_s(t)$ is a baseband direct-sequence spectrum-spreading signal with a bandwidth that is large compared to the data rate. A direct-sequence spectrum-spreading waveform itself is written as a baseband waveform

$$c_s(t) = \sum_{j=-\infty}^{\infty} b_j \sigma(t - jT_c)$$

where $\sigma(t)$, called a chip, is usually a time-limited rectangular pulse of duration T_c, and the sequence of b_j is a *pseudorandom* binary sequence known to both the transmitter and receiver. The term pseudorandom means that the sequence mimics a random sequence, although it may be generated deterministically as a function of a secret key. Figure 12.6 shows how any modulation waveform $c(t)$ can be spread and despread by $c_s(t)$, relying only on the property that $|c_s(t)|^2 = 1$.

The parameter T_c is called the *chip duration*, and the sequence

$$\boldsymbol{b} = \dots, b_{-1}, b_0, b_1, b_2, \dots$$

is called the *signature sequence* or the *spreading sequence*. The signature sequence is known to the transmitter and receiver, but not to the jammer. For practical reasons, the signature sequence is usually periodic, but for security, the period should be very long. For reasonable choices of signature sequence and chip pulse shape, $1/T_c$ is a rough estimate of the bandwidth of the spread-spectrum signal. For most direct-sequence systems, $T_c \ll T$, so the bandwidth of the spread-spectrum signal is much larger than

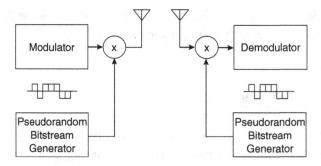

Figure 12.6. Direct-sequence spread-spectrum.

that of the data signal. For a direct-sequence spread-spectrum waveform, the processing gain can be written

$$\text{processing gain} = \frac{W}{R} = \frac{T}{T_c}.$$

Often, the duration T of the data pulse is an integer multiple of the chip duration and the symbol timing is synchronous with the chip timing, but this is not necessary. If $T = NT_c$ for some integer N there are N chips per data pulse, and the bandwidth of the spread-spectrum waveform is roughly N times the data rate.

To conclude this section, we rewrite the direct-sequence waveform in several other ways because the alternative descriptions suggest alternative implementations. If $s(t)$ is also a rectangular pulse, an alternative description of the direct-sequence waveform is given by redefining the data pulse $s(t)$ to absorb the signature sequence into the pulse. Then, for each ℓ, the data pulse is different. That is, for the ℓth data bit, let

$$s_\ell(t) = \left[\sum_{j=0}^{N-1} b_{\ell+j}\sigma(t - jT_c) \right] s(t).$$

Then,

$$w(t) = \sum_{\ell=-\infty}^{\infty} a_\ell s_\ell(t - \ell T)$$

is the direct-sequence spread-spectrum waveform. Now the ℓth bit is modulated as BPSK onto the wideband pulse $s_\ell(t)$. With this interpretation, we see that we may demodulate the ℓth bit as BPSK by using a filter matched to $s_\ell(t)$. This requires a programmable filter that can be reconfigured for each ℓ.

Figure 12.7 shows an implementation of the modulator and demodulator in which $s_\ell(t)$ is factored into the cascade of two filters, one an N-tap discrete-time filter whose

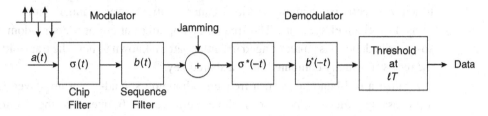

Figure 12.7. A filtering view of direct-sequence spread spectrum.

coefficients are given by the N coefficients of the signature sequence, and one a filter whose impulse response is the rectangular pulse $\sigma(t)$. The sequence filter

$$b(t) = \sum_{j=0}^{N-1} b_{j+\ell N}\delta(t - jT_c)$$

changes from bit to bit, and

$$a(t) = \sum_{\ell=-\infty}^{\infty} a_\ell\delta(t - \ell T).$$

Yet another way to write this is to allow the ℓth subsequence of length N of the signature sequence to be multiplied by a_ℓ. This leaves the sign of the signature bit unchanged if a_ℓ is positive and inverted if a_ℓ is negative. Thus, let

$$b'_{j+\ell} = a_\ell b_{j+\ell} \qquad \begin{matrix} j = 0,\ldots,N-1 \\ \ell = 0,\ldots \end{matrix}$$

and

$$w(t) = \sum_{j=-\infty}^{\infty} b'_j\sigma(t - jT_c).$$

With this view, $\sigma(t)$ can be chosen as a Nyquist pulse for time interval T_c. If $\sigma(t)$ is the pulse at the output of its matched filter, the chip matched-filter output samples $w(jT_c)$ are inverted or not inverted according to the value of b_j, then added together to synthesize a matched-filter output for the ℓth data bit.

12.5 Frequency-hopping spread spectrum

A frequency-hopping antijam waveform employs a large collection of specified carrier frequencies. The waveform rapidly and randomly moves from one carrier frequency to another, perhaps changing carrier frequency after every channel symbol or after every few channel symbols. The frequency-hopping pattern appears random and is so-described, but it is a prearranged irregular pattern known to both the transmitter and the receiver, usually computed from a shared key.

Against a full-band jammer, a frequency-hopping spread-spectrum waveform has a processing gain of N because, if there are N carrier frequencies, the jammer can only direct $1/N$ of the jamming power to any one frequency. However, even a fairly simple jammer will counter the frequency hopper by using a partial-band jammer. For example, suppose that a noncoherent, binary orthogonal communication system uses a randomly selected carrier frequency chosen from a set of $N = 1000$ carrier frequencies; the processing gain is 30 dB. The jammer will easily notice, however, that a single frequency of a typical waveform might be jammed with an E_b/J_0 of about 4 dB with a bit error rate of about 10^{-2}, and one bit out of a thousand is on this frequency. Hence by jamming only one frequency, the jammer produces a 10^{-5} bit error rate with an E_b/J_0 on that channel of 4 dB, whereas if there were no frequency hopping, it would need to ensure an E_b/J_0 of 9.6 dB to achieve a 10^{-5} bit error rate. This means that the frequency-hopping system has only a 5.6 dB advantage over an unprotected single-frequency system. The apparent 30 dB processing gain implied by using 1000 carrier frequencies turns out to be only a 5.6 dB gain. This frequency-hopping system is nearly worthless.

To remedy this vulnerability, the antijam waveform can use a diversity system by sending more than one copy of each symbol. Figure 12.8 shows a binary orthogonal signaling waveform with two-way diversity. Each bit is sent twice, with each copy of the bit sent on a randomly selected carrier frequency. In general, one can employ an M-ary orthogonal signaling waveform with L-ary diversity. Because it is difficult to build a frequency-hopping local oscillator that maintains coherence across frequency hops, most such applications use noncoherent demodulation with a noncoherent diversity combination. If, instead, coherence were maintained across frequency hops, there would be a small improvement. However, with M-ary orthogonal signaling with moderate values of M, noncoherent diversity combining is almost as good as coherent diversity combining.

Figure 12.9 shows an implementation of a demodulator for a frequency-hopping, noncoherent-combining, L-ary diversity, binary FSK signaling waveform for the case of perfect side information. The perfect side information is used to reject the jammed

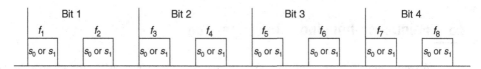

Figure 12.8. A binary frequency-hopping system with diversity.

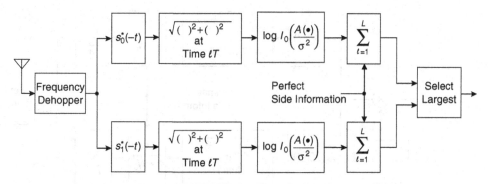

Figure 12.9. Noncoherent combining of dehopped diversity pulses.

copies of the signal. The remaining received copies of each symbol are noncoherently summed using log-Bessel-function combining as discussed in Section 7.5.

In practice, perfect side information is not available. Estimated side information must be generated by the receiver itself by observing the received signal. If the signal is small compared to the noise, as is often the case prior to the matched filter, then an estimate of the noise variance on each frequency is simply proportional to the sample power in the received signal prior to the matched filter:

$$\widehat{J}_0 \sim \int_{-T/2}^{T/2} |v(t)|^2 dt,$$

with an appropriate proportionality constant that does not concern us here. Indeed, if the jammer uses white gaussian noise and the signal is negligibly small, this is a maximum-likelihood estimate of the jammer power and the needed side information is provided.

Figure 12.10 shows the demodulator using estimated side information. Because the estimated variance of the jamming power is only an approximation of the true jamming power, the nonlinear $\log I_0(x)$ combining rule of Figure 12.9 may not be justified when the jammer power is not accurately known. Therefore, in Figure 12.10, the nonlinear combining has been replaced by a simple weighted average. Now this combining rule is given without formal justification. It may be satisfactory based on empirical observations.

A simple and robust technique to circumvent the need for side information is to insert a passband hardlimiter prior to the matched filter, as shown in Figure 12.11. This method works well whenever the signal is weaker than the noise at the input to the matched filter.

Recall from Section 11.8 that, if the input to the matched filter is hardlimited, then both the signal and the noise at the output of the matched filter are attenuated in such a way that the signal-to-noise ratio is $(2/\pi)E_p/J_{0\ell}$, and the output noise variance is independent of the input. Because of the hardlimiting, the noise at the output of the

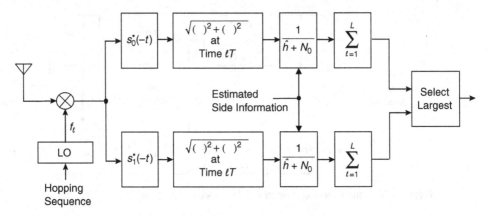

Figure 12.10. Noncoherent combining of dehopped diversity pulses.

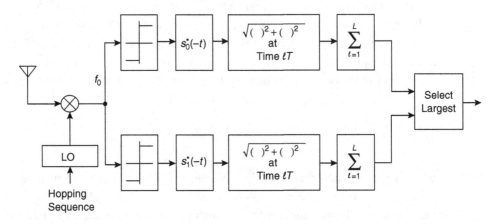

Figure 12.11. A robust demodulator using a hardlimiter.

matched filter is not gaussian, but we will approximate the output noise as gaussian. This approximation is good if $s(t)$ has a lot of structure because then the matched filter is required to sum many input samples suggesting the conclusion of the central limit theorem. Consequently, we can anticipate the approximate equation

$$p_e \approx Q\left(\sqrt{\left(\frac{2}{\pi}\right)\sum_{\ell=1}^{L}\frac{E_b/L}{N_0 + J_{0\ell}}}\right).$$

To maximize the right side, the sum on L should be minimized. A wise jammer constrained by $\sum_{\ell} J_{0\ell} = J$ will set $J_{0\ell} = J_0$ for all ℓ. Then

$$p_e \approx Q\left(\sqrt{\left(\frac{2}{\pi}\right)\frac{E_b}{N_0 + J_0}}\right).$$

The hardlimiter has effectively reduced E_b by $2/\pi$, or 2 dB. This energy penalty is the cost of using the hardlimiter to create a robust demodulator for a diversity waveform that counters partial-message jamming.

Problems for Chapter 12

12.1. (*High–low game.*) Suppose that an adversary randomly selects an integer between 0 and 127 inclusively, and you are to find that integer by a sequence of guesses. After each guess you are told whether you are high, low, or correct. You want to make the expected number of your guesses as small as possible, and the adversary wants to choose the integer to make it as large as possible.

    a. What strategy should you and your adversary each use?

    b. Now suppose that you must announce your strategy before the adversary selects the integer. (A spy divulges your design.) How does this change the strategies? (Both opponents have generators of random numbers available.)

12.2. a. Given L identical copies of a pulse with equal amplitudes, either $+a$ or $-a$, the ℓth copy is observed in independent, additive white gaussian noise of power density spectrum N_ℓ. Prove that the maximum-likelihood demodulator for the sign of the amplitude uses the sufficient statistic

$$u = \sum_{\ell=1}^{L} \frac{1}{N_\ell} u_\ell$$

    where u_ℓ is the matched-filter output for the ℓth pulse.

    b. Prove that the signal-to-noise ratio of this statistic is

$$\frac{S}{N} = E_p \left(\frac{1}{L} \sum_{\ell} \frac{1}{N_\ell} \right).$$

    c. Use a Lagrange multiplier to show that, given the constraint that

$$\frac{1}{L} \sum_{\ell=1}^{L} N_\ell = N_0,$$

    the signal-to-noise ratio is minimized by setting $N_\ell = N_0$ for $\ell = 1, \ldots, L$.

12.3. A diversity system for digital communication sends the same waveform through two additive noise channels so that if one channel is lost, the symbol can still be received correctly through the other. When both channels are working, it is best to combine the received signals prior to a threshold. A binary diversity system sends identical binary FSK signals on two carrier frequencies, f_0 and

f_0', each of which is received noncoherently with independent phase errors and independent additive gaussian noise.

a. At what point in the demodulator should the two signals be added together?

b. Sketch a block diagram of the demodulator.

c. Set up an equation for the probability of error.

d. How should the demodulator be redesigned if the phase errors on the two channels are always equal?

12.4. A Barker pulse is a pulse given by

$$s(t) = \sum_{\ell=0}^{n-1} c_\ell p(t - \ell T)$$

where $p(t)$ is a pulselet and c_ℓ for $\ell = 0, \ldots, n-1$ is a Barker sequence. Show that $\psi(\tau)$, the autocorrelation function of $s(t)$, can be written

$$\psi(\tau) = \sum_{\ell=-n+1}^{n-1} \gamma_\ell \pi(\tau - \ell T)$$

where γ_i is the autocorrelation function of the Barker sequence, and $\pi(\tau)$ is the autocorrelation function of the pulselet.

12.5. A communication channel is subjected to additive burst noise that occurs at random times with a ten-percent duty cycle. This means that the burst noise occupies ten percent of any time interval that is long compared to the duration of individual bursts. During the burst, the signal-to-noise ratio is 0 dB and the additive noise probability density function is unknown; otherwise, the signal-to-noise ratio is 30 dB and the noise is additive gaussian noise.

a. Specify a combination of a multilevel signaling scheme and a Reed–Solomon code that will give a good performance on this channel. What is the data rate?

b. An alternative signaling scheme is proposed that allegedly mitigates the effects of the burst noise. The proposal is to transform signal intervals of duration T into another orthogonal coordinate system. (A Fourier transform is one such transformation.) The alleged justification is that the burst noise is "spread out" in the new coordinate system. Is this proposal sound? What data rate may be expected by using this method?

c. To mitigate the effects of defects in an optical storage medium, a proposal is made to use the Fourier transform property of holography to "spread" the effect of the defects over all stored bits. Is this a sound technique?

12.6. Partial message jamming can be used against M-ary signaling, either coherent or noncoherent. On a graph of p_e versus E_b/N_0 for 32-ary orthogonal signaling, sketch the probability of demodulation error against an optimal partial-message jammer by arguing that, for small E_b/J_0, the duty factor must equal one, and otherwise p_e must behave as the reciprocal of E_b/J_0.

12.7. A digital communication system uses an eight-ary orthogonal signaling waveform with two-way diversity. The diversity receiver is capable of recognizing which symbols are jammed and demodulating using only the single unjammed pulse when it is available. The data is protected by a (15, 13) single-error-correcting Reed–Solomon code.

a. What is the minimum jammer duty factor that is meaningful to the jammer if the decoder uses an errors-only decoder?

b. What is the minimum jammer duty factor that is meaningful to the jammer if the decoder uses an errors-and-erasures decoder?

c. Set up all equations needed to compute p_e versus S/J for a fixed value of duty factor.

12.8. a. Given the primitive polynomial $p(x) = x^3 + x + 1$, construct an $M = 8$ simplex family of waveforms by using the seven m-sequences of length 7 and the all-zero sequence to form eight PSK waveforms.

b. Show that an $M = 2^m$ simplex pulse alphabet can be formed in this way for any m.

12.9. An antijam communication system uses a waveform $c(t)$ that is a superposition of sinc pulses

$$c(t) = \sum_{\ell=-\infty}^{\infty} a_\ell \mathrm{sinc}\left(\frac{t - \ell T_b}{T_b}\right),$$

and a signature sequence (b_k) modulating a sequence of rectangular pulses

$$g(t) = \sum_{k=-\infty}^{\infty} b_k \mathrm{rect}\left(\frac{t - kT_c}{T_c}\right)$$

where T_b/T_c is an integer. The received signal in additive white gaussian noise is

$$v(t) = g(t)c(t) + n(t).$$

a. Prove that passing the signal $g(t)v(t)$ through the filter $h(t) = \mathrm{sinc}(t/T_b)$ and sampling at time T_b maximizes the signal-to-noise ratio of the ℓth symbol. That is,

$$v(\ell T_b) = \int_{-\infty}^{\infty} g(\xi)v(\xi)\mathrm{sinc}\left(\frac{\xi - \ell T_b}{T_b}\right)d\xi$$

is a statistic with a maximum signal-to-noise ratio for the ℓth bit. Is there any intersymbol interference?

b. Repeat with the change that

$$g(t) = \sum_{k=-\infty}^{\infty} b_k \operatorname{sinc}\left(\frac{t - kT_c}{T_c}\right).$$

12.10. An antijam communication system uses a BPSK waveform $c(t)$, given by

$$c(t) = \sum_{\ell=-\infty}^{\infty} a_\ell s(t - \ell T_b),$$

where $s(t) * s(-t)$ is a Nyquist pulse, and a signature sequence composed of a sequence of rectangular pulses

$$g(t) = \sum_{k=-\infty}^{\infty} b_k \operatorname{rect}\left(\frac{t - kT_c}{T_c}\right)$$

where T_b/T_c is a large integer and b_k is a random sequence. The transmitted signal is $w(t) = g(t)c(t)$. The received signal is

$$v(t) = w(t) + e(t)$$

where $e(t)$ is a jamming signal of power J.

a. The receiver consists of the following steps,
1. $v'(t) = v(t)g(t)$
2. matched filter to $s(t)$ for $v'(t)$
3. sample at ℓT.

What (approximate) spectrum should the jammer use to minimize the signal-to-noise ratio of the samples?

b. What is the (approximate) signal-to-noise ratio if the jammer uses bandlimited white gaussian noise in the band $1/T_c$?

c. Given that the jammer is known to use the spectrum calculated in part b, can the receiver be redesigned to increase the signal-to-noise ratio?

12.11. A conventional QPSK waveform using the square signaling constellation

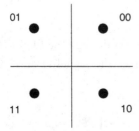

is jammed by a signal

$$e(t) = e_R(t) + je_I(t)$$

where $e_R(t)$ and $e_I(t)$ are white, independent, and gaussian, but with different variances

$$\text{var}[e_R(t)] = 4\text{var}[e_I(t)] = \frac{4}{5}J_0.$$

a. If the receiver is unaware that the variances are unequal, what is the probability of bit error as a function of E_b/J_0?

b. If the receiver (but not the transmitter) is aware of the distribution of jammer power, how should the decision regions be redefined? What is the probability of bit error as a function of E_b/J_0?

c. If the transmitter is also aware of the distribution of jammer power, how should the signal constellation be changed? What is the probability of bit error as a function of E_b/J_0?

12.12.　An M-ary orthogonal waveform

$$c(t) = \sum_{\ell=-\infty}^{\infty} s_{m\ell}(t - \ell T)\cos 2\pi f_0 t$$

is transmitted through a jamming channel in which the jammer elects to use a sinusoid as the jamming signal. The received passband signal is

$$v(t) = c(t) + n(t) + e(t)$$

where $n(t)$ is white gaussian noise, and $e(t) = E\cos 2\pi f_1 t$ and E is large. (See Problem 11.8.) The demodulator consists of a passband hardlimiter intended to suppress impulsive interference, f_0 allowed by a bank of matched filters, followed by a coherent threshold test. How does the jamming signal affect the probability of demodulation error?

12.13.　(**Voting Systems and Communication Diversity**) A large group of voters is to decide between two candidates. Each voter is to vote for either candidate A or candidate B. There are 50 million people who have decided to vote for candidate A and 49 million who have decided to vote for candidate B. An additional 100 million people have no preference what so ever and will each vote by making an independent random equiprobable choice. (All numbers are large enough so that the central limit theorem applies.)

a. What is the probability that candidate A will win in the popular vote?

b. An adversary is able to cast an additional 0.9 million false votes for candidate B by fraud. Can the adversary change the outcome? What now is the probability that candidate A will win? How does this change if the adversary can record one million (or more) false votes?

Suppose now that an electoral system for voting is used as follows: the voters are divided into fifty regions of equal size. Assume that each region has the three types of voters in the same proportion: one million voters for candidate A, 0.98 million voters for candidate B, and 2 million undecided. The majority popular vote in each region determines the single electoral vote from that region. The candidate with the largest number of electoral votes wins.

c. What is the probability that candidate A will win in the electoral vote?

d. Can the adversary with 0.9 million false votes change the outcome by fraud? With what strategy? Must the false voters be distributed over more than one region? How many? What now is the probability that candidate A will win? How does this change if the adversary can cast one million (or more) false votes?

Notes for Chapter 12

The method of spread-spectrum signaling has a long and interesting history that has been summarized in articles by Scholtz (1982) and by Price (1983). Tutorial treatments can be found in the article by Pickholtz, Schilling, and Milstein (1982), and in the book by Holmes (1982). An early use of frequency-hopping spread spectrum for radar appeared in a patent by Guanella (1938). A patent application for a direct-sequence spread-spectrum communication system was filed in 1953 by deRosa and Rogoff (1979).

The theory of m-sequences was developed by Gilbert, Golomb, Welch, and Zierler in the mid-1950s, and is described by Zierler (1959), Selmer (1966), and Golomb (1967). Sets of sequences with good crosscorrelation properties were discovered by Kasami (1966), and Gold (1967). The broad topic of signature sequences was surveyed by Sarwate and Pursley (1980), and by MacWilliams and Sloane (1976).

Bibliography

Aaron, M. R. and D. W. Tufts, "Intersymbol Interference and Error Probability," *IEEE Transactions on Information Theory*, vol. IT-12, pp. 24–36, 1966.

Amoroso, F., "Pulse and Spectrum Manipulation in the Minimum (Frequency) Shift Keying (MSK) Format," *IEEE Transactions on Communications*, vol. COM-24, pp. 381–384, 1976.

Amoroso, F. and J. A. Kivett, "Simplified MSK Signaling Technique," *IEEE Transactions on Communications*, vol. COM-25, pp. 433–441, 1977.

Anderson, J. B. and D. P. Taylor, "A Bandwidth-Efficient Class of Signal Space Codes," *IEEE Transactions on Information Theory*, vol. IT-24, pp. 703–712, 1978.

Arens, R., "Complex Process for Envelopes of Normal Noise," *IRE Transactions on Information Theory*, vol. IT-3, pp. 204–207, 1957.

Arimoto, S., "Encoding and Decoding of *p*-ary Group Codes and the Correction System," (in Japanese), *Information Processing in Japan*, vol. 2, pp. 321–325, 1961.

Armstrong, E. A., "A Method of Reducing Disturbances in Radio Signaling by a System of Frequency Modulation," *Proceedings of the IRE*, vol. 24, pp. 689–740, 1936.

Aschoff, V., "The Early History of the Binary Code," *IEEE Communications Magazine*, vol. 21, pp. 4–10, 1983.

Aulin, T., N. Rydbeck, and C.-E. Sundberg, "Continuous Phase Modulation. Part II: Partial Response Signaling," *IEEE Transactions on Communications*, vol. COM-29, pp. 210–225, 1981.

Aulin, T. and C.-E. Sundberg, "Continuous Phase Modulation. Part I: Full Response Signaling," *IEEE Transactions on Communications*, vol. COM-29, pp. 196–209, 1981.

Austin, M., "Decision-Feedback Equalization for Digital Communication over Dispersive Channels," *MIT Research Laboratory Electronics Technical Report 461*, 1967.

Bahl, L. R., J. Cocke, F. Jelinek, and J. Raviv, "Optimal Decoding of Linear Codes for Minimizing Symbol Error Rate," *IEEE Transactions on Information Theory*, vol. IT-20, pp. 284–287, 1974.

Balakrishnan, A. V., "A Contribution to the Sphere-Packing Problem of Communication Theory," *Journal of Mathematical Analysis and Applications*, vol. 3, pp. 485–506, 1961.

Barker, R. H., "Group Synchronizing of Binary Digital Systems," *Communication Theory*, London, Butterworth, 1953.

Baum, L. E. and T. Petrie, "Statistical Inference for Probabilistic Functions of Finite State Markov Chains," *Annals of Mathematical Statistics*, vol. 37, pp. 1554–1563, 1966.

Bennett, W. R. and S. O. Rice, "Spectral Density and Autocorrelation Functions Associated with Binary Frequency-Shift Keying," *Bell System Technical Journal*, vol. 42, pp. 2355–2385, 1963.

Berger, T., "Optimum PAM Compared with Information-Theoretic Bounds," *Northeast Electronics Research and Engineering Meeting Record*, pp. 242–243, 1966.

Berrou, C. and A. Glaveiux, "Near Optimum Error Correcting Coding and Decoding: Turbo Codes," *IEEE Transactions on Communications*, vol. COM-44, pp. 1261–1271, 1996.

Berrou, C., A. Glaveiux, and P. Thitimajshima, "Near Shannon Limit Error-Correcting Coding and Decoding," *Proceedings 1993 IEEE International Conference on Communications*, vol. 2, pp. 1064–1070, 1993.

Blachman, N. M., "Band-Pass Nonlinearities," *IEEE Transactions on Information Theory*, vol. IT-10, pp. 162–164, 1964.

Blachman, N. M., "Detectors, Band-Pass Nonlinearities, and Their Optimization: Inversion of the Chebychev Transform," *IEEE Transactions on Information Theory*, vol. IT-17, pp. 398–404, 1971.

Blahut, R. E., "Transform Techniques for Error-Control Codes," *IBM Journal of Research Development*, vol. 23, pp. 299–315, 1979.

Blahut, R. E., *Algebraic Codes for Data Transmission*, Cambridge University Press, 2003.

Blake, I. F., "The Leech Lattice as a Code for the Gaussian Channel," *Information and Control*, vol. 19, pp. 66–74, 1971.

Brennan, D. G., "On the Maximal Signal-to-Noise Ratio Realizable from Several Noisy Signals," *Proceedings of the IRE*, vol. 43, p. 1530, 1955.

Brennan, D. G., "Linear Diversity Combining Techniques," *Proceedings of the IRE*, vol. 47, pp. 1075–1102, 1959, reprinted *Proceedings of the IEEE*, vol. 91, pp. 331–356, 2003.

Bussgang, J. J., "Cross-Correlation Functions of Amplitude Distorted Gaussian Signals," *MIT Research Laboratory Electronics Technical Report 216*, 1952.

Cahn, C. R., "Performance of Digital Phase-Modulation Communication System," *IRE Transactions on Communication Systems*, vol. CS-7, pp. 3–6, 1959.

Cahn, C. R., "Combined Digital Phase and Amplitude Modulation Communication Systems," *IRE Transactions on Communication Systems*, vol. CS-8, pp. 150–154, 1960.

Cahn, C. R., "A Note on Signal-To-Noise Ratios in Bandpass Limiters," *IEEE Transactions on Information Theory*, vol. IT-7, pp. 39–43, 1961.

Calderbank, A. R. and J. E. Mazo, "A New Description of Trellis Codes," *IEEE Transactions on Information Theory*, vol. IT-30, pp. 784–791, 1984.

Calderbank, A. R. and L. H. Ozarow, "Nonequiprobable Signaling on the Gaussian Channel," *IEEE Transactions on Information Theory*," vol. IT-36, pp. 726–740, 1990.

Calderbank, A. R. and N. J. A. Sloane, "Four-Dimensional Modulation with an Eight-State Trellis Code," *AT&T Technical Journal*, vol. 64, pp. 1005–1018, 1985.

Calderbank, A. R. and N. J. A. Sloane, "An Eight-Dimensional Trellis Code," *Proceedings of the IEEE*, vol. 74, pp. 757–759, 1986.

Calderbank, A. R. and N. J. A. Sloane, "New Trellis Codes Based on Lattices and Cosets," *IEEE Transactions on Information Theory*, vol. IT-33, pp. 177–195, 1987.

Campopiano, C. N. and B. G. Glazer, "A Coherent Digital Amplitude and Phase Modulation Scheme," *IRE Transactions on Communication Systems*, vol. CS-10, pp. 90–95, 1962.

Carson, J. R., "Notes on the Theory of Modulation," *Proceedings of the IRE*, vol. 10, pp. 57–64, 1922.

Cheun, K., "Performance of Direct-Sequence Spread Spectrum Rake Receivers with Random Spreading," *IEEE Transactions on Communications*, vol. COM-45, pp. 1130–1143, 1997.

Conway, J. H. and N. J. A. Sloane, *Sphere Packings, Lattices and Groups*, New York, Springer, 1987.

Costas, J. P., "Synchronous Communications," *Proceedings of the IRE*, vol. 44, pp. 1713–1718, 1956.

Davenport, W. B., "Signal-To-Noise Ratios in Bandpass Limiters," *Journal of Applied Physics*, vol. 24, pp. 720–727, 1953.

Davisson, L. D. and L. B. Milstein, "On the Performance of Digital Communication Systems with Bandpass Limiters, Part 1: One Link Systems," *IEEE Transactions on Communications*, vol. COM-20, pp. 972–975, 1972.

de Buda, R., "Coherent Demodulation of Frequency-Shift Keying with Low Deviation Ratio," *IEEE Transactions on Communications*, vol. COM-20, pp. 429–435, 1972.

de Buda, R., "The Upper Bound of a New Near Optimal Code," *IEEE Transactions on Information Theory*, vol. IT-21, pp. 441–445, 1975.

deRosa, L. A. and M. Rogoff, "Secure Single Sideband Communication System Using Modulated Noise Subcarrier," U.S. Patent 4,176,316, 1979.

Doelz, M. L. and E. H. Heald, "Minimum-Shift Data Communication System," U.S. Patent 2,977,417, 1961.

Doelz, M. L., E. Heald, and D. Martin, "Binary Data Transmission Techniques for Linear Systems," *Proceedings of the IRE*, vol. 45, pp. 656–661, 1957.

Dolby, R. M., "An Audio Noise Reduction System," *Journal of the Audio Engineering Society*, vol. 15, pp. 383, 1967.

Dugundji, J., "Envelopes and Pre-envelopes of Real Wave-Forms," *IRE Transactions on Information Theory*, vol. IT-4, pp. 53–57, 1958.

Dwork, B. M., "Detection of a Pulse Superimposed on Fluctuation Noise," *Proceedings of the IRE*, vol. 38, pp. 771–774, 1950.

Elias, P., "Error-Free Coding," *IRE Transactions on Information Theory*, vol. IT-4, pp. 29–37, 1954.

Ericson, T., "Structure of Optimum Receiving Filters in Data Transmission Systems," *IEEE Transactions on Information Theory*, vol. IT-17, pp. 352–353, 1971.

Eyuboğlu, M. V., "Detection of Coded Modulation Signals on Linear Severely Distorted Channels Using Decision-Feedback Noise Prediction with Interleaving," *IEEE Transaction on Communications*, vol. COM-36, pp. 401–409, 1988.

Eyuboğlu, M. V., and G. D. Forney, Jr., "Trellis Precoding: Combined Coding, Precoding and Shaping for Intersymbol Interference Channels," *IEEE Transactions on Information Theory*, vol. IT-38, pp. 301–314, 1992.

Fisher, R. A., "On the Mathematical Foundations of Theoretical Statistics," *Philosophical Transactions of Royal Society*, A222, pp. 309–368, 1922.

Forney, G. D., Jr., "Convolutional Codes I: Algebraic Structure," *IEEE Transactions on Information Theory*, vol. IT-16, pp. 720-738, 1970.

Forney, G. D., Jr., "Lower Bounds on Error Probability in the Presence of Large Intersymbol Interference," *IEEE Transactions on Communications*, vol. COM-20, pp. 76–77, 1972.

Forney, G. D., Jr., "Maximum-Likelihood Sequence Estimation of Digital Sequences in the Presence of Intersymbol Interference," *IEEE Transactions on Information Theory*, vol. IT-18, pp. 363–378, 1972.

Forney, G. D., Jr., "The Viterbi Algorithm," *Proceedings of the IEEE*, vol. 61, pp. 268–278, 1973.

Forney, G. D., Jr., "Coset Codes I: Introduction and Geometrical Classification," *IEEE Transactions on Information Theory*, vol. IT-34, pp. 1123–1151, 1988.

Forney, G. D., Jr., "Coset Codes II: Binary Lattices and Related Codes," *IEEE Transactions on Information Theory*, vol. IT-34, pp. 1152–1187, 1988.

Forney, G. D., Jr., "The Forward–Backward Algorithm," *Proceedings of the 34th Annual Allerton Conference on Communication, Control, and Computing*, pp. 432–446, Monticello, IL, 1996.

Forney, G. D., Jr., "On Iterative Decoding and the Two-Way Algorithm," *Proceedings of the International Symposium on Turbo Codes and Related Topics*, Brest, France, 1997.

Forney, G. D., Jr., R. G. Gallager, G. R. Lang, F. M. Longstaff, and S. U. Qureshi, "Efficient Modulation for Band-Limited Channels," *IEEE Journal on Selected Areas Communications*, vol. SAC-2, pp. 632–647, 1984.

Forney, G. D., Jr., and G. Ungerboeck, "Modulation and Coding for Linear Gaussian Channels," *IEEE Transactions on Information Theory*, vol. IT-44, pp. 2384–2415, 1998.

Frank, R. L., "Polyphase Codes With Good Nonperiodic Correlation Properties," *IEEE Transactions on Information Theory*, vol. IT-9, pp. 43–45, 1963.

Frank, R. L., "Polyphase Complementary Codes," *IEEE Transactions on Information Theory*, vol. IT-26, pp. 641–647, 1980.

Franks, L. E., "Carrier and Bit Synchronization in Data Communication – A Tutorial Review," *IEEE Transactions on Communications*, vol. COM-28, pp. 1107–1121, 1980.

Franks, L. E., *Synchronization Subsystems: Analysis and Design*, in K. Feher (ed.), Digital Communications, Englewood Cliffs, NJ, Prentice-Hall, 1983.

Franks, L. E. and J. P. Bubrouski, "Statistical Properties of Timing Jitter in a PAM Timing Recovery Scheme," *IEEE Transactions on Communications*, vol. COM-22, pp. 913–920, 1974.

Gabor, D., "Theory of Communication," *Journal of the IEE*, vol. 93, (pt. 111), pp. 429–441, 1946.

George, T. S., "Fluctuation of Ground Clutter Return in Airborne Radar Equipment," *Philco Corporation Research Division Report 159*, 1950. Also, Institute of Electrical Engineering Monograph No. 22, 1952.

Golay, M. J. E., "Multislit Spectrometry," *Journal of the Optical Society of America*, vol. 39, p. 437, 1949.

Golay, M. J. E., "Complementary Series," *IRE Transactions on Information Theory*, vol. IT-7, pp. 82–87, 1961.

Gold, R., "Optimal Binary Sequences for Spread Spectrum Multiplexing," *IEEE Transactions on Information Theory*, vol. IT-13, pp. 619–621, 1967.

Golomb, S. W., *Shift Register Sequences*, San Francisco, Holden-Day, 1967.

Golomb, S. W., B. Gordon, and L. R. Welch, "Comma-Free Codes," *Canadian Journal of Mathematics*, vol. 10, pp. 202–209, 1958.

Golomb, S. W. and R. A. Scholtz, "Generalized Barker Sequences," *IEEE Transactions on Information Theory*, vol. IT-11, pp. 533–537, 1965.

Gruen, W. J., "Theory of AFC Synchronization," *Proceedings of the IRE*, vol. 41, pp. 1043–1048, 1953.

Guanella, G., *Distance Determining System*, Swiss Patent (U.S. Patent 2,253,975, 1941), 1938.

Guidoux, L., "Egaliseur Autoadaptif a Double Echantillonnage," *L'Onde Electrique*, vol. 55, pp. 9–13, 1975.

Hagenauer, J. and P. Hoeher, "A Viterbi Algorithm with Soft-Decision Outputs and Its Applications," *Proceedings of the IEEE 1989 Global Communications Conference*, Dallas, TX, pp. 47.1.1–47.1.7, 1989.

Hagenauer, J., E. Offer, and L. Papke, "Iterative Decoding of Binary Block and Convolutional Codes," *IEEE Transactions on Information Theory*, vol. IT-42, pp. 429–445, 1996.

Hamming, R. W., "Error Detecting and Error Correcting Codes," *Bell System Technical Journal*, vol. 29, pp. 147–160, 1950.

Hancock, J. C. and R. W. Lucky, "Performance of Combined Amplitude and Phase-Modulated Communication Systems," *IRE Transactions on Communication Systems*, vol. CS-8, pp. 232–237, 1960.

Harashima, H. and H. Miyakawa, "A Method of Code Conversion for a Digital Communication Channel with Intersymbol Interference," *Transactions of the Institute of Electronic Communication Engineering (Japan)*, vol. 52-A, pp. 272–273, 1969.

Harashima, H. and H. Miyakawa, "Matched-Transmission Technique for Channels with Intersymbol Interference," *IEEE Transactions on Communications*, vol. COM-20, pp. 774–780, 1972.

Hartley, R. V. L., "The Transmission of Information," *Bell System Technical Journal*, vol. 7, pp. 535–563, 1928.

Heller, J. A., *Short Constraint Length Convolutional Codes*, Jet Propulsion Laboratory Space Program Summary 37-54 III, pp. 171–177, 1968.

Heller, J. A. and J. M. Jacobs, "Viterbi Detection for Satellite and Space Communications," *IEEE Transactions on Communication Technology*, vol. COM-19, pp. 835–848, 1971.

Helstrom, C. W., "The Resolution of Signals in White Gaussian Noise," *Proceedings of the IRE*, vol. 43, pp. 1111–1118, 1955.

Holmes, J. K., *Coherent Spread Spectrum Systems*, New York, Wiley, 1982.

Jaffe, R. and E. Rechtin, "Design and Performance of Phase-Lock Circuits Capable of Near-Optimum Performance Over a Wide Range of Input Signal and Noise Levels," *IEEE Transactions on Information Theory*, vol. IT-1, pp. 66–76, 1955.

Jain, C. P., "Limiting of Signals in Random Noise," *IEEE Transactions on Information Theory*, vol. IT-18, pp. 332–340, 1972.

Jones, J. J., "Hard-Limiting of Two Signals in Random Noise," *IEEE Transactions on Information Theory*, vol. IT-9, pp. 34–42, 1963.

Kabal, P. and S. Pasupathy, "Partial Response Signaling," *IEEE Transactions on Communications*, vol. COM-23, pp. 921–934, 1975.

Kasami, T., *Weight Distribution Formula for Some Class for Cyclic Codes*, Coordinated Science Laboratory, University of Illinois, Urbana, *Technical Report No. R-285*, 1966.

Kassam, S. A. and H. V. Poor, "Robust Techniques for Signal Processing: A Survey," *Proceedings of the IEEE*, vol. 73, pp. 433–481, 1985.

Kobayashi, H., "Correlative Level Coding and Maximum-Likelihood Decoding," *IEEE Transactions on Information Theory*, vol. IT-17, pp. 586–594, 1971.

Kobayashi, H. and D. T. Tang, "Application of Partial Response Channel Coding to Magnetic Recording Systems," *IBM Journal of Research and Development*, vol. 14, pp. 368–375, 1970.

Koetter, R., A. C. Singer, and M. Tüchler, "Turbo Equalization," *IEEE Signal Processing Magazine*, vol. 21, pp. 67–80, 2004.

Kotel'nikov, V. A., *The Theory of Optimum Noise Immunity*. Translated by R. A. Silverman. New York, McGraw-Hill, 1959.

Kretzmer, E. R., "Generalization of a Technique for Binary Data Communication," *IEEE Transactions on Communications Technology*, vol. COM-14, pp. 67–68, 1966.

Landau, H. J., "On the Recovery of a Band-Limited Signal, After Instantaneous Companding and Subsequent Band Limiting," *Bell System Technical Journal*, vol. 39, pp. 351–364, 1960.

Lawton, J. G., "Comparison of Binary Data Transmission Systems," Presented at the *2nd National Conference on Military Electronics*, 1958.

Leech, J. and N. J. A. Sloane, "New Sphere Packings in Dimensions 9–15," *Bulletin of the American Mathematics Society*, vol. 76, pp. 1006–1010, 1970.

Lender, A., "Correlative Digital Communication Techniques," *IEEE Transactions on Communication Technology*, vol. COM-12, pp. 128–135, 1964.

Lindsey, W. C., *Synchronous Systems in Communications*, Englewood Cliffs, NJ, Prentice-Hall, 1972.

Lucky, R. W., "Automatic Equalization for Digital Communication," *Bell System Technical Journal*, vol. 44, pp. 547–588, 1965.

Lucky, R. W., "Techniques for Adaptive Equalization of Digital Communication," *Bell System Technical Journal*, vol. 45, pp. 255–286, 1966.

Lucky, R. W., "Signal Filtering with the Transversal Equalizer," *Proceedings of the 7th Annual Allerton Conference on Circuits and System Theory*, pp. 792–804, 1969.

Lucky, R. W. and J. C. Hancock, "On the Optimum Performance of N-ary Systems Having Two Degrees of Freedom," *IRE Transactions on Communication Systems*, vol. CS-10, pp. 185–192, 1962.

Lyon, D. L., "Envelope-Derived Timing Recovery in QAM and SQAM Systems," *IEEE Transactions on Communications*, vol. COM-23, pp. 1327–1331, 1975.

Lyon, D. L., "Timing Recovery in Synchronized Equalized Data Communication," *IEEE Transactions on Communications*, vol. COM-23, pp. 269–274, 1975.

MacWilliams, F. J. and N. J. A. Sloane, "Pseudo-Random Sequences and Arrays," *Proceedings of the IEEE*, vol. 64, pp. 1715–1729, 1976.

Manasse, R., R. Price, and R. M. Lerner, "Loss of Signal Detectability in Bandpass Limiters," *IRE Transactions on Information Theory*, vol. IT-4, pp. 34–38, 1958.

Massey, J. L., "Optimum Frame Synchronization," *IEEE Transactions on Communications*, vol. COM-20, pp. 115–119, 1972.

Massey, J. L. and M. K. Sain, "Inverses of Linear Sequential Circuits," *IEEE Transactions on Computers*, vol. C-17, pp. 330–337, 1968.

Mathes, R. C., U.S. Patent 1,295,553, 1919.

Maury, J. L. and F. J. Styles, *Development of Optimum Frame Synchronization Codes for Goddard Space Flight Center PCM Telemetry Standards*, in Proceedings of the 1965 National Telemetering Conference, Los Angeles, CA, 1964.

Mazo, J. E. and J. Salz, "On the Transmitted Power in Generalized Partial Response," *IEEE Transactions on Communications*, vol. COM-24, pp. 348–352, 1976.

Merkle, R. C., "Secure Communication over an Insecure Channel," *Communications of the Association for Computing Machinery*, vol. 21, pp. 294–299, 1978.

Messerschmitt, D. G., "Generalized Partial Response for Equalized Channels with Rational Spectra," *IEEE Transactions on Communications*, vol. COM-23, pp. 1251–1258, 1975.

Miyakawa, H. and H. Harashima, "Capacity of Channels with Matched Transmission Technique for Peak Transmitting Power Limitation," *National Convention Record IECE Japan*, pp. 1268–1269, 1969.

Miyakawa, H., H. Harashima, and Y. Tanaka, "A New Digital Modulation Scheme, Multimode CPFSK," *Proceedings, Third International Conference Digital Satellite Communications*, pp. 105–112, Kyoto, Japan, 1975.

Monsen, P., *Linear Equalization for Digital Transmission over Noisy Dispersive Channels*, Doctor of Engineering Science Dissertation, Columbia University, New York, 1970.

Mueller, M. S. and J. Salz, "A Unified Theory of Data-Aided Equalization," *Bell System Technical Journal*, vol. 60, pp. 2023–2038, 1981.

Murota, K. and K. Hirada, "GMSK Modulation for Digital Mobile Radio Telephony," *IEEE Transactions on Communications*, vol. COM-29, pp. 1044–1050, 1981.

Nielsen, P., "Some Optimum and Suboptimum Frame Synchronizers for Binary Data in Gaussian Noise," *IEEE Transactions on Communications*, vol. COM-21, pp. 770–772, 1973.

North, D. O., "An Analysis of the Factors Which Determine Signal/Noise Discrimination in Pulsed-Carrier Systems," *RCA Technical Report PTR-6C*, 1943.

Nuttall, A. H., "Error Probabilities for Equi-Correlated M-ary Signals Under Phase Coherent and Phase-Incoherent Reception," *IEEE Transactions on Information Theory*, vol. IT-8, pp. 305–314, 1962.

Nyquist, H., "Certain Factors Affecting Telegraph Speed," *Bell System Technical Journal*, vol. 3, pp. 324–346, 1924.

Nyquist, H., "Certain Topics in Telegraph Transmission Theory," *AIEE Transactions*, vol. 47, pp. 617–644, 1928.

Oliver, B. M., J. R. Pierce, and C. E. Shannon, "The Philosophy of PCM," *Proceedings of the IRE*, vol. 36, pp. 1324–1331, 1948.

Omura, J. K., "On Optimum Receivers for Channels with Intersymbol Interference," *IEEE International Symposium on Information Theory, Book of Abstracts*, Noordwijk, Holland, 1970.

Osborne, W. P. and M. B. Luntz, "Coherent and Noncoherent Detection of CPFSK," *IEEE Transactions on Communications*, vol. COM-22, pp. 1023–1036, 1974.

Pelchat, M. G., R. C. Davis, and M. B. Luntz, "Coherent Demodulation of Continuous Phase Binary FSK Signals," *Proceedings of the International Telemetry Conference*, Washington, DC, 1971.

Pickholtz, R. L., D. L. Schilling, and L. B. Milstein, "Theory of Spread-Spectrum Communications – A Tutorial," *IEEE Transactions on Communications*, vol. COM-30, pp. 855–884, 1982.

Pierce, J. N., "Theoretical Diversity Improvement in Frequency-Shift Keying," *Proceedings of the IRE*, vol. 46, pp. 903–910, 1958.

Price, R., "The Detection of Signals Perturbed by Scatter and Noise," *IRE Transactions on Information Theory*, vol. IT-4, pp. 163–170, 1954.

Price, R., "Optimum Detection of Random Signals in Noise with Applications to Scatter Multipath Communication," *IRE Transactions on Information Theory*, vol. IT-2, pp. 125–135, 1956.

Price, R., "A Useful Theorem for Nonlinear Devices Having Gaussian Inputs," *IRE Transactions on Information Theory*, vol. IT-4, pp. 69–72, 1958.

Price, R., "Nonlinearly Feedback Equalized PAM vs. Capacity for Noisy Filter Channels," *Proceedings of the 1972 IEEE International Conference on Communications*, pp. 22–12 to 22–17, 1972.

Price, R., "Further Notes and Anecdotes on Spread-Spectrum Origins," *IEEE Transactions on Communications*, vol. COM-31, pp. 85–97, 1983.

Price, R. and P. E. Green, Jr., "A Communication Technique for Multipath Channels," *Proceedings of the IRE*, vol. 46, pp. 555–569, 1958.

Proakis, J. G., *Advances in Equalization for Intersymbol Interference*. A. J. Viterbi (editor), Advances in Communication Systems Theory and Applications, vol. 4, New York, Academic, 1975.

Qureshi, S. U., "Adaptive Equalization," *IEEE Communications Magazine*, vol. 20, pp. 9–16, 1982.

Qureshi, S. U. and G. D. Forney, Jr., "Performance of a T/2 Equalizer," *National Telecommunications Conference Record*, 11.1-I–11.1-9, 1977.

Reed, I. S., "The Effect of a Limiter on the Relative Amplitudes of Two Signals in Noise," *Group Report 47-18, MIT Lincoln Lab*, Lexington, MA, 1958.

Reed, I. S. and G. Solomon, "Polynomial Codes over Certain Finite Fields," *Journal of the Society of Industrial and Applied Mathematics*, vol. 8, pp. 300–304, 1960.

Reiger, S., "Error Rates in Data Transmission," *Proceedings of the IRE*, vol. 46, pp. 919–920, 1958.

Rice, S. O., "Mathematical Analysis of Random Noise," *Bell System Technical Journal*, vol. 23, pp. 283–332, 1944; vol. 24, pp. 46–156, 1945.

Rice, S. O., "Envelope of Narrow-Band Signals," *Proceedings of the IEEE*, vol. 70, pp. 692–699, 1982.

Rimoldi, B. E., "A Decomposition Approach to CPM," *IEEE Transactions on Information Theory*, vol. IT-34, pp. 260–270, 1988.

Salz, J., "Optimum Mean-Square Decision-Feedback Equalization," *Bell System Technical Journal*, vol. 52, pp. 1341–1373, 1973.

Sarwate, D. V. and M. B. Pursley, "Crosscorrelation Properties of Pseudorandom and Related Sequences," *Proceedings of the IEEE*, vol. 68, pp. 593–619, 1980.

Scholtz, R. A., "Maximal and Variable Word-Length Comma-Free Codes," *IEEE Transactions on Information Theory*, vol. IT-15, pp. 300–306, 1969.

Scholtz, R. A., "Frame Synchronization Techniques," *IEEE Transactions on Communications*, vol. COM-28, pp. 1204–1213, 1980.

Scholtz, R. A., "Notes on Spread-Spectrum History," *IEEE Transactions on Communication*, vol. COM-31, pp. 82–84, 1982.

Schonhoff, T. A., "Symbol Error Probabilities for *M*-ary CPFSK: Coherent Detection," *IEEE Transactions on Communications*, vol. COM-24, pp. 644–652, 1976.

Selmer, E. S., "Linear Recurrence Relations over Finite Fields," *Ph.D. Thesis*, Department of Mathematics, University of Bergen, Norway, 1966.

Shannon, C. E., "A Mathematical Theory of Communications," *Bell System Technical Journal*, vol. 27, pp. 379–423, 1948 (Part I), pp. 623–656 (Part II) reprinted in book form with introduction by W. Weaver, University of Illinois Press, Urbana, IL, 1949, Anniversary Edition 1998.

Shannon, C. E., "Communication in the Presence of Noise," *Proceedings of the IRE*, vol. 27, pp. 10–21, 1949.

Slepian, D., "Permutation Modulation," *Proceedings of the IEEE*, vol. 53, pp. 228–236, 1965.

Sloane, N. J. A., "Tables of Sphere Packings and Spherical Codes," *IEEE Transactions on Information Theory*, vol. IT-27, pp. 327–338, 1981.

Stein, S., "Unified Analysis of Certain Coherent and Non-Coherent Binary Communications Systems," *IEEE Transactions on Information Theory*, vol. IT-10, pp. 43–51, 1964.

Stiffler, J. J., *Theory of Synchronous Communications*, Englewood Cliffs, NJ, Prentice-Hall, 1971.

Thitimajshima, P., "Les Codes Convolutifs Recursifs Systematiques et Leur Application á la Concaténation Paralléle", *Ph.D. Dissertation*, Université de Bretagne Occidentale, Brest, France, 1993.

Tomlinson, M., "New Automatic Equalizer Employing Modulo Arithmetic," *Electronic Letters*, vol. 7, pp. 138–139, 1971.

Tüchler, M., R. Koetter, and A. C. Singer, "Turbo Equalization: Principles and New Results," *IEEE Transactions on Communications*, vol. COM-50, pp. 754–767, 2002.

Tufts, D. W., "Nyquist's Problem: The Joint Optimization of Transmitter and Receiver in Pulse Amplitude Modulation," *Proceedings of the IEEE*, vol. 53, pp. 248–259, 1965.

Turin, G. L., "The Asymptotic Behavior of Ideal *M*-ary Systems," *Proceedings of the IRE*, vol. 47, pp. 93–94, 1959.

Turin, G. L., "An Introduction to Matched Filters," *IRE Transactions on Information Theory*, vol. IT-6, pp. 311–329, 1960.

Turyn, R. J., "Sequences with Small Correlation," in H. B. Mann (ed.), *Error Correcting Codes*, New York, Wiley, pp. 195–228, 1968.

Turyn, R. and J. Storer, "On Binary Sequences," *Proceedings of the American Mathematical Society*, vol. 12, pp. 394–399, 1961.

Ungerboeck, G., "Theory on the Speed of Convergence in Adaptive Equalizers for Digital Communication," *IBM Journal of Research and Development*, vol. 16, pp. 546–555, 1972.

Ungerboeck, G., "Adaptive Maximum-Likelihood Receiver for Carrier-Modulated Data-Transmission Systems," *IEEE Transactions on Communications*, vol. COM-22, pp. 624–636, 1974.

Ungerboeck, G., "Fractional Tap-Spacing Equalizer and Consequences for Clock Recovery in Data Modems," *IEEE Transactions on Communications*, vol. COM-24, pp. 856–864, 1976.

Ungerboeck, G., "Trellis Coding with Expanded Channel Signal Sets," *IEEE International Symposium on Information Theory, Book of Abstracts*, Ithaca, NY, 1977.

Ungerboeck, G., "Channel Coding with Multilevel Phase Signals," *IEEE Transactions on Information Theory*, vol. IT-28, pp. 55–67, 1982.

Van Vleck, J. H. and D. Middleton, "A Theoretical Comparison of Visual, Aural, and Meter Detection of Pulse Signals in the Presence of Noise," *Journal of Applied Physics*, vol. 171, pp. 940–971, 1946.

Van Vleck, J. H. and D. Middleton, "The Spectrum of Clipped Noise," *Proceedings of the IEEE*, vol. 54, pp. 2–19, 1966.

Verdu, S., "Maximum-Likelihood Sequence Detection for Intersymbol Interference Channels: A New Upper Bound on Error Probability," *IEEE Transactions on Information Theory*, vol. IT-33, pp. 62–68, 1987.

Viterbi, A. J., "On Coded Phase-Coherent Communication," *IRE Transactions on Space Electronics Telemetry*, vol. SET-7, pp. 3–12, 1961.

Viterbi, A. J., "Phase-Locked Loop Dynamics in the Presence of Noise by Fokker-Planck Techniques," *Proceedings of the IEEE*, vol. 51, pp. 1737–1753, 1963.

Viterbi, A. J., "Error Bounds for Convolutional Codes and an Asymptotically Optimum Decoding Algorithm," *IEEE Transactions on Information Theory*, vol. IT-13, pp. 260–269, 1967.

Voronoi, G., "Recherches sur les Paralleloedres Primitives," *Journal Reine Angew Mathematique*, vol. 134, pp. 198–287, 1908.

Warrier, D. and U. Madhow, "Spectrally Efficient Noncoherent Communication," *IEEE Transactions on Information Theory*, vol. IT-48, pp. 651–668, 2002.

Wei, L. F., "Rotationally Invariant Convolutional Channel Coding with Expanded Signal Space. Part 1: 180°. Part 11: Nonlinear Codes," *IEEE Journal on Selected Areas Communication*, vol. SAC-2, pp. 659–686, 1984.

Wei, L. F., "Trellis-Coded Modulation with Multidimensional Constellations," *IEEE Transactions on Information Theory*, vol. IT-33, pp. 483–501, 1987.

Welti, G. R., "Quaternary Codes for Pulsed Radar," *IRE Transactions on Information Theory*, vol. IT-6, pp. 400–408, 1960.

Welti, G. and J. Lee, "Digital Transmission with Coherent Four Dimensional Modulation," *IEEE Transactions on Information Theory*, vol. IT-20, pp. 497–502, 1974.

Widrow, B., "Adaptive Filter, I: Fundamentals," *Technical Report No. 6764-6*, Stanford Electronics Laboratory, Stanford University, Stanford, CA, 1966.

Wilson, S. G., H. A. Sleeper, and N. K. Srinath, "Four Dimensional Modulation and Coding: An Alternative to Frequency-Reuse," *Proceedings of the 1984 IEEE International Conference on Communications*, Amsterdam, 1984.

Woodward, P. M., *Probability and Information Theory with Applications to Radar*, New York, McGraw-Hill, 1954.

Wozencraft, J. M., "Sequential Decoding for Reliable Communications," *1957 National IRE Convention Record*, vol. 5, Part 2, pp. 11–25, 1957.

Ziemer, R. E. and W. H. Tranter, *Principles of Communications: Systems, Modulation, and Noise*, 5th edn, New York, Wiley, 2001.

Zierler, N., "Linear Recurring Sequences," *Journal of the Society of Industrial and Applied Mathematics*, vol. 7, pp. 31–48, 1959.

Index

Printed in the United States
By Bookmasters